AF512846

Recent Advances in Legume Crop Protection

Recent Advances in Legume Crop Protection

Guest Editors

Kai Li

Yu Gao

Zhaofeng Huang

MDPI

Basel • Beijing • Wuhan • Barcelona • Belgrade • Novi Sad • Cluj • Manchester

Guest Editors
Kai Li
College of Agriculture
Nanjing Agricultural University
Nanjing
China

Yu Gao
College of Plant Protection
Jilin Agricultural University
Changchun
China

Zhaofeng Huang
Institute of Plant Protection
Chinese Academy of Agricultural Sciences
Beijing
China

Editorial Office
MDPI AG
Grosspeteranlage 5
4052 Basel, Switzerland

This is a reprint of the Special Issue, published open access by the journal *Agronomy* (ISSN 2073-4395), freely accessible at: https://www.mdpi.com/journal/agronomy/special_issues/31W3D34GRL.

For citation purposes, cite each article independently as indicated on the article page online and as indicated below:

Lastname, A.A.; Lastname, B.B. Article Title. *Journal Name* **Year**, *Volume Number*, Page Range.

ISBN 978-3-7258-5443-1 (Hbk)
ISBN 978-3-7258-5444-8 (PDF)
https://doi.org/10.3390/books978-3-7258-5444-8

© 2025 by the authors. Articles in this book are Open Access and distributed under the Creative Commons Attribution (CC BY) license. The book as a whole is distributed by MDPI under the terms and conditions of the Creative Commons Attribution-NonCommercial-NoDerivs (CC BY-NC-ND) license (https://creativecommons.org/licenses/by-nc-nd/4.0/).

Contents

About the Editors

Kai Li

He is an associate professor of College of Agriculture, Nanjing Agricultural University, Ph. D. graduated in Nanjing Agricultural University. He serves as a scientist in the virus disease prevention and control position of the national modern agricultural industrial technology system. He engages in the identification, classification and dynamic monitoring of soybean mosaic virus strains in China, the identification and dynamic monitoring of resistance of new soybean varieties (lines) to soybean mosaic virus, the creation of soybean germplasm resistance, resistance genetic analysis, resistance gene mining and breeding of soybean varieties against disease. At present, he has published more than 30 papers as the first or corresponding author; participated in witting 2 books.

Yu Gao

He is an associate professor of College of Plant Protection, Jilin Agricultural University, Ph. D. graduated in Chinese Academy of Agricultural Sciences. He serves as a post scientist of the National Modern Industrial Technology System. He serves concurrently as: Deputy Director of the Key Laboratory for Soybean Pest and Disease Control, Ministry of Agriculture and Rural Affairs; External Technical Expert, National Engineering Research Center for Tea Quality and Safety, Ministry of Science and Technology; Deputy Secretary-General and Executive Council Member, Jilin Provincial Plant Protection Society; Science and Technology Special Envoy, Jilin Province; Member, Soybean Specialized Committee, 9th Crop Variety Approval Committee, Jilin Province; Editorial Board Member, Biology and Agrochemicals; Young Editorial Board Member, Journal of Jilin Agricultural University and Soybean Science and Technology. He engages in the research of agricultural pest ecology, pesticide residue detection, and edible insects. At present, he has published more than 70 papers as the first or corresponding author, including one ESI 'highly cited paper'; participated in witting six books.

Zhaofeng Huang

He is a professor of Chinese Academy of Agricultural Sciences, Ph. D. graduated from the Chinese Academy of Agricultural Sciences. He engaged in the research of herbicide resistance mechanism, prevention and control of weeds in agricultural fields, focusing on the malignant weeds in crop fields, carried out the monitoring and early warning of the resistance of malignant weeds such as amaranth, field spinosa, quinoa, etc., to the main herbicide, and elucidated the mechanism of target resistance to herbicides, Based on this, he developed the detection kit for the PPO-resistant amaranth inhibitor, which provides the theoretical basis for the green prevention and control of agricultural weeds. He has published more than 30 papers in Weed Science, Crop Protection, Scientific reports, Pesticide Biochemistry and Physiology, Journal of Plant Protection, Plant Protection and other journals.

Preface

Leguminous crops occupy a central place in the global food system, supplying critical calories in the form of starch, high-quality protein for both direct human consumption and animal feed, edible and industrial oils, and an array of fresh and processed vegetables. Yet their genetic yield potential is continually eroded by a complex of fungal, bacterial, viral and oomycete pathogens, by insect and mite pests with piercing-sucking or chewing mouthparts, and by highly competitive annual and perennial weeds. Climate change accelerates pathogen and pest life cycles, enables range expansion, and disrupts the synchrony between host phenology and natural-enemy activity, while shifts in planting patterns and intensified monocropping create ecological niches that favor new biotypes and invasive species. To counteract these pressures, the legume-production sector is embracing an integrated management paradigm that fuses traditional agronomic wisdom with cutting-edge advances in biotechnology, molecular biology and genomics. Resistance-gene discovery pipelines now exploit pangenomes, pangenomics and machine-learning algorithms to mine nucleotide-binding leucine-rich repeat immune receptors and pattern-recognition receptors that confer durable, broad-spectrum resistance. Gene-editing platforms such as CRISPR/Cas and base-editing systems are deployed to stack multiple defense pathways without linkage drag, while marker-assisted and genomic-selection breeding accelerate the introgression of quantitative resistance loci into elite cultivars. Simultaneously, microbiome engineering identifies and formulates consortia of endophytic bacteria, mycorrhizal fungi and entomopathogenic viruses that outcompete pathogens, prime systemic acquired resistance and deter herbivores through volatile or proteinaceous effectors. On the landscape scale, precision-agriculture technologies—drone-based multispectral imaging, AI-driven decision support systems and variable-rate smart sprayers—enable site-specific application of biological or biorational pesticides, reducing chemical load and preserving beneficial arthropods. Cultural tactics such as diversified rotations, strip intercropping with non-host species, conservation tillage and mulching suppress soil-borne inoculum, disrupt pest mating and suppress weed seedbanks. By integrating these complementary tactics, farmers can stabilize legume yields, cut production costs, protect environmental quality and enhance the resilience of agroecosystems under a rapidly changing climate. This Reprint therefore invited original research, reviews and methodological papers that advance our understanding and practical deployment of integrated strategies for managing diseases, insect pests and weeds in legume crops.

Kai Li, Yu Gao, and Zhaofeng Huang
Guest Editors

agronomy

MDPI

Editorial

Recent Advances in Legume Crop Protection

Yu Gao [1,2], Zhaofeng Huang [3] and Kai Li [4,*]

1 College of Plant Protection, Jilin Agricultural University, Changchun 130118, China
2 Key Laboratory of Soybean Disease and Pest Control, Ministry of Agriculture and Rural Affairs, Changchun 130118, China
3 State Key Laboratory for Biology of Plant Diseases and Insect Pests, Institute of Plant Protection, Chinese Academy of Agricultural Sciences, Beijing 100193, China
4 Soybean Research Institute & MARA National Center for Soybean Improvement & MARA Key Laboratory of Biology and Genetic Improvement of Soybean & State Innovation Platform for Integrated Production and Education in Soybean Bio-Breeding & State Key Laboratory for Crop Genetics and Germplasm Enhancement and Utilization & Jiangsu Collaborative Innovation Center for Modern Crop Production, Nanjing Agricultural University, Nanjing 210095, China
* Correspondence: kail@njau.edu.cn

The legume family is economically important and is one of the most important sources of starch, protein, oil, and vegetables for human food around the world, playing an irreplaceable role in guaranteeing human food security. However, the frequent occurrence of diseases, pests, and weeds is one of the important factors restricting the high and stable yield of legume crops [1]. Major diseases of legumes include root rots, bacterial blights, and mosaic diseases. In addition, damage caused by malignant weeds, nematodes, and sap-sucking/-chewing insects is also included as a constraint on legume production [2]. This Special Issue is focused on 13 selected topics (12 articles and 1 editorial) from different universities and research institutes. The research fields covered include mechanisms of resistance to pests and diseases, identification and monitoring of diseases, alternative control strategies, weed management, and drone-mounted multispectral imaging technology. Current problems in the prevention and control of legume crop diseases, pests, and weeds are discussed with a view to providing subsequent research and applications of key pest- and disease-monitoring and -control technologies with a reference point.

Disease and insect resistance of soybean cultivars are extremely important and effective measures for controlling soybean diseases and pests [3]. The following four papers cover relevant topics:

Contribution 1: Raza et al. from Junyi Gai's team used resistant (Kefeng-1) and susceptible (NN1138-2) soybean cultivars inoculated with SMV-SC4, and transcriptional analyses at 0, 6, 24, and 48 h post-inoculation identified 1,201 core differentially expressed genes (DEGs). Most DEGs were activated early in the resistant cultivar. Gene ontology analysis revealed enriched DEGs in three key functional categories contributing to resistance: signal transduction, oxidoreductase activity, and response to auxin. These DEGs exhibited significantly higher differential expressions in Kefeng-1 versus NN1138-2. This study elucidates molecular networks underlying soybean resistance to SMV, providing crucial insights for developing virus-resistant soybean cultivars.

Contribution 2: Antwi-Boasiako et al. utilized two RIL populations, namely ZM6 and MN, to identify quantitative trait loci (QTLs) associated with red crown root resistance in soybeans. In total, 15 and 14 QTLs were found to be related to RCR resistance in ZM6 and MN populations, respectively. Six 'QTL hotspots' for resistance to RCR from the ZM6 and MN RIL populations were detected on chromosomes 1, 7, 10, 11, 13, and 18. Through gene annotations, gene ontology enhancement, and RNA sequencing assessment, 23 genes located within six 'QTL Hotspots' were found as potential candidate genes that could provide resistance to red crown root in soybeans.

Contribution 3: Murodova et al. used 96 chickpea accessions from ICARDA infected with fusarium wilt (FW). Six pathogenic species (*Neocosmospora solani*, *N. nelsonii*, *N. falciformis*, *N. brevis*, *Fusarium brachygibbosum*, and *F. gossypinum*) were identified by molecular analysis (targeting ITS, *tef1-α*, and tub2 regions). Genetic diversity assessment using 69 polymorphic SSR markers revealed 191 alleles across all markers. Association mapping (employing GLM and MLM approaches) identified five consistent marker–trait associations for FW resistance. This study represents the first association mapping for FW resistance in ICARDA chickpeas, thus identifying key genomic regions for targeted resistance breeding to enhance global chickpea sustainability.

Contribution 4: Soybean 'Zhengqing' (or namely stay-green syndrome) is a major issue that causes soybean yield reduction in North China [4]. Wang D. et al. investigated differentially expressed genes (DEGs) in the sensitive soybean variety (HD0702) impacted by/without 'Zhengqing' (1,858 DEGs in the pods and 2,814 DEGs in the leaves). The chlorophyll content of the pods increased, soluble sugar levels significantly increased, whereas indole-3-acetic acid and abscisic acid decreased.

The identification and monitoring of soybean diseases have always presented a challenge to producers [5]. Three papers cover topics relevant to this subject, including rot, seed, and foliar diseases.

Contribution 5: As a soil-borne disorder, soybean root rot is driven by a complex of pathogens that threaten soybean production worldwide. The pathogenic fungal strains from Heilongjiang Province were identified as *Fusarium oxysporum*, *F. graminearum*, *F. asiaticum*, *Pythium macrosporum*, and *Rhizoctonia solani*, through morphological and molecular identification. Among them, *F. oxysporum* was the dominant species. The application of fludioxonil and pyraclostrobin had better control effects for *F. asiaticum*.

Contribution 6: Wang Y. et al. sampled 14 soybean cultivars and profiled their seed-borne microbiota with both conventional assays and high-throughput sequencing. The results indicated that seeds sourced from Jilin Province carried the greatest abundance of fungi and bacteria, followed by those from Liaoning and Heilongjiang Province. Characterizing these microbial communities establishes a foundation for seed quarantine protocols and disease-management strategies.

Contribution 7: Meng et al. found that the green normalized difference vegetation index could be used to monitor the soybean bacterial blight disease using drone-mounted multispectral imaging. The soybean yield loss was significantly higher at disease grade four for this disease. This approach leverages a random-forest model to survey disease classes and predict yield loss, laying the groundwork for future precision plant-protection strategies.

This Special Issue contains five papers that explore selected issues related to the integrated management of soybean diseases, pests, and weeds.

Contribution 8: For the sustained control of soybean root rot, Liu et al. have made breakthroughs. The researchers developed an innovative strategy, where a wheat–straw-fiber-based mulch film coated with carbendazim and chitosan mixture was prepared through a bar-coating technology. This special film achieved the desired physical properties and could effectively inhibit the growth of *F. solani* and boost soybean growth.

Contribution 9: In the coming period, alternative solutions of chemical control will be investigated, moving from artificially synthesized chemical insecticides to plant-based insecticides. The study by Niu et al. proposed four alternative essential oils (marjoram oil, clary sage oil, perilla leaf oil, and spearmint oil) against an emerging pest, *Thrips flavus*, in soybean fields in Northeast China. Linalool, isopropyl myristate, limonene, and carvone were the primary chemical constituents of these essential oils. The spearmint oil was significantly attractive to female adults in the olfactory test, indicating its potential for developing a thrips attractant.

Contribution 10: The whitefly (*Bemisia tabaci*) represents a cryptic species complex and ranks among the world's most damaging pests, owing to its extensive host range and global reach. Alvarez et al. assessed the development and survival of the Mediterranean biotype on soybean, common bean, cotton, bell pepper, and tomato across multiple temperatures. Their findings show that conditions suitable for this biotype prevail throughout much of Brazil, especially in cropping systems marked by heavy insecticide use. These outcomes are pivotal for forecasting the Mediterranean biotype's establishment and spread across Brazil's varied climatic zones.

Contribution 11: Yangui et al. explored the use of discrete choice experiments to analyze farmers' decision-making preferences in integrated *Orobanche* management and their willingness to pay for different control measures. It highlights significant variability in farmer acceptance of these measures, influenced by factors such as their financial situation and the severity of *Orobanche* infestation.

Contribution 12: Pamungkas et al. evaluated the efficacy and selectivity of three pre-emergence herbicides (pendimethalin, s-metolachlor, and flumioxazin) on dominant soybean weeds in Northeast Thailand across rainy (2023) and dry (2024/25) seasons, using two soybean varieties. They proposed that season-specific herbicide selection is critical. CM60 is better adapted for herbicide use under Thai conditions. Manual weeding remains the gold standard, but herbicides like s-metolachlor offer viable alternatives. *Cyperus rotundus* requires targeted management due to its persistence.

In summary, the 12 articles in this Special Issue collectively underscore the critical need for integrated management of disease, pests, and weeds in legume crops. These findings will provide valuable strategies to improve the productivity of legume crops, from elucidating the identification and monitoring of diseases, mining for resistance genes, to revealing crop resistance mechanisms to pests and diseases, as well as multiple control measures for pests, diseases, and weeds. Given the intensification of crop diseases and pests caused by global climate change and changes in planting structures, the adoption of integrated management theories and methods for diseases, pests, and weeds remains crucial for ensuring food security and the sustainable development of agriculture [6,7].

Author Contributions: Conceptualization, Y.G., K.L. and Z.H.; writing—original draft preparation, Y.G., K.L. and Z.H.; writing—review and editing, Y.G., K.L. and Z.H. All authors have read and agreed to the published version of the manuscript.

Funding: This research received no external funding.

Institutional Review Board Statement: Not applicable.

Data Availability Statement: All original data and findings reported herein are contained within the article. Any additional questions should be addressed to the corresponding authors.

Acknowledgments: The Guest Editors of the Special Issue 'Recent Advances in Legume Crop Protection' sincerely appreciate all authors and reviewers who contributed their valuable work to the Special Issue, making this edition of the journal a great success.

Conflicts of Interest: The authors declare no conflicts of interest.

List of Contributions:

1. Raza, M.M.; Jia, H.; Gu, S.; Gai, J.; Li, K. Transcriptome Insights into Resistance Mechanisms Against Soybean Mosaic Virus Strain SC4 in Soybean. *Agronomy* **2025**, *15*, 906. https://doi.org/10.3390/agronomy15040906.

2. Antwi-Boasiako, A.; Zhang, C.; Almakas, A.; Liu, J.; Jia, S.; Guo, N.; Chen, C.; Zhao, T.; Feng, J. Identification of QTLs and Candidate Genes for Red Crown Rot Resistance in Two Recombinant Inbred Line Populations of Soybean [*Glycine max* (L.) Merr.]. *Agronomy* **2024**, *14*, 1693. https://doi.org/10.3390/agronomy14081693.
3. Murodova, S.M.; Bozorov, T.A.; Aytenov, I.S.; Ochilov, B.O.; Qulmamatova, D.E.; Salakhutdinov, I.B.; Isokulov, M.Z.; Khalillaeva, G.O.; Azimova, L.A.; Meliev, S.K. Uncovering Fusarium Species Associated with Fusarium Wilt in Chickpeas (*Cicer arietinum* L.) and the Identification of Significant Marker–Trait Associations for Resistance in the International Center for Agricultural Research in the Dry Areas' Chickpea Collection Using SSR Markers. *Agronomy* **2024**, *14*, 1943. https://doi.org/10.3390/agronomy14091943.
4. Wang, D.; Wang, Y.; Sun, R.; Yang, Y.; Zhao, W.; Yu, G.; Wang, Y.; Wang, F.; Zhou, L.; Huang, Z. Transcriptomics and Physiological Analyses of Soybean Stay-Green Syndrome. *Agronomy* **2025**, *15*, 82. https://doi.org/10.3390/agronomy15010082.
5. Liu, J.; Cui, W.; Zhao, Q.; Ren, Z.; Li, L.; Li, Y.; Sun, L.; Ding, J. Identification, Characterization, and Chemical Management of *Fusarium asiaticum* Causing Soybean Root Rot in Northeast China. *Agronomy* **2025**, *15*, 388. https://doi.org/10.3390/agronomy15020388.
6. Wang, Y.; Bai, Q.; Meng, F.; Dong, W.; Fan, H.; Zhu, X.; Duan, Y.; Chen, L. High-Throughput Sequence Analysis of Microbial Communities of Soybean in Northeast China. *Agronomy* **2025**, *15*, 436. https://doi.org/10.3390/agronomy15020436.
7. Meng, W.; Li, X.; Zhang, J.; Pei, T.; Zhang, J. Monitoring of Soybean Bacterial Blight Disease Using Drone-Mounted Multispectral Imaging: A Case Study in Northeast China. *Agronomy* **2025**, *15*, 921. https://doi.org/10.3390/agronomy15040921.
8. Liu, S.; Jin, Z.; Zhou, P.; Shang, H.; Yang, H.; Li, L.; Li, R.; Zhang, Y.; Chen, H. Preparation of Wheat-Straw-Fiber-Based Degradable Mulch Film for Sustained Release of Carbendazim and Its Application for Soybean Root Rot Control. *Agronomy* **2025**, *15*, 71. https://doi.org/10.3390/agronomy15010071.
9. Niu, Y.; Pei, T.; Zhao, Y.; Zhou, C.; Liu, B.; Shi, S.; Xu, M.-L.; Gao, Y. Exploring the Efficacy of Four Essential Oils as Potential Insecticides against *Thrips flavus*. *Agronomy* **2024**, *14*, 1212. https://doi.org/10.3390/agronomy14061212.
10. Alvarez, D.d.L.; Hayashida, R.; Santos, D.M.; Silva, F.B.d.; Müller, C.; Krause-Sakate, R.; Hoback, W.W.; Oliveira, R.C.d. Thermal Tolerance and Host Plant Suitability of *Bemisia tabaci* MED (Gennadius) in Brazilian Legume Crops. *Agronomy* **2025**, *15*, 1622. https://doi.org/10.3390/agronomy15071622.
11. Yangui, A.; Mlayeh, T.; Abbes, Z.; Kharrat, M. Towards an Integrated Orobanche Management: Understanding Farmers' Decision-Making Processes Using a Discrete Choice Experiment. *Agronomy* **2025**, *15*, 219. https://doi.org/10.3390/agronomy15010219.
12. Pamungkas, U.R.R.; Chankaew, S.; Jongrungklang, N.; Monkham, T.; Gonkhamdee, S. The Efficacy of Pre-Emergence Herbicides against Dominant Soybean Weeds in Northeast Thailand. *Agronomy* **2025**, *15*, 1725. https://doi.org/10.3390/agronomy15071725.

References

1. Xia, Y.; Sun, G.; Xiao, J.; He, X.; Jiang, H.; Zhang, Z.; Zhang, Q.; Li, K.; Zhang, S.; Shi, X.; et al. AlphaFold-guided redesign of a plant pectin methylesterase inhibitor for broad-spectrum disease resistance. *Mol. Plant* **2024**, *17*, 1344–1368. [CrossRef] [PubMed]
2. Xu, H.; Chen, S.W.; Wang, Y.Y.; Pan, J.Z.; Liu, X.Z.; Wang, C.W.; Wang, X.X.; Cui, X.Y.; Chen, X.; Li, J.B.; et al. A Faboideae-specific floral scent betrays seeds to an important granivore pest. *J. Agric. Food Chem.* **2023**, *71*, 12668–12677. [CrossRef] [PubMed]
3. Jiang, H.; Qu, S.; Liu, F.; Sun, H.; Li, H.; Teng, W.; Zhan, Y.; Li, Y.; Han, Y.; Zhao, X. Multi-omics analysis identified the GmUGT88A1 gene, which coordinately regulates soybean resistance to cyst nematode and isoflavone content. *Plant Biotechnol. J.* **2025**, *23*, 1291–1307. [CrossRef] [PubMed]
4. Cheng, R.X.; Mei, R.X.; Yan, R.; Chen, H.Y.; Miao, D.; Cai, L.N.; Fan, J.Y.; Li, G.R.; Xu, R.; Ye, W.G.; et al. A new distinct geminivirus causes soybean stay-green disease. *Mol. Plant* **2022**, *15*, 927–930. [CrossRef] [PubMed]
5. Warpechowski, L.F.; Steinhaus, E.A.; Moreira, R.P.; Godoy, D.N.; Preto, V.E.; Braga, L.E.; Wendt, A.D.; Reis, A.C.; Lima, E.F.B.; Farias, J.R.; et al. Why does identification matter? Thrips species (Thysanoptera: Thripidae) found in soybean in southern Brazil show great geographical and interspecific variation in susceptibility to insecticides. *Crop Prot.* **2024**, *178*, 106592. [CrossRef]

6. Chen, J.; Jiang, K.; Li, Y.; Wang, S.; Bu, W. Climate change effects on the diversity and distribution of soybean true bugs pests. *Pest Manag. Sci.* **2024**, *80*, 5157. [CrossRef] [PubMed]
7. Ruiu, L. Microbial biopesticides in agroecosystems. *Agronomy* **2018**, *8*, 235. [CrossRef]

Disclaimer/Publisher's Note: The statements, opinions and data contained in all publications are solely those of the individual author(s) and contributor(s) and not of MDPI and/or the editor(s). MDPI and/or the editor(s) disclaim responsibility for any injury to people or property resulting from any ideas, methods, instructions or products referred to in the content.

agronomy

MDPI

Article

Transcriptome Insights into Resistance Mechanisms Against Soybean Mosaic Virus Strain SC4 in Soybean

Muhammad Muzzafar Raza †, Huiying Jia †, Shengyu Gu, Junyi Gai * and Kai Li *

Soybean Research Institute & MARA National Center for Soybean Improvement & MARA Key Laboratory of Biology and Genetic Improvement of Soybean & State Innovation Platform for Integrated Production and Education in Soybean Bio-Breeding & State Key Laboratory for Crop Genetics and Germplasm Enhancement and Utilization & Jiangsu Collaborative Innovation Center for Modern Crop Production, Nanjing Agricultural University, Nanjing 210095, China; 2017201099@njau.edu.cn (M.M.R.); 2016101068@njau.edu.cn (H.J.); 2018201036@njau.edu.cn (S.G.)

* Correspondence: sri@njau.edu.cn (J.G.); kail@njau.edu.cn (K.L.)

† These authors contributed equally to this work.

Abstract: Soybean, an economically valuable oil and protein crop, is vulnerable to numerous biotic stresses throughout its growth period. Soybean mosaic virus (SMV), a destructive plant pathogen, induces substantial yield reduction and seed quality deterioration globally. In China, a total of 22 distinct SMV strains have been documented, with SMV-SC4 being a widely spread strain. The Chinese cultivar Kefeng-1 (KF) is resistant to this strain. To investigate the resistance mechanism, transcriptional analysis was performed at 0, 6, 24, and 48 h post-inoculation of SC4 in KF (Resistant) and NN1138-2 (NN) (Susceptible). A total of 1201 core differentially expressed genes (DEGs) were identified as active ones against SC4 infection, with most originating from the resistant cultivar at the early infection stages. Gene ontology enrichment analysis indicated that the DEGs directly involved in signal transduction and those related to plant stress response contributed to KF resistance indirectly, including protein phosphorylation, protein kinase activity, oxidation–reduction, oxidoreductase activity, catalytic activity, metal ion transport, and response to auxin. A total of 27 genes in "Signal transduction" with most of them were disease resistance conserved domains, 52 genes active in oxidoreductase activity involving in removing ROS from SMV attack, and 8 genes in "Response to auxin", a phytohormone that plays a role in biotic stress response in addition to growth and development. These genes expressed more differentially in the resistant versus susceptible cultivar. Our findings provide insights into the molecular networks related to soybean response to SMV, which may be relevant in understanding soybean resistance against the viral infections.

Keywords: gene ontology (GO); Kyoto encyclopedia of genes and genomes (KEGG); deferentially expressed genes (DEGs); signal transduction

1. Introduction

Plant viruses are significant threats for crop yield and quality with their impact becoming a global issue. Legumes, particularly soybeans [*Glycine max* (L.) Merr.], have been severely affected by viruses, especially the soybean mosaic virus (SMV) of the genus *Potyvirus* [1]. Fortunately, researchers have isolated, classified, and categorized SMV strains based on their differential responses in different resistant soybean lines. In China, 22 SMV strains (SC1-SC22) have been identified with SC4 being a widely spread and moderately virulent pathogenic strain [2,3], while seven strains (G1–G7) have been reported in the United States [4,5] and five strains (A–E) in Japan [6].

 https://doi.org/10.3390/agronomy15040906

Naturally, plants possess defense systems to combat pathogenic attacks, and so is for soybeans. Recognition receptors present on cell surface detect strain specific associated molecules or internal signals produced during infection which initiate the first line of defense known as pattern-triggered immunity (PTI), which cause a basic level of resistance. Inside the host cells, pathogen effectors are recognized by the receptors, activating the second line of defense called effector-triggered immunity (ETI) which offers a stronger defensive response [7]. The interplay between PTI and ETI, constituting a complex regulatory network, activates coordinated signaling cascades. These biochemical events encompass reactive oxygen species (ROS) burst and oscillatory calcium ion fluxes, which function as pivotal second messengers in plant immune signaling pathways [8,9]. Transmembrane transport and metal ion transport are involved in the movement of secondary metabolites. Similarly, protein kinase activity and signal transduction play roles in natural resistance against biotic and abiotic stresses in plants [10–13]. The RNA interference (RNAi) pathway serves as a fundamental component of plant antiviral defense mechanisms, executing sequence-specific degradation of viral RNAs through RNA silencing. This mechanism likely contributes to the observed differences in gene expression between the resistant and susceptible soybean varieties. Critical molecular determinants, including genes encoding pattern recognition receptors, signal transduction components, and resistance (R) proteins, have been identified as key mediators orchestrating SMV resistance mechanisms.

The plant defense system is also mediated by several hormones, including auxin, salicylic acid (SA), cytokinin, ethylene, jasmonic acid (JA), and abscisic acid (ABA). Auxin not only directly participates in signaling pathways for resistance but also regulates these hormones through signal transduction [14]. Although auxin primarily governs plant growth and development under various environmental conditions, its low concentration can control gene expression via precise transcription factors and proteins modified for biotic responses in signaling networks [15].

In China, resistance to SMV is present in many cultivars, such as RN-9, Qihuang-1, and Dabaima, especially Kefeng-1 [16–18]. However, the specific pathways and mechanisms leading to defense against SMV in Kefeng-1 remain unclear. The present study aims to investigate the molecular mechanism of resistance against SC4 in Kefeng-1. Transcriptomic analysis was conducted on two extreme cultivars—Kefeng-1 (resistant to SC4) and Nannong1138-2 (susceptible to SC4)—to identify transcriptomic patterns after SC4 inoculation. Through this analysis, we expect to identify gene networks and pathways that may control the resistance mechanism against SMV infection.

2. Materials and Methods

2.1. Plant Materials, SMV Strain, and SMV Inoculation

Kefeng-1 (KF) is a soybean cultivar that has been verified to be resistant to most of SMV strains, while Nannong1138-2 (NN) is a susceptible cultivar that is susceptible to all known SMV strains in China. The SMV strain SC4, sourced by the National Center for Soybean Improvement (Nanjing, China), was collected, purified, and thoroughly identified. The inoculum was prepared by grinding leaf tissue containing the virus with 0.01 M phosphate-buffered saline (pH 7.2) using a mortar and pestle, as described in a previous study [2]. When the first true leaf unfolded, the inoculum was gently and quickly brushed onto the leaves, and the plants were cultivated in an incubator with alternating day and night cycles. Three biological replicates of leaves were collected at 0, 6, 24, and 48 h post-inoculation (hpi), frozen in liquid nitrogen, and stored at −80 °C.

2.2. Serological Determination of Sequenced Materials

A double antibody sandwich enzyme-linked immunosorbent assay (DAS-ELISA) was used to detect the SMV content in Kefeng-1 and NN1138-2 after inoculation with SMV SC4 strain. We followed an antibody diagnostic kit (V094-R2, Nanodiaincs, Fayetteville, AR, USA) manufacturer's instructions and read the absorption value at 405 nm with the Infinite 200PRO (TECAN, Männedorf, Switzerland). The positive criterion was that the OD_{405} value was significantly higher than twice the negative control value.

2.3. Total RNA Extraction and cDNA Library Construction for Transcriptome Analysis

RNA was isolated from 24 leave samples using the Trizol reagent kit (Invitrogen, Carlsbad, CA, USA) as per manufacturer's instructions. The RNA quality was measured with an Agilent 2100 Bioanalyzer (Agilent Technologies, Palo Alto, CA, USA) and confirmed by RNase-free agarose gel electrophoresis. Subsequently eukaryotic mRNA was enriched with Oligo (dT) beads, fragmented, and converted into cDNA using the NEBNext Ultra RNA Library Prep Kit for Illumina (NEB #7530, New England Biolabs, Ipswich, MA, USA). The purified double-stranded cDNA fragments were end-repaired, A -tailed, and ligated to Illumina sequencing adapters. The ligation reaction was purified using AMPure XP Beads (1.0×) and PCR-amplified. The generated cDNA library was subsequently sequenced on an Illumina Novaseq6000 platform by Gene Denovo Biotechnology Co., Ltd. (Guangzhou, China).

2.4. Principal Component Analysis

A principal component analysis (PCA) was carried out with the R package (version 2.19.1) g models (http://www.r-project.org/, accessed on 28 February 2025). PCA is a statistical technique that transforms hundreds of thousands of correlated variables, such as gene expression data, into a set of linearly uncorrelated variables known as principal components. This method is extensively used to uncover the underlying structure and relationships within a dataset.

2.5. Differentially Expressed Genes (DEGs)

Differential expression analysis of RNA was conducted with DESeq2 [19] for comparison between two different groups and edgeR [20] for pairwise sample comparisons. Genes or transcripts with the false discovery rate (FDR) below 0.05 and an absolute fold change of ≥2 was classified as differentially expressed.

2.6. Gene Ontology Analysis of DEGs

Differentially expressed genes (DEGs) were assigned to the corresponding terms in the gene ontology (GO) database (http://www.geneontology.org/, accessed on 28 February 2025), the count of DEGs for each term was determined to generate a list of genes with specific GO functions. A hypergeometric test was then used to find GO terms that were significantly enriched ($p \leq 0.05$) enriched in DEGs compared to background.

2.7. Pathway Enrichment Analysis

Pathway-based analysis provides deeper insights into biological functions of genes. KEGG [21], a chief public database pathway for information, was used for this purpose. In our study, pathway enrichment analysis identified metabolic and signal transduction pathways that were significantly enriched among the differentially expressed genes to whole genome background. The calculated p-values were adjusted with an FDR of threshold of ≤0.05. Pathways meeting this criterion were deemed significantly enriched in the DEGs.

3. Results

3.1. Transcriptome or RNA-Seq Analysis

DAS-ELISA results showed that KF with robust resistance and NN with susceptibility were suitable materials for exploring the mechanism of resistance by transcriptomics (Table S1). A total of 1.13×10^9 base pairs of Raw Data (deposited in the NCBI Sequence Read Archive (SRA) under accession number PRJNA1062726) were obtained for 24 samples, with an average clean data coverage of 99.73% (Table S2). The GC content after filtration ranged between 41.95% and 44.50%, and the Q30 of each sample was above 93.24% (Table S3). These results indicate high randomness and reliability of the sequencing fragments. Mapping with the reference genome *G. max* Wm82.a2. v1 showed an average total coverage of 95.62% (Table S4), indicating high assembly integrity. For quality control, PCA was performed on expression files from all samples to ensure that samples from the same group were similar (Figure S1). The figure shows that samples from the same treatment time and material clustered together, while samples from different materials at the same treatment time were relatively close, indicating similar expression patterns. However, one sample from NN at 24 hpi deviated from normal clustering and was excluded from further analysis.

3.2. Identification of DEGs at Four-Time Stages After SC4-Inoculation

RNA-seq analysis revealed that at each time point (0, 6, 24, and 48 hpi) after virus infection, the number of DEGs between the two cultivars ranged from 1148 to 2425, with a balance between up-regulated and down-regulated genes (Figure 1A). Significantly more DEGs were observed between soybean cultivars NN and KF at the KF-0h-vs.-KF-6h and NN-0h-vs.-NN-6h stages compared to other time points, particularly for up-regulated genes. As time progressed, the number of DEGs gradually returned to a lower level compared to untreated plants (Figure 1B), suggesting that genes active in the early stages of infection may play pivotal roles in responding to SMV infection.

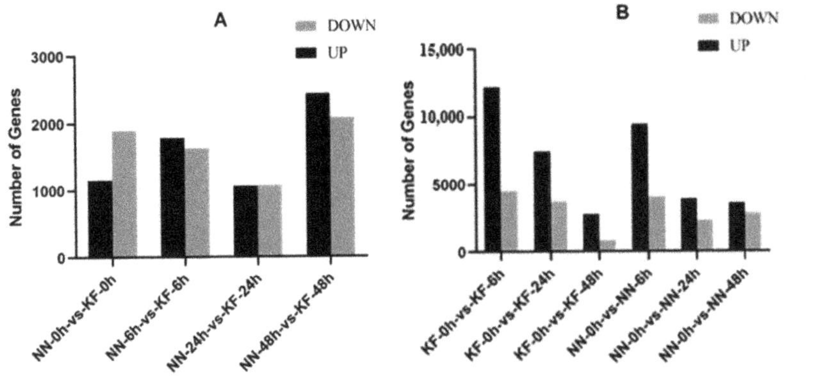

Figure 1. Comparative analysis of differentially expressed genes (DEGs) across four distinct stages following SC4 infection on soybean. (**A**) Comparative analysis of DEGs between soybean cultivars NN1138-2 (NN) and Kefeng-1 (KF) at each of the four stages. (**B**) Stage-specific comparative analysis of DEGs between NN1138-2 (NN) and Kefeng-1 (KF) after SC4 inoculation.

To understand how plants respond to invading pathogens, we analyzed all DEGs in both cultivars by comparing four groups: KF0h vs. KF6h, NN0h vs. NN6h, KF0h vs. NN0h and KF6h vs. NN6h. The DEGs in groups KF0h vs. KF6h and NN0h vs. NN6h were 5731 (464 + 4466 + 190 + 611) and 2504 (308 + 157 + 1787 + 252), respectively, with an additional 10,994 shared DEGs (270 + 629 + 576 + 9519). The DEGs between the other two groups (KF0h vs. NN0h and KF6h vs. NN6h) were 1791 (499 + 464 + 576 + 252) and 2154 (606 + 308 + 611 + 629), with 1250 shared DEGs (633 + 157 + 190 + 270). After accounting for shared genes, the total DEGs associated with resistance and susceptibility were 9340 [(5731 + 2504) + (499 + 606)] or [(5731 + 2504) + (1791 + 2154)] − [(464 + 576 + 629 + 611 + 308 + 252)] (Figure S2).

Similarly, at 24 hpi, the DEGs in groups KF0h vs. KF24h and NN0h vs. NN24h) were 6133 (594 + 5017 + 328 + 194) and 1185 (132 + 80 + 180 + 793), respectively, with an additional 5019 shared DEGs (173 + 195 + 436 + 4215). The DEGs between the other two groups (KF0h vs. NN0h and KF24h vs. NN24h) were 1934 (724 + 594 + 180 + 436) and 1039 (384 + 132 + 328 + 195), with 1107 shared DEGs (660 + 194 + 173 + 80). After excluding shared DEGs, the total DEGs linked to resistance and susceptibility were 8426 [(6133 + 1185) + (724 + 384)] or [(6133 + 1185) + (1934 + 1039)] − [(594 + 436 + 195 + 328 + 132 + 180)] (Figure S3).

At 48 hpi, the DEGs in groups KF0h vs. KF48h and NN0h vs. NN48h were 1331 (287 + 773 + 173 + 98) and 4174 (1364 + 194 + 324 + 2292), respectively, with 2277 shared DEGs (226 + 560 + 1357 + 134). The DEGs between the other two groups (KF0h vs. NN0h and KF48h vs. NN48h) were 1535 (790 + 287 + 324 + 134) and 3002 (905 + 1364 + 173 + 560), with 1506 shared DEGs (988 + 98 + 226 + 194). After excluding shared genes, the total DEGs associated with resistance and susceptibility were 7200 [(1331 + 4174) + (790 + 905)] or [(1331 + 4174) + (1535 + 3002)] − [(287 + 134 + 506 + 173 + 324 + 1364)] (Figure S4).

The Venn diagram and group-wise calculation of DEGs indicate that nearly double the number of DEGs (5731) were present in KF cultivar compared to NN cultivar at 6 hpi, increasing to 6133 at 24 hpi. This suggests that most genes in active pathways originate from the resistant cultivar at the early stages of infection, peaking at 24 hpi. However, at 48 hpi, the total number of DEGs decreased in KF cultivar while increasing in NN cultivar. The reduced activity of genes in the resistant cultivar at 48 hpi compared to 6 and 24 hpi suggests that most genes related to the defense mechanism are active during the initial to 24 hpi of infection. Once the pathogen is controlled, these genes return to normal levels, whereas in the susceptible cultivar, the infection continues, involving more genes.

To further narrow down the most relevant DEGs involved in the resistance process, total DEGs at all three time points were assessed to identify shared DEGs. A total of 18,221 DEGs were identified, with 5154, 4034, and 3489 DEGs uniquely expressed at 6 hpi, 24 hpi, and 48 hpi, respectively. Additionally, 1833 DEGs were shared between 6 hpi and 24 hpi, 1152 between 6 hpi and 48 hpi, 1358 between 24 hpi and 48 hpi, and 1201 DEGs were shared across all three time points (Figure 2). These 1201 genes, expressed at all time points after SMV inoculation, may be involved in the resistance in both cultivars. To explore and compare the functions of these 1201 shared genes and the total 18,221 DEGs, GO enrichment analyses were performed separately.

3.3. GO Enrichment Analysis of the Total 18,221 DEGs and 1201 Joint DEGs Among Stages

GO enrichment analysis was divided into three major categories: "Biological process", "Cellular component", and "Molecular function", with 30 subcategories (Figure 3). Nearly half of the total DEGs were found in the "molecular function" category, primarily distributed among protein, DNA, ion, and ADP binding, which are mainly associated with regular growth and development activities. Additionally, 869 genes were active in protein kinase activity, 370 in oxidoreductase activity, and 270 in catalytic activity, which are related to diseases resistance signaling in plants. The other half of the DEGs were primarily found in the "biological process" category, distributed among protein phosphorylation, oxidation-reduction Process, transport, transmembrane transport, and signal transduction. Very few DEGs were found in the "Cellular Component" category. Protein phosphorylation and oxidation-reduction processes were the major categories, with 1637 DEGs associated with disease resistance, supported by transmembrane transport, transport and signal transduction, which contained 337, 191, and 166 DEGs, respectively.

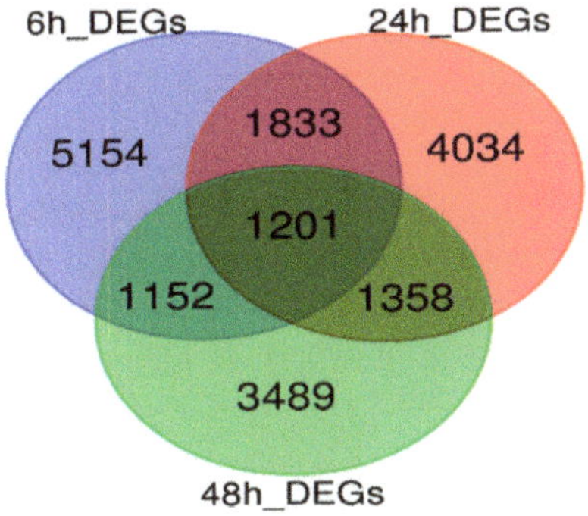

Figure 2. Venn diagram illustrating the unique and overlapping differentially expressed genes (DEGs) at 6, 24, and 48 h post-inoculation of SMV SC4 on soybean.

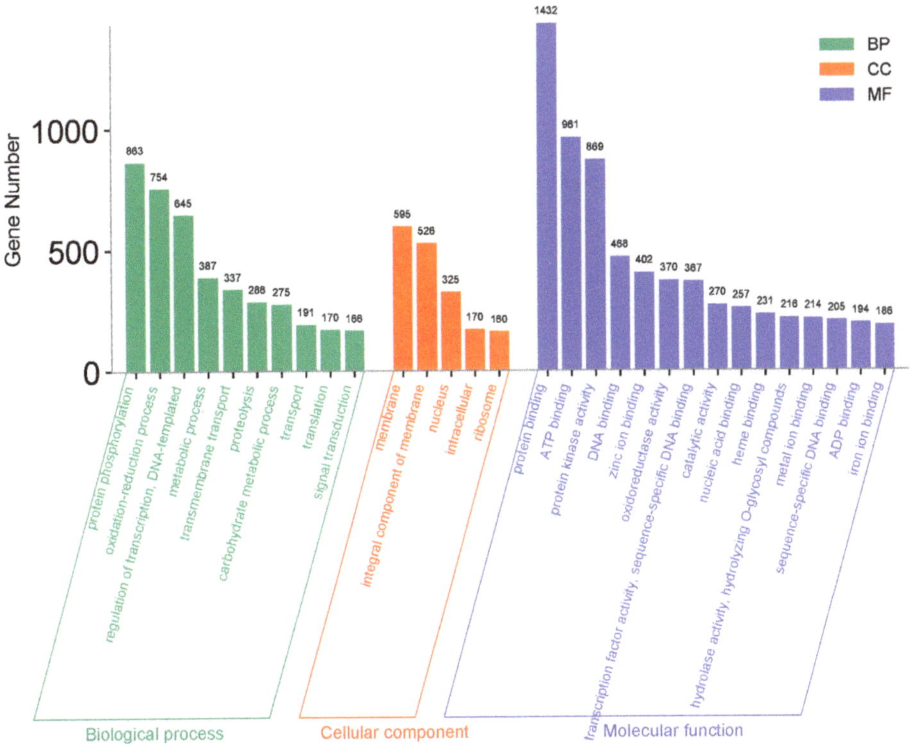

Figure 3. GO functional analysis histogram for totally identified 18,221 DEGs over four time points of soybean after SMV SC4 inoculation. Biological processes (BP) highlighted in dark cyan, cellular components (CC) in sienna, and molecular functions (MF) in steel blue.

To further narrow down the DEGs and identify the most likely mechanisms against SMV-SC4, a GO function analysis of 1201 joint DEGs was performed (Figure 4). The results revealed that a larger portion of the DEGs was associated with GO term "Biological process", with the majority falling into the subclass of protein phosphorylation, which included

90 DEGs. This was followed by oxidation-reduction processes (55 DEGs), regulation of transcription and DNA-templated processes (37 DEGs), transmembrane transport (30 DEGs), and signal transduction (27 DEGs). All these processes are related to disease resistance mechanisms. Additionally, transport, metal ion transport, and response to auxin contained 15, 12, and 8 DEGs, respectively, which are supportive activities for signal transduction. Other processes such as proteolysis, metabolic process, carbohydrate metabolic process, lipid metabolic process, protein folding, recognition of pollen, and microtubule-based movement also had DEGs, but these are typically involved in regular cellular functions. In the "Cellular component" category, a total of 103 genes were distributed across subclasses: integral components of the membrane (42 DEGs), Membrane (32 DEGs), Nucleus (12 DEGs), cell wall (9 DEGs), and kinesin complex (8 DEGs). Similarly, the "Molecular function" category had DEGs distributed across subclasses: protein binding (108 DEGs), protein kinase activity (92 DEGs), ATP binding (79 DEGs), oxidoreductase activity (52 DEGs), DNA binding (29 DEGs), zinc ion binding (27 DEGs), hydrolase activity (26 DEGs), catalytic activity (24 DEGs), transcription factor activity (24 DEGs), and ADP binding (23 DEGs) (Figure 4). The joint DEGs active during the resistance process in protein phosphorylation, Transmembrane transport, oxidation-reduction processes, and signal transduction in the biological process, supported by protein kinase activity and oxidoreductase activity in molecular function, contribute to resistance against SMV.

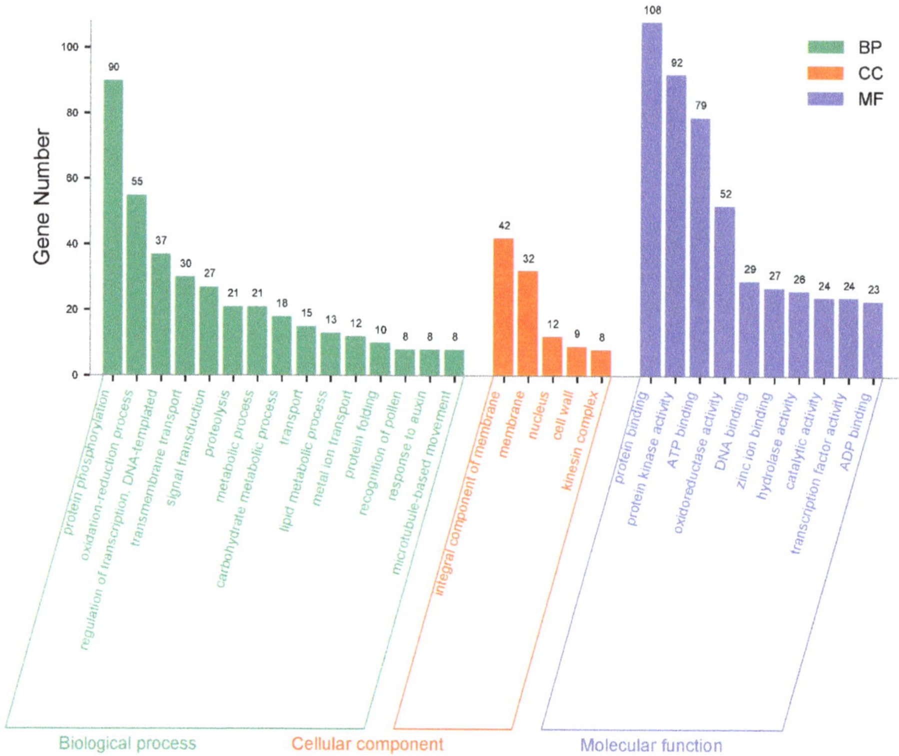

Figure 4. GO functional analysis histogram for 1201 jointly identified DEGs over four time points of soybean after SMV SC4 inoculation. Biological processes (BP) represented in dark cyan, cellular components (CC) in sienna, and molecular functions (MF) in steel blue.

3.4. Protein Phosphorylation, Signal Transduction and Protein Kinase Activity in Early Defense Response to SMV

Plants protect themselves from pathogen attacks by initiating a complex natural defense response. Plant-pathogen interactions trigger signal transduction cascades that mobilize defense mechanisms, ultimately leading to disease resistance responses [10]. Signal transduction, linked with protein phosphorylation and protein kinase activity, collectively activates the plant's defense system against pathogen attacks. Metal ions play a vital role in plant signal transmission as secondary messengers, and the accumulation and movement of secondary metabolites, facilitated by transmembrane transport, are also crucial for plant growth and defense functions [10–13].

In the Venn diagrams, as DEGs in KF cultivar increased at 6 and 24 hpi, genes active in signal transduction also increased during the early hours of pathogen attack. By 48 hpi, their number decreased in KF cultivar as the infection was controlled, but increased in NN cultivar as pathogen damage continued. Similarly, in the GO functional analysis of total DEGs and joint DEGs, signal transduction and protein kinase activity play a vital role in disease signaling response. Signal transduction serves as a channel through which other processes like transmembrane transport, metal ion transport, and response to auxin participate in the defense mechanism. Genes involved in metal ion transport and response to auxin were only observed when narrowing down to key genes. These genes are indirectly associated with the signal transduction process, as auxin is a key hormone that not only regulates growth and development but also modulates other hormones in signaling pathways. A few DEGs were also active in the cell wall under "Molecular function", which may be involved in the defense mechanism, as the cell wall is related to transport and communication functions (Figure 5).

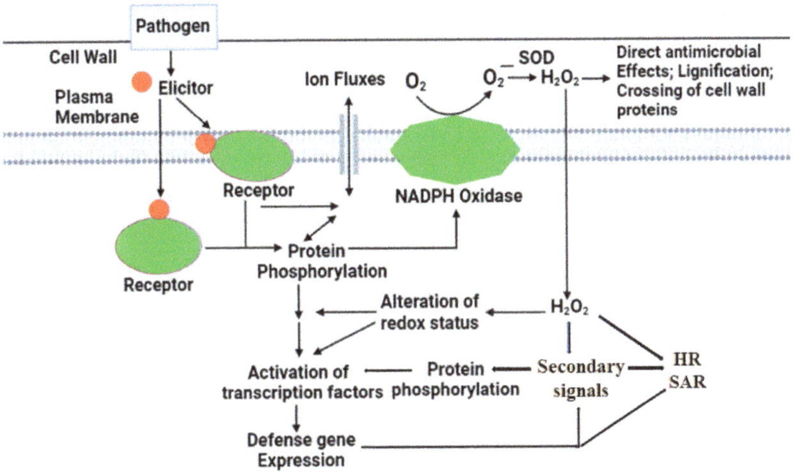

Figure 5. Model of signal transduction in response to pathogen (SMV SC4) attack on soybean plant. Illustrating the involvement of cell wall, plasma membrane, protein phosphorylation, redox reactions, and secondary signaling pathways in regulating resistance mechanisms.

3.5. Oxidation-Reduction and Oxidoreductase Activity in Resistance

When a pathogen invades plant tissue, the production of reactive oxygen species (ROS) is observed. Although ROS is produced during regular growth and development, its concentration increases during stress, causing redox imbalance. Therefore, it is essential to reduce excessive ROS levels to mitigate their toxic effects. However, moderate ROS levels act as signaling molecules that interact with other signaling pathways to activate the plant's defense system [22]. In the GO functional analysis histograms of 1201 joint DEGs and total 18,221 DEGs, genes were observed in these two terms under "Biological process" and "Molecular function". The second-highest number of active genes (754) in the total DEGs histogram and 55 genes in the joint DEGs histogram were associated with the oxidation-reduction process, indicating its significant contribution to the defense mechanism of KF cultivar. Similarly, 370 and 52 genes in the total and joint DEGs histograms, respectively, were active in oxidoreductase activity (Figures 3 and 4). Genes active in these redox activities are involved in removing high concentrations of ROS generated in response to SMV attack. Another important term under molecular function is catalytic activity.

Many catalase enzymes in plants convert H_2O_2 into water and oxygen when plants face environmental stresses [23]. These enzymes are present at major sites where H_2O_2 is produced, such as cytosol, mitochondria, and chloroplasts. Various catalases play versatile role in the crop plant system. The conversion of H_2O_2 by catalases within specific cells or organelles at specific times during growth and development directly or indirectly affects signal transduction in plants. In summary, signal transduction is the key process through which all other defense mechanisms are linked to produce the final output.

3.6. Genes Related to Signal Transduction and Response to Auxin in Soybean

A considerable number of resistance genes in plants have been identified and classified into eight fundamental classes, which can be further divided into various super families based on their functional protein domains. Most of the genes cloned to date belong to the NBS-LRR kinase super families, which provide resistance against a wide range of pathogens, including viruses, bacteria, and pests [24]. In the GO enrichment analysis of the current study, 27 genes related to signal transduction (Table S5) were identified, most of which belong to specific disease-resistant groups and super families, such as Casitas B-lineage lymphoma (CBL)- interacting serine/threonine-protein kinase. Five genes from this family showed deferential expression in both cultivars at different time points. *Glyma13G370000*, *Glyma07G023500* and *Glyma08G218400* were upregulated in KF cultivar, while *Glyma03G260200* and *Glyma18G212200* were upregulated in NN (Figure 6A). Eight genes with predicted disease resistance functions showed expression in the resistant cultivar at almost all three points, except for *Glyma19G054900*, which was expressed in the susceptible cultivar (Figure 6B). A few genes were also found with predicted functions related to tobacco mosaic virus (TMV) resistance. Another group of eight genes was found in the "Response to auxin" subclass (Table S6). Their differential expression, calculated using FPKM values from RNA-seq data, showed that *Glyma04G006300*, *Glyma07G043500*, *Glyma09G219900*, and *Glyma16G011900* were more highly expressed in KF cultivar, while *Glyma09G220400*, *Glyma12G034800*, *Glyma12G035800*, and *Glyma19G258800* were more highly expressed in NN cultivar (Figure 6C).

In conclusion, the transcriptomic study revealed that genes involved in signal transduction directly, and genes active in other categories related to plant stress response, such as protein phosphorylation, protein kinase activity, oxidation-reduction, oxidoreductase activity, catalytic activity, metal ion transport, and response to auxin, indirectly contribute to the resistance mechanism in KF cultivar through signal transduction.

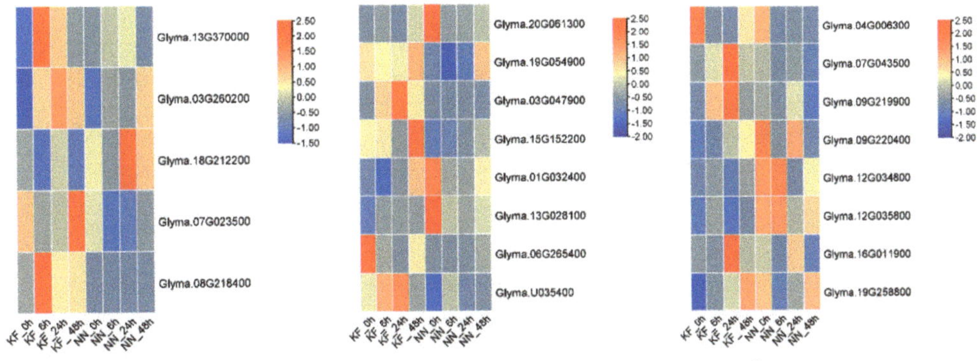

Figure 6. Heatmap generated from FPKM values derived from RNA-seq data of interaction of soybean with SMV SC4 strain. Depicting the expression profiles of genes involved in: (**A**) (CBL)-interacting serine/threonine-protein kinase, (**B**) disease resistance proteins, (**C**) response to auxin.

4. Discussion

4.1. Transcriptome Study on the Defense Mechanism Against Soybean Mosaic Virus

In the plant defense system against viruses, there are often different defense mechanisms. One common defense mechanism, part of the innate immune system, involves pattern-triggered immunity (PTI), where recognition factors are usually localized on the cell surface [8,9,25]. Another, more powerful defense mechanism is effector-triggered immunity (ETI), which is often specific to certain strains and can mediate rapid necrosis of virus-infected areas, preventing further spread. The genes inducing this response are often referred to as resistance (R) genes.

The activation of PTI and ETI involves different types of receptors—Pattern Recognition Receptors (PRRs) for PTI and NOD-like Receptors (NLRs) for ETI—and different early signal transduction processes. However, there is significant overlap in downstream outputs, such as Calcium Flux, Reactive Oxygen Species (ROS) Burst, Transcriptional Reprograming, and Phytohormone Signaling [26]. In the present study, we identified genes for example *Glyma13G370000* (CBL-interacting serine/threonine-protein kinase 5), *Glyma08G218400* (CBL-interacting serine/threonine-protein kinase 5-like), *Glyma03G047900* (Disease resistance protein RML1A-like), *GlymaU035400* (R1protein) (Table S5) in sub-categories signal transduction with their higher expression in the resistant cultivar compare to susceptible cultivar (Figure 6A,B). It can be speculated that these genes may be involved in ETI mechanism removal ROS by translating into specific amino acids to encounter the attack of specific SMV strain, in our case SC4on soybean cultivar Kefeng-1 (KF) which also give signals to other genes to cooperate with them to get rid of the infection at early stage. In short, ETI is the main defense system which makes KF cultivar resistant against SC4.

4.2. Resistance to SMV Is Mediated by a Network of Processes and Signaling Pathways

Soybean plants defend against SMV invasion by initiating a multicomponent defense response. Proteins coded by the disease resistance genes recognize specific SMV strain attacks and bind to specific pathogen (virus)-derived virulence (Avr) proteins. This initiates an internal signaling cascade, activating the host plant's defense arsenal, leading to localized cell death at the infection site and systemic acquired resistance (SAR) across the plant, preventing the further infection spread and causing a hypersensitive response (HR) [24]. Interactions between pathogens and plants, signal initiation, and signal transduction exhibit resistance in many plants against various pathogens [27]. In the current study, 27 genes active in signal transduction under the biological process were identified as involved in resistance against virus attacks. Most of these genes have predicted disease resistance functions and belong to groups specialized for disease resistance, such as CBL-interacting serine/threonine-protein kinase, and were more highly expressed in the resistant cultivar compared to the susceptible cultivar in the RNA seq dataset. Previously, genes with specific conserved domains belonging to the NBS–LRR or LRR kinase super families have been cloned for resistance against bacteria, nematodes, viruses, and fungi [27–30].

In addition to signal transduction, DEGs were active in Oxidation-reduction, Protein kinase activity, oxidoreductase activity, metal ion transport, and transmembrane transport, all of which are part of the resistance mechanism. Protein kinase activity involves the removal of a phosphate group from an ATP molecule and its attachment to an acceptor molecule with a free hydroxyl group. Kinase enzymes in this activity regulate many cellular functions, including signal transduction, cell division, differentiation, growth, and development. Metal ions play a vital role in plant signal transmission as secondary messengers, and the accumulation and movement of secondary metabolites, facilitated by transmembrane transport, are also crucial for plant growth and defense functions [12,13].

Moreover, 55 genes in oxidation-reduction and 52 in oxidoreductase activity were expressed in the GO enrichment analysis histogram. Oxidation-reduction induced by pathogens is usually accompanied by a large amount of (ROS), which signals successful infection identification and activates plant defenses [31]. Production of ROS during ETI is essential for eliciting a more robust defense response. A defining characteristic of ETI is an intense and sustained burst of ROS. When high levels of ROS reached, they are associated with the hypersensitive response. In the initial stage, an oxidative burst usually occurs at the virus invasion site, sending signals to activate the entire defense system. Subsequently, ROS accumulate to higher levels, playing an important signaling role in activating plant defenses and triggering hypersensitive cell death (HR) [32]. These findings demonstrate the role of oxidation-reduction, signal transduction, and ROS production in the initial resistance of KF cultivar.

4.3. Role of Hormones in the Defense Mechanism

Plant hormones significantly influence the soybean plant defense system. Jasmonic acid, ethylene, Salicylic acid and auxins are the essential phytohormones that regulate various aspects of plant immunity, including pathogen (virus) recognition, signal transduction, and initiation of genes related to defense. Additionally, plant growth regulating hormones such as cytokinin and gibberellic acid modulate immune responses through intricate regulatory networks [33]. Auxin, naturally existing as IAA, is synthesized from pyruvic acid through various pathways. It rapidly modifies the expression of many genes by eliminating the inhibitory function of AUX/IAA proteins. Its role as a fundamental part of the hormone signaling system in plants under biotic stress has been described by many researchers [34–38]. For example, auxin-sensitive factors in tomatoes play an intermediate role between auxin action and biotic response. In two viral diseases of tomato—Tomato spotted wilt virus (TSWV) and Tomato brown rugose fruit virus (ToBRFV)—auxin plays a signaling role in the defense system [39,40].

In the current transcriptomic study on SMV resistance, eight genes expressed in the "Response to auxin" subclass under the "Biological Process" category may be involved in the resistance mechanism of KF cultivar, as their differential expression was higher in KF cultivar. These genes were active during the middle stage of infection in the GO term enrichment analysis, indicating auxin's involvement in the defense system against the virus. The movement of auxin into, out of, or within cells, facilitated by transporter or pores, and the auxin-mediated signaling pathway, are crucial for plant growth and development.

However, auxin's involvement in defense responses has also been suggested. Ghanashyam and Jain (2009) studied the function of auxin-inducible genes under biotic stress conditions and observed upregulated differential expression in rice [38]. Auxin is involved in the TOR (target of rapamycin) signaling pathway, which responds to environmental stress and acts as a signal inducer in translation in plants [41,42].

Recent transcriptomic studies on soybean mosaic virus (SMV) resistance across different resistant cultivars and strains indicate that three major pathways are involved in the resistance process: plant-pathogen interaction pathways, hormone signal transduction pathways, and mitogen-activated protein kinase (MAPK) signaling pathways. These pathways are regulated by clusters of genes [43,44]. A study by Zhu et al. [44] investigated SMV strain N1 resistance in the soybean cultivar Dongnong 93-046 using transcriptomic analysis. They identified 41,189 differentially expressed genes (DEGs) associated with the resistance mechanism, with 9196 DEGs showing FPKM values above 10. Their study highlighted 894 core genes involved in plant-pathogen interaction, linoleic acid metabolism, and plant hormone signaling transduction.

In our current study, we analyzed transcriptomic responses to SMV strain SC4 using two soybean cultivars with contrasting resistance profiles: Kefeng-1 (resistant) and NN1138-2 (susceptible). We identified 18,221 DEGs across both varieties, with 1201 core DEGs expressed at 0, 6, 24, and 48 h after SMV inoculation. These core DEGs were consistently active throughout the resistance process, from Virus recognition to defense activation. Some genes were preferentially expressed in the resistant cultivar (KF) during early infection stages but remained inactive in the susceptible cultivar (NN). Conversely, certain genes were highly upregulated in the susceptible cultivar at later infection stages but not in the resistant cultivar, providing valuable insights into the genetic basis of resistance and susceptibility. Additionally, we identified 27 genes involved in signal transduction, most of which contained specific domains associated with disease resistance, including tobacco mosaic virus (TMV) resistance domains. These findings suggest that our study provides a more comprehensive view of SMV resistance mechanisms compared to previous studies.

5. Conclusions

Signal transduction, linked with protein phosphorylation and protein kinase activity, collectively activates the plant's defense system against pathogen attacks. Metal ions play a vital role in plant signal transmission as secondary messengers, and the accumulation and movement of secondary metabolites, facilitated by transmembrane transport, are also crucial for plant growth and defense functions. Many catalase enzymes in plants convert H_2O_2 into water and oxygen when plants face environmental stresses. These enzymes are present at major sites where H_2O_2 is produced, such as cytosol, mitochondria, and chloroplasts. Various catalases play versatile role in the crop plant system. The conversion of H_2O_2 by catalases within specific cells or organelles at specific times during growth and development directly or indirectly affects signal transduction in plants. In summary, signal transduction is the key process through which all other defense mechanisms are linked to produce the final output. Our transcriptomic study reveals the similar results; genes involved in signal transduction directly, and genes active in other categories related to plant stress response such as protein phosphorylation, protein kinase activity, oxidation–reduction, oxidoreductase activity, catalytic activity, metal ion transport, and response to auxin indirectly contribute to the resistance mechanism in KF cultivar through signal transduction.

Supplementary Materials: The following supporting information can be downloaded at: https://www.mdpi.com/article/10.3390/agronomy15040906/s1, Figure S1. Principal component analysis (PCA) plot; Figure S2. Venn diagram of Differentially expressed genes (DEGs) in resistant and susceptible cultivars at 6 h post-inoculation (hpi), across four comparison groups, including shared DEGs; Figure S3. Venn diagram of Differentially expressed genes (DEGs) in resistant and susceptible cultivars at 24 h post-inoculation (hpi), across four comparison groups, including shared DEGs; Figure S4. Venn diagram of Differentially expressed genes (DEGs) in resistant and susceptible cultivars at 48 h post-inoculation (hpi), across four comparison groups, including shared DEGs; Table S1. DAS-ELISA data of Kefeng-1 and NN-1138 inoculated with SMV-SC4; Table S2. The average coverage of clean data across 24 samples reached; Table S3. The GC content after filtration ranged from 41.95% to 44.50%, and the Q30 score for each sample exceeded 93.24%; Table S4. The average coverage across all samples; Table S5. Twenty-seven genes associated with signal transduction; Table S6. Eight genes related to "Response to auxin" from GO.

Author Contributions: J.G. and K.L. planned the experiment, M.M.R. and H.J. executed the experiment and drafted the whole manuscript, while S.G. assisted in execution and write up. All authors have read and agreed to the published version of the manuscript.

Funding: This work was supported by grants from the National Key Research and Development Program of China (2024YFD1201404), China Agriculture Research System of MOF and MARA (No. CARS-04), Jiangsu Collaborative Innovation Center for Modern Crop Production (JCIC-MCP), and Collaborative Innovation Center for Modern Crop Production co-sponsored by Province and Ministry (CIC-MCP).

Data Availability Statement: The original contributions presented in this study are included in the article/Supplementary Materials. Further inquiries can be directed to the corresponding author(s).

Conflicts of Interest: All authors declared that they have no conflicts of interest.

References

1. Adams, M.J.; Antoniw, J.F.; Beaudoin, F. Overview and Analysis of the Polyprotein Cleavage Sites in the Family Potyviridae. *Mol. Plant Pathol.* **2005**, *6*, 471–487. [CrossRef] [PubMed]
2. Li, K.; Yang, Q.H.; Zhi, H.J.; Gai, J.Y. Identification and Distribution of Soybean Mosaic Virus Strains in Southern China. *Plant Dis.* **2010**, *94*, 351–357. [CrossRef] [PubMed]
3. Wang, D.; Tian, Z.; Li, K.; Li, H.; Huang, Z.; Hu, G.; Zhi, H. Identification and Variation Analysis of Soybean Mosaic Virus Strains in Shandong, Henan and Anhui Provinces of China. *Soybean Sci.* **2013**, *32*, 806–809.
4. Cho, E.K.; Goodman, R.M. Strains of Soybean Mosaic Virus: Classification Based on Virulence in Resistant Soybean Cultivars. *Phytopathology* **1979**, *69*, 467–470.
5. Cho, E.K.; Goodman, R.M. Evaluation of Resistance in Soybeans to Soybean Mosaic Virus Strains 1. *Crop Sci.* **1982**, *22*, 1133–1136. [CrossRef]
6. Takahashi, K.; Tanaka, T.; Iida, W.; Tsuda, Y. Studies on Virus Diseases and Causal Viruses of Soybean in Japan. *Bull. Tohoku Natl. Agric. Exp. Stn.* **1980**, *62*, 1–130.
7. Yu, X.-Q.; Niu, H.-Q.; Liu, C.; Wang, H.-L.; Yin, W.; Xia, X. PTI-ETI Synergistic Signal Mechanisms in Plant Immunity. *Plant Biotechnol. J.* **2024**, *22*, 2113–2128. [CrossRef]
8. Lin, H.; Wang, M.; Chen, Y.; Nomura, K.; Hui, S.; Gui, J.; Zhang, X.; Wu, Y.; Liu, J.; Li, Q.; et al. An MKP-MAPK Protein Phosphorylation Cascade Controls Vascular Immunity in Plants. *Sci. Adv.* **2022**, *8*, eabg8723. [CrossRef]
9. Li, P.; Zhao, L.; Qi, F.; Htwe, N.M.P.S.; Li, Q.; Zhang, D.; Lin, F.; Shang-Guan, K.; Liang, Y. The Receptor-like Cytoplasmic Kinase RIPK Regulates Broad-Spectrum ROS Signaling in Multiple Layers of Plant Immune System. *Mol. Plant* **2021**, *14*, 1652–1667. [CrossRef]
10. Ding, L.N.; Li, Y.T.; Wu, Y.Z.; Li, T.; Geng, R.; Cao, J.; Zhang, W.; Tan, X.L. Plant Disease Resistance-Related Signaling Pathways: Recent Progress and Future Prospects. *Int. J. Mol. Sci.* **2022**, *23*, 16200. [CrossRef]
11. Liu, J.Z.; Lam, H.M. Signal Transduction Pathways in Plants for Resistance against Pathogens. *Int. J. Mol. Sci.* **2019**, *20*, 2335. [CrossRef] [PubMed]
12. Nogia, P.; Pati, P.K. Plant Secondary Metabolite Transporters: Diversity, Functionality, and Their Modulation. *Front. Plant Sci.* **2021**, *12*, 758202. [CrossRef] [PubMed]

13. Shi, X.; Bao, J.; Lu, X.; Ma, L.; Zhao, Y.; Lan, S.; Cao, J.; Ma, S.; Li, S. The Mechanism of Ca^{2+} Signal Transduction in Plants Responding to Abiotic Stresses. *Environ. Exp. Bot.* **2023**, *216*, 105514. [CrossRef]
14. Mazzoni-Putman, S.M.; Brumos, J.; Zhao, C.; Alonso, J.M.; Stepanova, A.N. Auxin Interactions with Other Hormones in Plant Development. Cold Spring Harb. *Perspect. Biol.* **2021**, *13*, a039990. [CrossRef]
15. Gomes, G.L.B.; Scortecci, K.C. Auxin and Its Role in Plant Development: Structure, Signalling, Regulation and Response Mechanisms. *Plant Biol.* **2021**, *23*, 894–904. [CrossRef]
16. Wang, D.; Ma, Y.; Liu, N.; Yang, Z.; Zheng, G.; Zhi, H. Fine Mapping and Identification of the Soybean R SC4 Resistance Candidate Gene to Soybean Mosaic Virus. *Plant Breed.* **2011**, *130*, 653–659. [CrossRef]
17. Rui, R.; Liu, S.; Karthikeyan, A.; Wang, T.; Niu, H.; Yin, J.; Yang, Y.; Wang, L.; Yang, Q.; Zhi, H.; et al. Fine-Mapping and Identification of a Novel Locus Rsc15 Underlying Soybean Resistance to Soybean Mosaic Virus. *Theor. Appl. Genet.* **2017**, *130*, 2395–2410. [CrossRef]
18. Zheng, G.; Yang, Y.; MA, Y.; Yang, X.; Chen, S.; Ren, R.; Wang, D.; Yang, Z.; Zhi, H. Fine Mapping and Candidate Gene Analysis of Resistance Gene R to Soybean Mosaic Virus in Qihuang 1. *J. Integr. Agric.* **2014**, *13*, 2608–2615. [CrossRef]
19. Love, M.I.; Huber, W.; Anders, S. Moderated Estimation of Fold Change and Dispersion for RNA-Seq Data with DESeq2. *Genome Biol.* **2014**, *15*, 550. [CrossRef]
20. Robinson, M.D.; McCarthy, D.J.; Smyth, G.K. EdgeR: A Bioconductor Package for Differential Expression Analysis of Digital Gene Expression Data. *Bioinformatics* **2010**, *26*, 139–140. [CrossRef]
21. Kanehisa, M. KEGG: Kyoto Encyclopedia of Genes and Genomes. *Nucleic Acids Res.* **2000**, *28*, 27–30. [CrossRef] [PubMed]
22. Suman, S.; Bagal, D.; Jain, D.; Singh, R.; Singh, I.K.; Singh, A. Biotic Stresses on Plants: Reactive Oxygen Species Generation and Antioxidant Mechanism. In *Frontiers in Plant-Soil Interaction*; Aftab, T., Hakeem, R.K., Eds.; Academic Press: London, UK, 2021; pp. 381–411.
23. Sharma, I.; Ahmad, P. Catalase. In *Oxidative Damage to Plants*; Academic Press: London, UK, 2014; pp. 131–148.
24. Padder, B.A. Plant Disease Resistance Genes: From Perception to Signal Transduction. In *Plant Signaling: Understanding the Molecular Crosstalk*; Hakeem, K., Rehman, R., Tahir, I., Eds.; Springer: New Delhi, India, 2014; pp. 345–354.
25. Bigeard, J.; Colcombet, J.; Hirt, H. Signaling Mechanisms in Pattern-Triggered Immunity (PTI). *Mol. Plant* **2015**, *8*, 521–539. [CrossRef] [PubMed]
26. Yuan, M.; Ngou, B.P.M.; Ding, P.; Xin, X.F. PTI-ETI Crosstalk: An Integrative View of Plant Immunity. *Curr. Opin. Plant Biol.* **2021**, *62*, 102030. [CrossRef]
27. Ijaz, S.; Haq, I.U.; Babar, M.; Nasir, B. Disease Resistance Genes' Identification, Cloning, and Characterization in Plants. In *Cereal Diseases: Nanobiotechnological Approaches for Diagnosis and Management*; Springer Nature: Singapore, 2022; pp. 249–269.
28. Ameline Torregrosa, C.; Wang, B.B.; O'Bleness, M.S.; Deshpande, S.; Zhu, H.; Roe, B.; Young, N.D.; Cannon, S.B. Identification and Characterization of Nucleotide-Binding Site-Leucine-Rich Repeat Genes in the Model Plant Medicago Truncatula. *Plant Physiol.* **2008**, *146*, 5–21. [CrossRef]
29. Bai, S.; Liu, J.; Chang, C.; Zhang, L.; Maekawa, T.; Wang, Q.; Xiao, W.; Liu, Y.; Chai, J.; Takken, F.L.W.; et al. Structure-Function Analysis of Barley NLR Immune Receptor MLA10 Reveals Its Cell Compartment Specific Activity in Cell Death and Disease Resistance. *PLoS Pathog.* **2012**, *8*, e1002752. [CrossRef]
30. Cao, Y.; Mo, W.; Li, Y.; Xiong, Y.; Wang, H.; Zhang, Y.; Lin, M.; Zhang, L.; Li, X. Functional Characterization of NBS-LRR Genes Reveals an NBS-LRR Gene That Mediates Resistance against Fusarium Wilt. *BMC Biol.* **2024**, *22*, 45. [CrossRef]
31. Liu, C.; Liu, Q.; Mou, Z. Redox Signaling and Oxidative Stress in Systemic Acquired Resistance. *J. Exp. Bot.* **2024**, *75*, 4535–4548. [CrossRef]
32. Weralupitiya, C.; Eccersall, S.; Meisrimler, C.-N. Shared Signals, Different Fates: Calcium and ROS in Plant PRR and NLR Immunity. *Cell Rep.* **2024**, *43*, 114910. [CrossRef]
33. Shafqat, A.; Abbas, S.; Ambreen, M.; Siddiqa Bhatti, A.; Kausar, H.; Gull, T. Exploring the Vital Role of Phytohormones and Plant Growth Regulators in Orchestrating Plant Immunity. *Physiol. Mol. Plant Pathol.* **2024**, *133*, 102359. [CrossRef]
34. Kalsi, H.S.; Karkhanis, A.A.; Natarajan, B.; Bhide, A.J.; Banerjee, A.K. AUXIN RESPONSE FACTOR 16 (StARF16) Regulates Defense Gene StNPR1 upon Infection with Necrotrophic Pathogen in Potato. *Plant Mol. Biol.* **2022**, *109*, 13–28. [CrossRef]
35. Liu, L.; Yahaya, B.S.; Li, J.; Wu, F. Enigmatic Role of Auxin Response Factors in Plant Growth and Stress Tolerance. *Front. Plant Sci.* **2024**, *15*, 1398818. [CrossRef]
36. Bouzroud, S.; Gouiaa, S.; Hu, N.; Bernadac, A.; Mila, I.; Bendaou, N.; Smouni, A.; Bouzayen, M.; Zouine, M. Auxin Response Factors (ARFs) Are Potential Mediators of Auxin Action in Tomato Response to Biotic and Abiotic Stress (*Solanum lycopersicum*). *PLoS ONE* **2018**, *13*, e0193517. [CrossRef]
37. Ghanashyam, C.; Jain, M. Role of Auxin-Responsive Genes in Biotic Stress Responses. *Plant Signal. Behav.* **2009**, *4*, 846–848. [CrossRef] [PubMed]
38. Guilfoyle, T.J.; Hagen, G. Auxin Response Factors. *Curr. Opin. Plant Biol.* **2007**, *10*, 453–460. [CrossRef]

39. Werghi, S.; Herrero, F.A.; Fakhfakh, H.; Gorsane, F. Auxin Drives Tomato Spotted Wilt Virus (TSWV) Resistance through Epigenetic Regulation of Auxin Response Factor ARF8 Expression in Tomato. *Gene* **2021**, *804*, 145905.
40. Vaisman, M.; Hak, H.; Arazi, T.; Spiegelman, Z. The Impact of Tobamovirus Infection on Root Development Involves Induction of Auxin Response Factor 10a in Tomato. *Plant Cell Physiol.* **2023**, *63*, 1980–1993. [CrossRef]
41. Schepetilnikov, M.; Ryabova, L.A. Auxin Signaling in Regulation of Plant Translation Reinitiation. *Front. Plant Sci.* **2017**, *8*, 1014. [CrossRef]
42. Schepetilnikov, M.; Ryabova, L.A. Recent Discoveries on the Role of TOR (Target of Rapamycin) Signaling in Translation in Plants. *Plant Physiol.* **2018**, *176*, 1095–1105. [CrossRef]
43. Niu, J.; Zhao, J.; Guo, Q.; Zhang, H.; Yue, A.; Zhao, J.; Yin, C.; Wang, M.; Du, W. WGCNA Reveals Hub Genes and Key Gene Regulatory Pathways of the Response of Soybean to Infection by Soybean Mosaic Virus. *Genes* **2024**, *15*, 566. [CrossRef]
44. Zhu, H.; Li, R.; Fang, Y.; Zhao, X.; Teng, W.; Li, H.; Han, Y. Weighted Gene Co-Expression Network Analysis Uncovers Critical Genes and Pathways Involved in Soybean Response to Soybean Mosaic Virus. *Agronomy* **2024**, *14*, 2455. [CrossRef]

Disclaimer/Publisher's Note: The statements, opinions and data contained in all publications are solely those of the individual author(s) and contributor(s) and not of MDPI and/or the editor(s). MDPI and/or the editor(s) disclaim responsibility for any injury to people or property resulting from any ideas, methods, instructions or products referred to in the content.

agronomy

MDPI

Article

Identification of QTLs and Candidate Genes for Red Crown Rot Resistance in Two Recombinant Inbred Line Populations of Soybean [*Glycine max* (L.) Merr.]

Augustine Antwi-Boasiako [1,2], Chunting Zhang [1], Aisha Almakas [1], Jiale Liu [1], Shihao Jia [1], Na Guo [1], Changjun Chen [3], Tuanjie Zhao [1,*] and Jianying Feng [1,*]

1 Key Laboratory of Biology and Genetics Improvement of Soybean, Ministry of Agriculture, Zhongshan Biological Breeding Laboratory (ZSBBL), National Innovation Platform for Soybean Breeding and Industry-Education Integration, State Key Laboratory of Crop Genetics & Germplasm Enhancement and Utilization, College of Agriculture, Nanjing Agricultural University, Nanjing 210095, China; bbbantwi@yahoo.com (A.A.-B.); 2018801229@njau.edu.cn (C.Z.); aisha90@126.com (A.A.); ljl15256082024@163.com (J.L.); chu1feng@163.com (S.J.); guona@njau.edu.cn (N.G.)

2 Council for Scientific and Industrial Research-Crops Research Institute (CSIR-CRI), Fumesua, Kumasi P.O. Box 3785, Ghana

3 College of Plant Protection, Nanjing Agricultural University, Nanjing 210095, China; changjun-chen@njau.edu.cn

* Correspondence: tjzhao@njau.edu.cn (T.Z.); fengjianying@njau.edu.cn (J.F.)

Abstract: With the rapid emergence and distribution of red crown rot (RCR) across countries, durable sources of resistance against *Calonectria ilicicola* in soybean [*Glycine max* (L.) Merrill] is required to control the disease. We employed two RIL populations for the experiment. We identified 15 and 14 QTLs associated with RCR resistance in ZM6 and MN populations, respectively, totaling 29 QTLs. Six and eight QTLs had phenotypic variation above 10% in ZM6 and MN populations, respectively. We identified six (6) "QTL hotspots" for resistance to RCR from the ZM6 and MN RIL populations on chromosomes 1, 7, 10, 11, 13, and 18. Gene annotations, gene ontology enhancement, and RNA sequencing assessment detected 23 genes located within six "QTL Hotspots" as potential candidate genes that could govern RCR resistance in soybeans. Our data will generally assist breeders in rapidly and effectively incorporating RCR resistance into high-yielding accession through marker-assisted selection.

Keywords: soybean; red crown rot; resistance; candidate gene; QTL mapping

1. Introduction

Soybean [*Glycine max* (L.) Merr.] is a significant legume crop globally because of its high nutritional values [1]. In China, soybeans are essential legumes for vegetable oil and animal feed are concerned, ranking fourth in its production globally in 2022 (FAOSTAT, https://www.fao.org/faostat, accessed on 31 December 2023). Economic analysis of soybean production shows a high return contributing to the economic and social impact, especially on smallholder farmers [2,3]. Hence, some researchers suggest the need to establish interventions to protect China's economic, environmental, and sustainable practices for soybean production [4].

Unfortunately, soybean yield parameters are usually confronted with a number of stresses ranging from biotic (pests, diseases, and weeds) to abiotic (drought, salinity, and heavy metals), regulating the quantity and quality of soybean seeds and their nutritional components [5,6]. One of the key constraints impeding soybean production arises from fungal pathogens, which cause 26 different diseases [7]. Particularly in China, *Calonectria ilicicola* Boedijn and Reitsma are among the economically important pathogens as they infect several plants such as soybean, blueberry, stout camphor tree, groundnuts, alfalfa, sassafras, and ginger [8–10]. Now, soybean red crown rot (RCR) disease caused by *C. ilicicola* is a threat to major soybean regions, especially in southern Asia. Soybean RCR negatively impacts the quantity of seed in a pod, its seed weight, and quality [11–13], thus limiting yield by 25–30% [14–16], and it is projected to reach 50–100% disease incidence on susceptible cultivars [11,14,17]. Morphologically, strains of *C. ilicicola* are characterized by anamorphic structures (macroconidia, vesicles, conidiophores, and conidia) and teleomorphic structures (perithecia, asci, and ascospores) [18,19]. *C. ilicicola* has also been identified via internal transcribed spacer (ITS) regions, calmodulin, histone3, and translation elongation factor 1-α [19,20]. Recently, novel *Calonectria* species and new hosts have been recorded in Southern China [21]. The fungus *C. ilicicola* infects the roots of soybeans, resulting in root rot advancing to pencil-like roots, damping off of younger seedlings, and untimely defoliation [22–24]. Currently, neither an effective fungicide nor a resistant cultivar controls RCR; however, agronomic practices are adopted to manage the disease [15,25,26]. The generation of RCR-resistant varieties continues as the foundation for disease management.

Research on soybean diseases is of growing interest owing to its consequences. Hence, several researchers have investigated the inherent features of disease traits via quantitative trait loci (QTL) through linkage analysis and quantitative trait nucleotides (QTNs) via genome-wide association studies (GWAS) [27,28]. The disease traits in soybeans are associated with the growing season, isolates and their diversity, and the soybean cultivar [7,29]. Thus, disease traits are complex and are controlled by several genes [30,31]. Therefore, it is crucial to identify key genes that could provide resistance to soybean crops. Molecular breeding provides an inclusive comprehension of the genetic basis underlying disease traits. Previous researchers have identified QTLs/QTNs utilizing either GWAS or linkage mapping analysis to unravel the genetic basis of soybean disease traits [31,32]. The success of QTL/QTN identification largely depends on the mapping strategies used. However, few soybean germplasm accessions show RCR resistance [25]. Hence, the application of GWAS becomes comparatively difficult as rare variants could be omitted due to the minor allele frequency (MAF) cut-off, leading to skewed phenotypic distribution toward the susceptibility genotypes [33]. On the contrary, biparental population RCR-resistance sources could be studied by utilizing the genetic background of the resistant parent. Linkage mapping analysis has proven fruitful in detecting markers linked with various soybean disease traits. For instance, linkage mapping has successfully been applied in soybean diseases such as rust [34,35], bacterial leaf pustule [32,36], mold white [37,38], *Phytophthora* root rot [39–41], and frog eye leafspot [42,43], resulting in detecting QTLs. Conversely, bi-parental mapping populations have yet to be employed in the search for QTL related to RCR resistance in soybeans. Therefore, the current investigation used a high-density linkage map of two RIL populations, namely ZM6 and MN, and evaluated two different times in an environment to identify QTLs and mine for potential candidate genes for RCR resistance in soybeans. We sought to identify QTLs and potential genes governing regulating RCR resistance in soybeans. This study deepens our understanding of the genetics of RCR resistance in soybeans, thus contributing to the identification, development, and application of markers related to RCR resistance for marker-assisted selection toward developing resistant cultivars.

2. Materials and Methods

2.1. Plant Material

After preliminary RCR resistance evaluation of parents of six recombinant inbred line (RIL) populations (LM6, ZM6, M6T, M6T, MN, ZN), the parents of ZM6 and MN RIL populations showed significant resistance differences, so these two populations were selected for mapping RCR resistance QTL. Both RIL populations were developed through the single-seed descent (SSD) method by advancing to the seventh generation. The ZM6, consisting of 122 lines, was developed by crossing a disease-sensitive variety, Zhengyang, and a disease-tolerant cultivar, Meng8206 to *C. ilicicola*. Also, MN, consisting of 98 lines, was developed using the soybean cvs. M8180 and NN1138-2, which are resistant and susceptible to *C. ilicicola* strain Y62, respectively. All the RIL population accessions were sourced from the Soybean Germplasm Gene Bank, located at the National Centre for Soybean Improvement (Ministry of Agriculture), Nanjing Agricultural University, Nanjing, China.

2.2. Pathogen Culture and Inoculation

The *C. ilicicola* strain Y62 was provided by the College of Plant Protection, Nanjing Agricultural University, Nanjing, China. The fungi were maintained on V8 media plates (90 mm) at 26 °C for short-term use by subculturing. Pathogen inoculation was carried out following the protocol of Jiang et al. (2020) [25].

2.3. Growth Conditions, Experimental Conditions, and Design

The planting media consisted of a mixture of vermiculite and nutritive soil in equal volumes (1:1, v/v). The media was autoclaved. After cooling at ambient temperature for two days, the media was combined with an inoculum–soil mixture to achieve a concentration of 2% (w/v) and then filled into the plastic pot as the growing media as described by Jiang et al. (2020) [25]. The seeds for planting consist of the two RIL populations, along with their parents. Each pot was planted with ten seeds, and the surface was then covered with a two-millimeter layer of the growing medium. Pots were positioned within a container, and water was directed to the pots through their drainage openings to make sure they were completely saturated in a greenhouse maintained at a temperature of 26 °C and a relative humidity of 50%. Water was supplied to the container to maintain soil moisture throughout the assay. To mitigate against potential location-related impacts, the plastic pots in the greenhouse were changed every two days. The study used a completely randomized design (CRD) with three replications.

2.4. Phenotypic Data

Phenotypic data were taken on the emergence rate (ER), survival rate (SR), and disease severity (DS). The data on the traits were taken twice. The first data set was taken between February and April 2023, termed as the first (1st) data set. The second time data set was taken between June and August 2023 as the second (2nd) data set.

2.4.1. Assessment of the Rate of Emergence and Survival for Resistance to *Calonectria ilicicola*

The soybean accessions were scored for emergence rate on the 5th day after planting (DAP). The emergence rate was expressed as the total number of seeds that emerged out of the total number of seeds planted expressed in percentage. The survival rate was taken on the 12th DAP. Survival rate was calculated as the total number of plants alive out of the total number of plants that emerged expressed in percentage.

2.4.2. RIL Populations Lines Response to *Calonectria ilicicola* Infection

The seedlings were evaluated for *C. ilicicola* resistance on the 14 DAP according to Jiang et al. (2020) [25]. The seedlings were gently removed from the pot, and the roots were washed to ensure that the roots were free from soil for visual examination. The resistance level of each RIL was determined using DS in addition to their SR and ER traits. A threshold of SR > 0.90 and ER > 0.85 was used as a standard to identify accessions with resistance. The SR helped to determine the mortality rate of the seedling, which was incorporated into the disease resistance rating scale. The DS employs a numerical scale ranging from 0 to 5, as indicated in Table 1. These were employed to categorize RILs according to their responses to *C. ilicicola* infection. The average of the three repeats of each RIL was further used to classify them using $y \leq 1.5$, $1.5 \geq y \leq 2.5$, $2.5 \geq y \leq 3.5$, $3.5 \geq y \leq 4.5$, and $y > 4.5$ as highly resistant, resistant, moderately susceptible, susceptible, and highly susceptible to *C. ilicicola*, respectively.

Table 1. Classification for the disease severity of red crown rot.

Scale	Damage Degree	Resistance Degree
0	No visible sign of necrotic lesions on the root	Immune
1	Only the tap root exhibits small necrotic lesions without obvious changes in its form and shape	Highly Resistant
2	Necrotic lesions spread to the crown and root system and seedling mortality less than 10%	Resistant
3	Roots show serious necrotic lesions with less than 50% loss by rot and seedling mortality of 10–20%	Moderately Susceptible
4	Roots show serious necrotic lesions with more than 50% root loss by rot and seedling mortality of 21–50%	Susceptible
5	Over 50% of root loss by rot with seedling mortality of more than 50%	Highly Susceptible

2.4.3. Analysis of Phenotypic Data

The GenStat software, version 12 (VSN International Ltd., Hemel Hempstead, UK), was utilized to generate descriptive statistics for the ER, SR, and DS traits in both RIL populations as well as their parents. These statistics included the mean, range, degree of variation (CV, %), skewness, and kurtosis. A Pearson correlational study was conducted to examine the relationship between ER, SR, and DS. The results were shown using the Corrplot package in R (version 0.84, Auckland, New Zealand). The level of significance was set at $p < 0.05$ [44].

2.5. SNP Genotyping and Linkage Map Construction for ZM6 and MN RIL Populations

We extracted the genomic DNA using young and fresh leaves of the entire populations of ZM6 and MN RILs in addition to the parents, following the standard established by [45]. The DNA library construction, sequencing, SNP acquisition, and integration of Bin/SLAF (specific-locus amplified fragment) markers were performed for the ZM6 and MN populations. For the ZM6 population, bin markers were used for genotyping [46,47]. Also, the MN population was genotyped using SLAF markers [46,47]. The high genetic maps of the ZM6 and MN populations comprised 2601 bin markers and 2062 SLAF markers, respectively. The combined length of the ZM6 and MN maps was 2630.22 cM and 2054.5 cM, having a mean distance of 1.02 cM and 1.36 cM between neighboring markers, respectively (Table S1). The mean length of each linkage group was 131.51 cM for the ZM6 linkage map and 102.73 cM for the MN linkage map. The average number of markers per chromosome was 130.05 and 103.1 from the ZM6 and MN populations, respectively (Table S1).

2.6. QTL Mapping for Emergence Rate, Survival Rate, and Disease Severity

The WinQTLCart (version 2.5, Raleigh, NC, USA) application was used to locate QTLs via composite interval mapping (CIM) [48]. The window size, working speed, and control were configured to be 10 cM, 1 cM, and less than 5 cM, respectively [49]. The LOD thresholds for identifying significant QTLs were established at 2.5, using 1000 permutations for each test and a type I error rate of 5% [50]. The precise physical positions of individual QTLs on all chromosomes were determined using the MapChart (version 2.1, Wageningen, Netherlands) tool [51]. Names were assigned to the QTLs by adhering to the accepted nomenclature system [52]. For instance, for the QTL denoted as *qER-7-1zm6*, the letter *q* stands for QTL preceded every QTL name, which was followed by a two-letter descriptor of the ER trait, *7* refers to the specific chromosome on which the QTL was identified, *1* displays the number of each QTL trait on the chromosome, and *zm6* represents the ZM6-RIL population where the QTL existed.

2.7. Mining of Candidate Genes for Major QTLs

We obtained the complete genomic data from the main "QTL hotspots" by utilizing the internet-based resource libraries from Phytozome (http://phytozome.jgi.doe.gov (accessed on 13 January 2024)) and SoyBase (http://www.soybase.org (accessed on 13 January 2024)), which also assisted in candidate gene identification along with the data obtained from previously published publications. We used the Gene Ontology (GO) online platform to analyze the identified potential genes. GO enrichment analysis was performed for all the QTL hotspots using agriGo (version 2.0, Beijing, China) [53].

The available RNA sequencing dataset by Kobayashi et al. (2022) [54] and our unpublished data were applied to investigate the transcription of identified candidate genes in different soybean tissues and phases of development. A heat map was generated using TBtools version 1.6 to visually represent the changes in gene expression trends for the anticipated candidate genes [55].

3. Results

3.1. Variability Characteristics of the Mapping Traits for the RIL Populations

The use of parents with varied phenotypes is necessary for the creation of mapping populations, and this is crucial for maximizing the effectiveness of QTL discovery [56]. For this investigation, two mapping RIL populations were utilized: ZM6 RIL, which originated from the cross between Zhengyang (♀) and Meng8206 (♂), and MN, which originated from the cross between M8180 (♀) and NN1138-2 (♂) (Table 2). The ZM6 and MN RIL populations display the mean, range (minimum and maximum value), standard deviation, skewness, and kurtosis of the ER, SR, and DS for the two RIL populations (ZM6 and MN) and their parents (Table 2). These measurements were examined for two conservative times (first and second). The parental genotypes exhibited relatively large variations for ER, SR, and DS (Table 2). The RILs mostly showed a transgressive segregation. The skewness and kurtosis were mostly less than 1, which is typical for quantitative attributes, implying that the populations are suitable for conducting QTL mapping (Table 2). Furthermore, the measured variables (ER, SR, and DS) exhibited a distinct range of distribution across both RIL populations (Figure 1). Based on the three traits screened, there were significant positive correlation coefficients between SR and ER in both mapping populations, while significant negative correlation coefficients existed between DS against ER and SR (Figure 2).

Table 2. Response of two soybean RIL populations and their parents to RCR.

Population/ Environment [a]	Trait [b]	Mean of Parent (%) [c]		RIL Populations [d]				
		♀	♂	Mean ± SE	Min.	Max.	Kurtosis	Skewness
ZM6/2	ER	73.33 ± 3.33	83.33 ± 3.33	83.11 ± 0.00	63.33	100.00	−0.86	0.13
	SR	82.14 ± 3.57	100 ± 0.00	86.07 ± 0.00	54.76	100.00	−0.09	−0.63
	DS	3.33 ± 0.33	1.00 ± 0.00	2.83 ± 1.00	1.00	4.00	−0.50	−0.10
ZM6/1	ER	76.67 ± 3.33	93.33 ± 3.33	80.44 ± 1.00	56.67	100.00	−0.30	−0.32
	SR	82.74 ± 3.90	100 ± 0.00	81.79 ± 1.00	39.68	100.00	−0.05	−0.59
	DS	4.33 ± 0.33	2.00 ± 0.00	3.46 ± 1.00	1.00	4.33	−0.50	−0.10
MN/2	ER	100.00 ± 0.00	80.00 ± 0.00	82.50 ± 5.20	35.67	100.00	1.31	−1.08
	SR	100.00 ± 0.00	83.33 ± 6.93	96.45 ± 0.04	74.17	100.00	3.86	−1.93
	DS	1.67 ± 34.64	3.67 ± 5.75	2.46 ± 6.09	1.00	4.00	−0.66	0.30
MN/1	ER	96.67 ± 5.97	70.00 ± 0.00	78.44 ± 3.92	60.00	90.00	3.67	−1.36
	SR	100.00 ± 0.00	76.67 ± 8.00	87.86 ± 3.00	52.86	100.00	1.28	−1.03
	DS	2.00 ± 0.00	3.6 ± 6.00	3.00 ± 6.00	1.33	4.00	−0.53	−0.52

[a] Environment: the first and second time screenings (1 and 2) in each RIL population. [b] ER-emergence rate, SR—survival rate, and DS—disease severity. [c] Zhengyang (♀)and Meng8206 (♂) for ZM6 RILs, while M8180 (♀)and NN1138-2 (♂) for MN RILs. [d] SE—standard error of mean, min—minimum, and max—maximum.

3.2. QTL Detected in the ZM6 Population for Red Crown Rot Resistance in Soybean

The composite interval mapping (CIM) strategy was employed using the 2601 bin markers obtained among the ZM6 RIL population and was polymorphic between the parents. A total of 15 QTL for the three traits in ZM6 RIL population were located on eight different chromosomes (Chr06, Chr07, Chr08, Chr10, Chr11, Chr13, Chr17, and Chr18), of which 5, 6, and 4 were for ER, SR, and DS, respectively, and each QTL had phenotypic variance explained (PVE) ranging between 8.40 and 14.87% (Table 3; Figure 3). The 5 QTL for ER consisted of *qER-7-1zm6* (Chr07), *qER-8-1zm6* (Chr08), *qER-10-1zm6* and *qER-10-2zm6* (Chr10), and (*qER-11-1zm6*) (Chr11) with their PVE ranging from 8.40 to 10.70%. With the exception of the 2 QTL on Chr10 (*qER-10-1zm6* and *qER-10-2zm6*), which inherited its alleles from the male parent, Zhengyang, the remaining three QTLs (*qER-7-1zm6*, *qER-8-1zm6*, and *qER-11-1zm6*) had their alleles from the female parent, Meng8206 (Table 3).

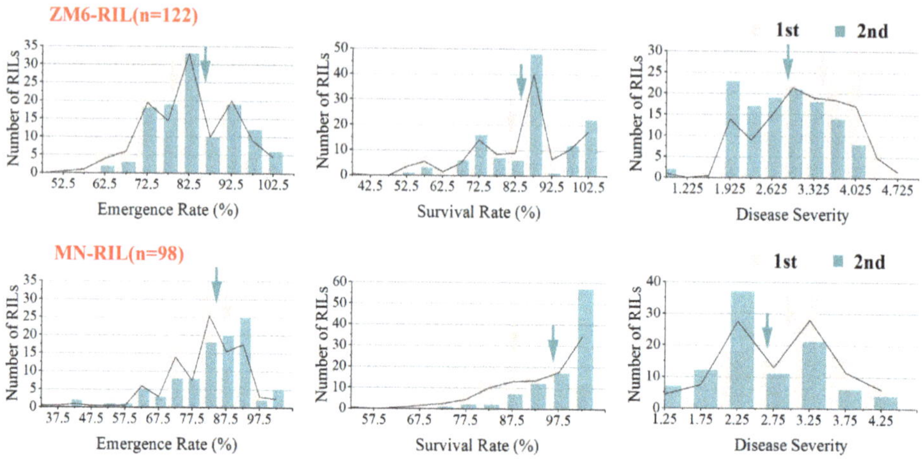

Figure 1. Frequency distribution of emergence rate, survival rate, and disease severity in ZM6 and MN RIL populations. The phenotype was averaged from the first and second time screenings in each RIL population. The trend shows the moving average. The arrows show the mean value of corresponding. The horizontal and vertical axis represent trait value and number of genotypes, respectively.

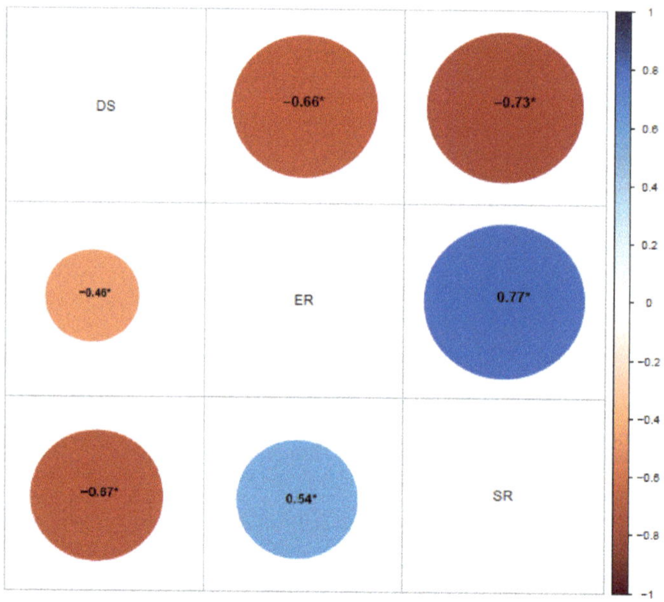

Figure 2. Pearson correlation matrix among the emergence rate (ER), survival rate (SR), and disease severity (DS) used to screen for red crown rot resistance in ZM6 and MN RIL populations (ZM6, upper part, and MN, lower part of the matrix). The phenotype was averaged from the first and second time screenings of the study in each RIL population. The asterisk shows significance at $p < 0.05$. The color scale of each correlation coefficient is shown in the color legend on the right-hand side of the correlation matrix.

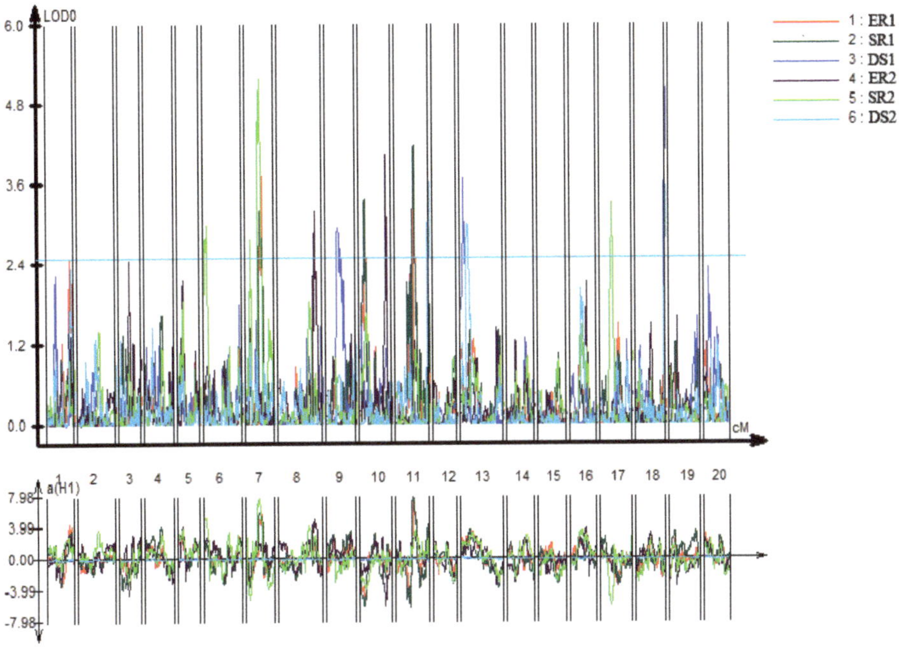

Figure 3. Genome-wide view of QTL detected in the ZM6 RIL population for soybean root crown rot disease via the CIM strategy. The phenotypic data were obtained from two time screenings (first time and second time). The disease resistance was screened by three traits: emergence rate (ER), survival rate (SR), and disease severity (DS). The y-axis of the top figure represents a logarithm of odd value with a threshold of 2.5 (sea blue line), while the x-axis represents the chromosome (Chr) number in the soybean genome. The bottom figure denotes the additivity of alleles and their origin from either of the parents: positive values represent allele from Meng8206 (♀), while negative values emanate from Zhengyang (♂). The color legend is shown on the right-hand side of the top figure.

Table 3. QTLs detected for RCR disease resistance via ER, SR, and DS in the ZM6 RIL population via the CIM strategy.

QTL [a]	Pos (cM) [b]	LOD [c]	Add [d]	PVE (%) [e]	CI (cM) [f]	Physical Region (bp)	Env [g]
qER-7-1zm6	64.51	3.82	5.81	10.37	61.4–69.7	15,375,767–17,747,973	1
qER-8-1zm6	150.81	3.22	5.04	8.40	147.7–154.7	41,485,100–42,915,255	2
qER-10-1zm6	16.61	3.25	−5.24	8.56	12.8–19.6	1,694,367–2,679,837	1
qER-10-2zm6	106.41	4.04	−6.00	10.70	105.1–109.9	43,900,754–44,741,960	2
qER-11-1zm6	73.81	3.86	7.20	10.40	71.2–75.5	14,962,695–15,949,296	1
qSR-6-1zm6	9.71	2.97	5.32	7.78	6.7–15.6	1,813,130–3,196,555	2
qSR-7-1zm6	48.81	4.55	7.08	12.71	46.2–50.1	9,304,376–10,428,532	2
qSR-7-2zm6	56.01	5.19	7.71	14.87	53–62	14,134,797–15,903,280	2
	59.21	3.21	6.00	8.62	52.7–61.4	14,134,797–15,452,798	1
qSR-10-1zm6	16.61	3.37	−6.00	8.88	10.3–19.6	1,603,735–2,732,880	1
qSR-11-1zm6	76.41	4.19	7.66	11.34	75.5–81.8	15,676,274–16,816,800	1
qSR-17-1zm6	46.41	3.32	−5.93	8.74	32.8–53.7	6,777,393–9,645,325	2
qDS-11-1zm6	139.11	3.81	−0.26	9.55	136–140.8	37,603,249–38,850,696	2
qDS-13-1zm6	8.91	3.69	0.27	9.26	3.1–11.2	4,552,834–5,592,448	1
qDS-18-1zm6	120.11	3.63	−0.28	9.57	118–122.7	59,218,992–60,685,675	1
qDS-18-2zm6	128.51	5.04	−0.32	12.95	124.8–129.5	61,300,197–62,014,706	1
	128.51	3.57	−0.25	8.89	124.2–129.5	60,909,812–62,014,706	2

[a] QTLs detected in different environments at the same, adjacent, or overlapping marker intervals were considered the same QTL. [b] Pos—Position of QTL in centiMorgan. [c] LOD—Logarithm of odds. [d] Add—Addictive, indicating the origin of beneficial alleles; positive values represent allele from Meng8206 (♀), while negative values emanated from Zhengyang (♂)). [e] PEV—phenotypic variance explained (%) expressed by the QTL. [f] CI—Confidence interval. [g] Env—Environment, the phenotypic data obtained from two time screenings (first time and first time).

Six QTLs were mapped for SR across five chromosomes: Chr06 (*qSR-6-1zm6*), Chr07 (*qSR-7-1zm6*, and *qSR-7-2zm6*), Chr10 (*qSR-10-1zm6*), Chr11 (*qSR-11-1zm6*), and Chr17 (*qSR-17-1zm6*) (Table 3). Out of these, *qSR-7-2zm6* had the most prominent QTL with a LOD value of 5.19, accounting for 14.87% of the phenotypic variation, which was followed by qSR-7-1zm6 with a LOD score of 4.55 and a difference in phenotype of 12.71% (Table 3). Four QTLs influencing DS were detected across three chromosomes (Chr11, Chr13, and Chr18), accounting for an average of 10.04% of PVE (Table 3). The Chr011 and Chr13 harbored one QTL each (*qDS-11-1zm6* and *qDS-13-1zm6*, respectively), while Chr18 had two QTL (*qDS-12-1zm6* and *qDS-12-2zm6*). Figure 4 shows the location of QTLs on the linkage genetic map of the ZM6 RIL population.

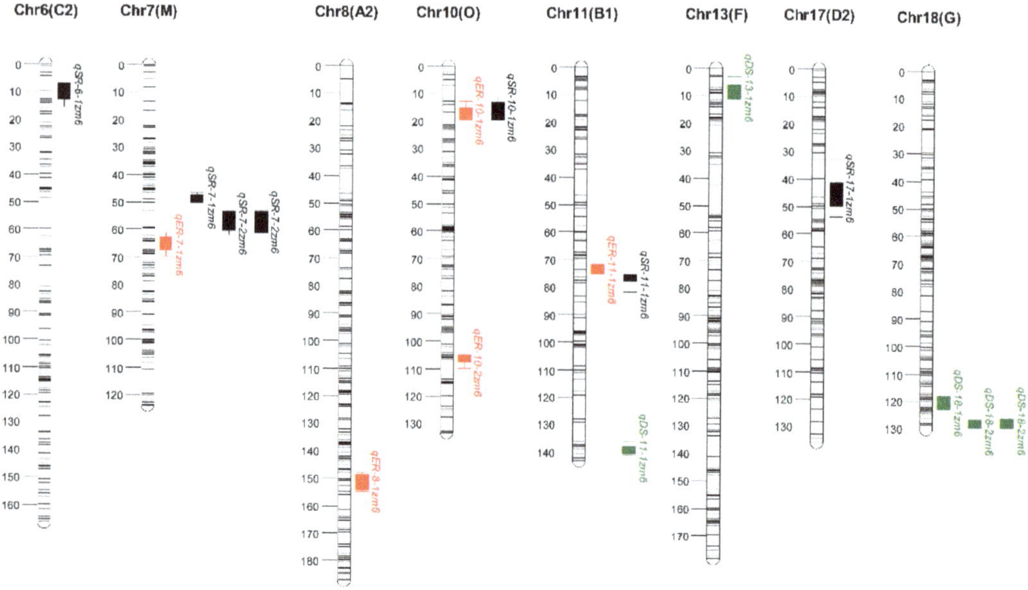

Figure 4. Location of QTLs on the genetic linkage map of ZM6 RIL population. Distances among markers are indicated using the physical location to the left of the populations and the QTL names located at their position on the right side. Chr-Chromosome. Colored bars represent different QTLs.

3.3. *QTL Detected in MN Population for Red Crown Rot Resistance in Soybean*

The MN RIL population detected 14 QTLs across eight chromosomes (Chr01, Chr02, Chr04, Chr08, Chr10, Chr13, Chr15, and Chr17) mapped for ER, SR, and DS comprising 5, 4, and 5 QTLs, respectively, with their phenotypic variation ranging from 8.69% to 25.24% (Figure 5; Table 4). For the ER, the five QTLs, namely, *qER-1-1mn*, *qER-10-1mn*, *qER-8-1mn*, *qER-8-2mn*, and *qER-15-1mn* had their LOD values in the range of 2.63 to 7.61 (Table 4). The QTLs for the SR were located on Chr01 (*qSR-1-1mn*, and *qSR-1-2mn*), Chr02 (*qSR-2-1mn*), and Chr17 (*qSR-17-1mn*) and their PVE ranged from 8.68% to 10.7% and their alleles were inherited from NN1138-2 (♂) except *qSR-2-1mn* (Table 4). Five QTLs, namely, *qDS-1-1mn*, *qDS-1-2mn*, *qDS-1-3mn*, *qDS-4-1mn*, and *qDS-13-1mn*, were mapped for DS with PVE ranging from 10.51% to 12.28% (Table 4). Two QTLs (*qER-1-1mn* and *qSR-1-2mn*) were detected on Chr01 across the second environment, accounting for 33.93% of the phenotypic variance, suggesting it is relative stability (Table 4). These two QTLs colocalized within the genomic interval 37,147,226–38,075,381 bp. Figure 6 shows the location of QTLs on the genetic linkage map of the MN RIL population.

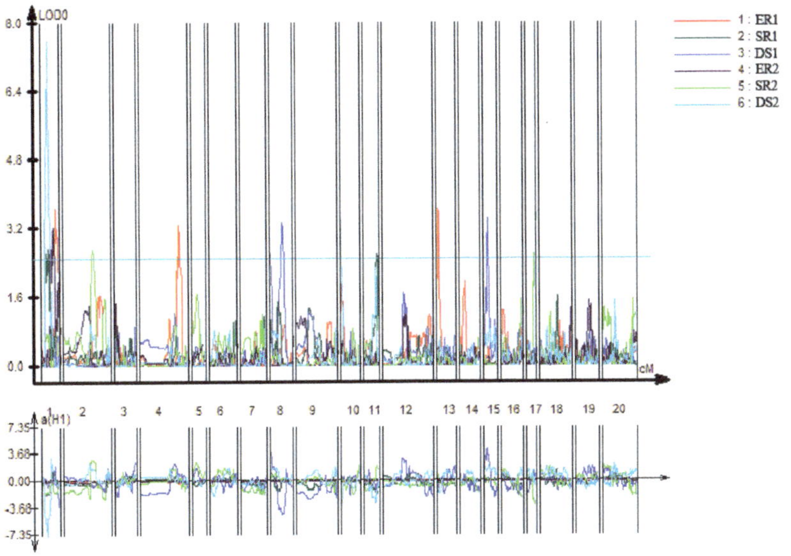

Figure 5. Genome-wide view of QTL detected in MN RIL population for root crown rot resistance via the CIM strategy. The phenotypic data were obtained from two time screenings (first time and second time). The disease resistance was screened by three traits: emergence rate (ER), survival rate (SR), and disease severity (DS). The y-axis of the top figure represents a logarithm of odd value with a threshold of 2.5 (sea blue line), while the x-axis represents chromosome (Chr) number in the soybean genome. The bottom figure denotes the additivity of alleles and their origin from either of the parents: positive values represent alleles from M8180 (♀), while negative values emanate from NN1138-2 (♂). The color legend is shown on the right-hand side of the top figure.

Table 4. QTLs detected for RCR disease resistance via ER, SR, and DS in the MN RIL population via the CIM strategy.

QTL [a]	Pos(cM) [b]	LOD [c]	Add [d]	PVE (%) [e]	CI (cM) [f]	Physical Region (bp)	Env [g]
qER-1-1mn	27.11	7.61	−7.24	25.24	25–29	8,823,531–22,021,358	1
qER-10-1mn	0.01	3.38	−3.81	10.72	0–8	2,275,280–4,174,791	1
qER-8-1mn	0.01	2.63	3.88	9.59	0–10	14,650,727–11,805,246	2
qER-8-2mn	48.71	3.35	−4.51	12.92	33.2–57.9	8,218,976–18,160,078	2
qER-15-1mn	16.11	3.44	4.31	12.66	13.5–22.3	5,519,255–113,832,93	2
qSR-2-1mn	111.51	2.67	2.97	9.32	105.5–125.6	7,361,306–15,293,225	1
qSR-17-1mn	34.01	2.91	−3.04	9.83	31.4–34.6	18,508,753–33,427,203	1
qSR-1-1mn	19.81	2.64	−1.83	10.7	14.2–23.3	3,617,559–8,823,788	2
qSR-1-2mn	29.01	2.73	−1.68	8.69	27.1–35	21,174,218–44,479,895	2
qDS-1-1mn	37.91	2.74	0.25	10.6	32.8–41.3	10,404,837–43,932,907	1
qDS-1-2mn	46.61	3.22	0.27	11.7	44.8–52.6	38,197,263–49,324,405	1
qDS-1-3mn	56.71	3.67	−0.26	11.84	51.4–58.2	41,313,930–50,832,292	2
qDS-4-1mn	154.41	3.28	−0.25	10.51	152.4–164	11,631,171–36,956,769	2
qDS-13-1mn	3.41	3.65	−0.27	12.28	0–7.8	688,713–5,592,448	2

[a] QTLs detected in different environments at the same, adjacent, or overlapping marker intervals were considered the same QTL. [b] Pos—Position of QTL in centiMorgan. [c] LOD—Logarithm of odds. [d] Add—Addictive, indicating the origin of beneficial alleles; positive values represent alleles from Meng8206 (♀), while negative values emanate from Zhengyang (♂). [e] phenotypic variance explained (%) expressed by the QTL. [f] CI—Confidence interval of the QTL. [g] Env—Environment.

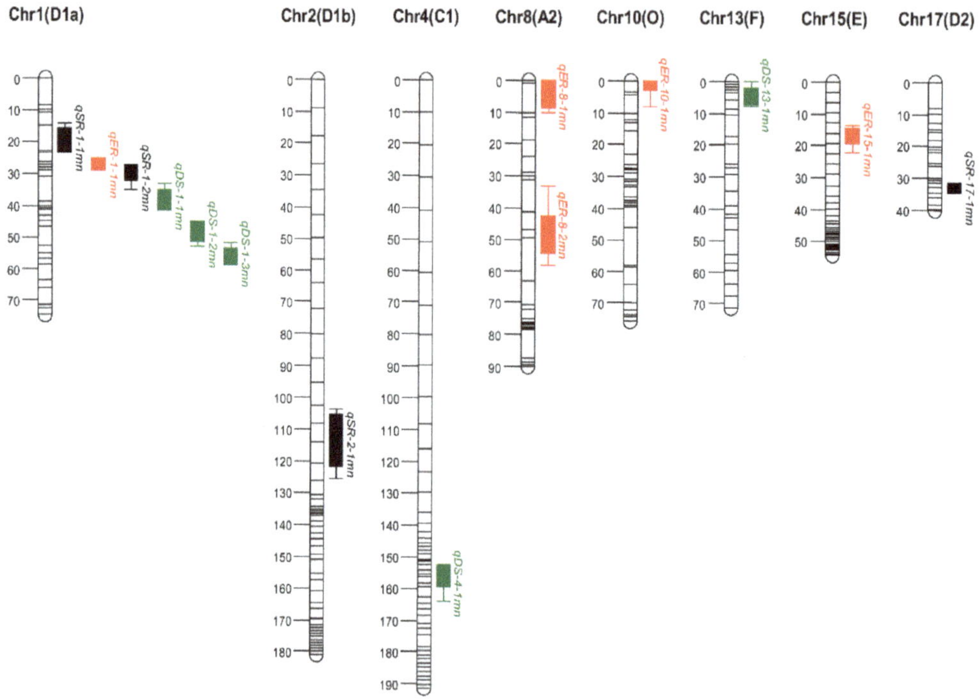

Figure 6. Location of the QTLs on the genetic linkage map of MN RIL populations. Distances among markers are indicated using the physical location to the left of the populations and the QTL names located at their position on the right side. Colored bars represent different QTLs.

3.4. Colocalization of QTLs in the QTL Hotspot

The QTL hotspot refers to the concentrated area on a chromosome that hosts many QTLs related to distinct traits. The stability of QTL across mapping populations and environments is essential for their use in practical plant breeding [57]. Hence, our stable QTLs were detected either linked to at least two of the traits (ER, SR, or DS) across an environment (first and second time of screening) within a population or across a population (found in ZM6 and MN). The QTLs *qER-1-1mn* and *qSR-1-2mn* were detected on Chr01 for two conservative times during the screening; hence, the stable QTL was termed as Hotspot A (Table 5). The QTLs were stable and are colocalized within the genomic interval of 8,823,531–44,479,895 bp. The mapping results of the ZM6 population found two pairs of QTLs, namely, *qSR-7-2zm6* colocalized within the genomic region of 14,134,797–15,903,280 bp across the two environments and renamed as QTL Hotspot B (Table 5). Also, *qER-10-1zm6* and *qSR-10-1zm6* colocalized within genomic interval 1,603,735–2,732,880 bp, termed as QTL Hotspot C. Hotspot D refers to QTL *qER-11-1zm6* in addition to *qSR-11-1zm6* colocalized within the genomic interval of 14,962,695–16,816,800. Interestingly, a stable QTL *qDS-13-1zm6* and *qDS-13-1mn* colocalized within genomic interval 68,8713-5,375,644 bp was detected across the MN and ZM6 population and named QTL Hotspot E (Tables 3–5). The QTLs *qSR-7-2zm6* and *qDs-18-2zm6* were constantly found in ZM6 in the first and second screenings, indicating that they were stable; they were thereafter classified as QTL hotspots B and F, respectively (Table 5).

Table 5. The six QTL hotspots identified in the ZM6 and MN RIL populations.

QTL Hotspot Name	QTL Name	LOD [a]	Add [b]	PVE (%) [c]	Physical Region (bp)
Hotspot A	*qER-1-1mn*	7.61	−7.24	25.24	8,823,531–44,479,895
	qSR-1-2mn	2.73	−1.68	8.69	
Hotspot B	*qSR-7-2zm6*	5.19	7.71	14.87	14,134,797–15,903,280
		3.20	6.00	8.62	
Hotspot C	*qER-10-1zm6*	3.25	−5.24	8.56	1,603,735–2,732,880
	qSR-10-1zm6	3.37	−6.00	8.88	
Hotspot D	*qER-11-1zm6*	3.86	7.20	10.40	14,962,695–16,816,800
	qSR-11-1zm6	4.19	7.66	11.34	
Hotspot E	*qDS-13-1zm6*	3.81	−0.26	9.55	688,713–5,592,448
	qDS-13-1mn	3.65	−0.27	12.28	
Hotspot F	*qDs-18-2zm6*	5.04	−0.32	12.95	60,909,812–62,014,706
		3.57	−0.25	8.89	

[a] LOD—Logarithm of odds. [b] Add—Adductive indicating the origin of beneficial alleles; positive values represent alleles from Meng8206 (♀), while negative values emanate from Zhengyang (♂). [c] PVE—phenotypic variance explained (%) explained by the QTL.

3.5. Candidate Gene Mining within Major "QTL Hotspots"

One of the goals of QTL mapping research is to identify possible candidate genes. This is made possible by the presence of complete genetic information and detailed gene descriptions. Therefore, model genes within the QTL hotspot and their corresponding gene annotations were obtained via SoyBase (https://www.soybase.org, retrieved on 26 December 2023) using soybean genome version 2. A total of 1279 gene models were found in the genomic region for each of the significant "QTL hotspots," of which 529, 116, 118, 122, 274, and 120 genes were found within hotspots of A, B, C, D, E, and F, respectively (Table S3).

The gene ontology (GO) analysis of enrichment was conducted using the web-based toolbox agriGO (version 2.0, Beijing, China) [53]. It enabled the depiction of the main categorizations of biological process (BP), molecular function (MF), and cellular component (CC) (Figures S1–S3). We observed the highest number of genes in CC followed by BP, and MF was the least for all the QTL hotspots, as expected for QTL hotspot C, for which the BP was higher than the CC, and the MF recorded the lowest number of genes (Figures S1–S3). Analysis of the GO showed that the majority of the genes associated with the six "QTL hotspots" in CC engage in the cell, cell part, and organelle processes; the BP partake largely in processes involving cellular, metabolic, and single-organism processes; and the MF genes mainly engage in the activities of catalytic, binding, and transport (Figures S1–S3).

Using gene annotation descriptions, existing research findings, and GO enrichment examination, our study identified 32, 12, 15, 12, 25, and 15 potential genes from "QTL Hotspot A", "QTL Hotspot B", "QTL Hotspot C", "QTL Hotspot D", "QTL Hotspot E", and "QTL Hotspot F", totaling 111 predicted genes (Table S4). These genes play a role in controlling disease infection development, including developing resistance and innate immune response. They are involved in various processes such as signal transduction; biosynthetic processes involving salicylic acid, lignin, and flavonoid; phenylpropanoid metabolic processes; and cell death. To enhance the accuracy of the earlier-predicted list of candidate genes, we used an available online database by Kobayashi et al. (2022) [54] (Table S5) and that of our RNA-Seq data (unpublished, though is available upon request). According to the RNA-seq data, 23 genes from a total of 111 projected candidate genes exhibited a significantly greater level of transcription log2 (FC) of ≥ 1 in the root, stem, R2-1d, R2-2d, S-1d, and S-2d (Figure S4A–F; Table 6). These include five genes (*Glyma.01G127100*, *Glyma.01G126600*, *Glyma.01G112300*, *Glyma.01G127200*, and *Glyma.01G127700*); two genes (*Glyma.07G133900*, and *Glyma.07G134100*); five genes (*Glyma.10G019900*, *Glyma.10G023300*, *Glyma.10G023400*, *Glyma.10G028200*, and *Glyma.10G025700*); three genes (*Glyma.11G181200*, *Glyma.11G192300*, and *Glyma.11G193600*); three genes (*Glyma.13G066100*, *Glyma.13G069200*, and *Glyma.13G076200*); and five genes (*Glyma.18G287000*, *Glyma.18G289100*, *Glyma.18G289600*,

Glyma.18G293200, and *Glyma.18G293300*) from QTL Hotspots A, B, C, D, E, and F, respectively (Figure S4; Table 6). Therefore, the 23 genes are potential candidates for regulating *C. ilicicola* resistance in soybeans. Nevertheless, additional functional validation is required to verify their specific functions in modulating resistance to soybean RCR disease.

Table 6. Predictive gene annotation information within the six QTL hotspots identified in the ZM6 and MN RIL populations.

QTL Hotspot	Gene Mapped IDs [a]	Annotation Descriptions [b]
A	*Glyma.01G127100*	Disease resistance-responsive; Dirigent-like protein
	Glyma.01G126600	Disease resistance-responsive; Dirigent-like protein
	Glyma.01G112300	Signal transduction; Leucine Rich Repeat
	Glyma.01G127200	Disease resistance-responsive; Dirigent-like protein
	Glyma.01G127700	Signal transduction; Defense response
B	*Glyma.07G133900*	Lignin catabolic process; multicopper oxidase
	Glyma.07G134100	Lignin catabolic process; multicopper oxidase
C	*Glyma.10G019900*	Glutathione metabolic process; glutathione S-transferase
	Glyma.10G023300	Protein phosphorylation; serine/threonine protein kinase
	Glyma.10G023400	Protein phosphorylation; serine/threonine protein kinase
	Glyma.10G028200	Bifunctional inhibitor/lipid-transfer protein/seed storage 2S albumin superfamily protein
	Glyma.10G025700	F-box family protein; protein binding
D	*Glyma.11G181200*	F-box family protein; protein binding
	Glyma.11G192300	Oxidation-reduction process; pheophorbide an oxygenase
	Glyma.11G193600	Cellular glucan metabolic process; cell wall biogenesis
E	*Glyma.13G066100*	DNA repair; ATP binding
	Glyma.13G069200	Zinc finger (AN1-like) family protein
	Glyma.13G076200	Defense response; Signal transduction; Leucine Rich Repeat
F	*Glyma.18G287000*	Defense response; signal transduction
	Glyma.18G289100	ATP binding, hsp70 protein
	Glyma.18G289600	ATP binding, hsp70 protein
	Glyma.18G293200	Drug transmembrane transport; multidrug resistance protein
	Glyma.18G293300	Drug transmembrane transport; multidrug resistance protein

[a] model gene retrieved on reference genome V2 from SoyBase (https://www.soybase.org/ accessed on 4 April 2024). [b] Gene ontology of the retrieved genes from SoyBase (https://www.soybase.org/ accessed on 3 April 2024).

4. Discussion

The objective of this study was to employ a high level of intraspecific mapping of ZM6 and MN populations, analyzed for two different periods, to find stable QTLs (QTL Hotspots) and to determine probable candidate genes associated with soybean's resistance to *C. ilicicola*. We found six QTL hotspots and predicted 23 potential candidate genes (PCGs).

4.1. Examining the Resistance to RCR in Two RIL Populations

Due to its fast spreading and severity in major production regions, soybean pathologists and breeders have aimed to identify and incorporate RCR resistance into high-quality soybean varieties. Earlier works have developed screening protocols for evaluating soybean germplasm for resistant accession identification [25,58–60]. There are a few sources of incomplete resistance, primarily found in wild relatives or non-elite germplasm. This requires a process of crossing and careful selection, which is made more difficult by environmental factors that affect the pathogen's lifestyle. The ANOVA analysis showed a substantial variation across all the RILs in the ZM6 and MN populations for all three traits, with the environment playing a substantial role (Table S2). The research found that the paternal accessions Zhengyang and NN1138-2 showed a significant resistance level, while Meng8206 and M8180 revealed some level of susceptibility. The disease severity index analysis revealed that the resistant accessions Zhengyang and NN1138-2 potentially harbor distinct resistance genes. The high correlation coefficients among the traits used for screening RCR resistance imply an independent and interdependent relationship among the traits, suggesting some common loci may be responsible for at least two of the traits studied (Figure 2).

4.2. QTL Detected for Red Crown Rot Resistance in ZM6 and MN Population

The availability of genetic markers is crucial to assist QTL identification. Furthermore, combining QTLs that resist *C. ilicicola* infection within a cultivar is an additional advantage of identifying QTLs connected with soybean RCR resistance. Although some progress has been made in establishing resistance to RCR, data regarding DNA markers associated with resistance to RCR are scarce. Linkage mapping has the potential to facilitate mutual verification and enhance the accuracy of produced results, promoting its application in several studies involving disease resistance gene identification in soybeans using RIL populations [34]. The high-density genetic maps for ZM6 were genotyped by RAD-seq technology [32], and the MN population was also genotyped by SLAF technology [61]. The markers on both linkage maps, ZM6 and MN, were incorporated into all the 20 linkage groups and covered a length of 2630.22 cM and 2054.50 cM, respectively. In ZM6, the average distance among adjacent markers was 1.01 cM, while in MN, it was 1.00 cM. By combining these high-density genetic maps, there is potential to accurately locate QTL that has significant and subtle effects on resistance to RCR in the mapping populations.

The capacity to detect QTL determines the extent of the genomic region, which is decided by the gene variation between parental individuals and is influenced by population structure and size [62,63]. The two RIL populations identified 15 and 14 QTLs associated with RCR resistance in ZM6 and MN populations, respectively, totaling 29 QTLs. For the ZM6, the most substantial effect was on chromosome 7, *qSR-7-2zm6*, with LOD 5.19, while for MN, the highest was on chromosome 1, *qER-1-1mn* with a LOD of 7.61 (Tables 3 and 4). This confirms that the phenotypic data gathered in different settings exhibit statistical differences [64]. The identification of QTL for resistance to RCR provides evidence that quantitative traits are impacted by multiple QTLs, each with relatively small individual effects. Hence, we hypothesize that resistance to RCR is a complex genetic characteristic regulated by several resistance gene loci. For instance, in soybeans, several QTLs are identified for *S. sclerotiorum* and *P. sojae resistance*, confirming that their genetic architecture is influenced by several genes [7,65,66].

The QTLs *qER-7-1zm6*, *qER-10-2zm6*, *qER-11-1zm6*, *qSR-7-1zm6*, *qSR-7-2zm6*, and *qSR-11-1zm6* in the ZM6 accounted for phenotypic variation above 10% and were on Chr07, Chr10, or Chr11. Therefore, the desired alleles of the genetic markers can significantly decrease the disease traits of RIL progenies and alter the response characteristics of *C. ilicicola*. Also, the MN population had eight QTLs with a PVE of more than 10%, of which *qER-1-1mn* had its PVE of 25.24% on Chr01. Thus, our findings demonstrated the dependability of QTL mapping. Furthermore, QTLs could be utilized as main targets for discovering candidate genes and implementing marker-assisted selection in future studies. The results prove that RIL has segregated homozygous alternative alleles and can detect QTL due to the presence of half alleles in the RIL [67].

4.3. Identification of QTL Hotspot for Resistance to RCR in Soybean

The durability of the QTL is essential for its application in a breeding strategy. This study additionally showed the localization of QTLs on chromosomes for many variables (ER, SR, and DS) associated with RCR resistance. For instance, in the two environments, we confirmed the presence of six (6) QTL hotspots for resistance to RCR from the ZM6 and MN RIL populations on chromosomes 1, 7, 10, 11, 13, and 18 (Table 5). Hotspot E consisted of *qDS-13-1zm6* and *qDS-13-1mn*, which were obtained from both populations colocalized in the same physical interval. Our investigation found that the *qSR-7-2zm6* and *qDS-18-2zm6* genes in the ZM6 population were localized in the same area on chromosome 7 for two consecutive years (Table 5). These findings indicate that there may be a shared resistance mechanism to several RCRs in soybeans, and a specific gene on chromosome 7 could be responsible for providing resistance to RCR. We have a firm conviction that these QTLs have the potential to be significant QTLs for either gene cloning or marker-assisted selection.

Determining the precise gene(s) accountable for resistance loci is an essential stage in the ongoing endeavor to examine the molecular processes and biological basis of quantitative resistance to diseases. The gene ontology (GO) study revealed that the majority of genes in the six hotspots were mainly involved in cellular functions such as the cell, cell component, organelle, cellular process, metabolism process, catalytic activity, binding, and transporter activity. These substances are crucial for the growth and development of plants, and multiple studies have verified their role in safeguarding plants from diseases caused by pathogens. To illustrate, the cells, cell parts, and organelles of plants participate in the hypersensitive response in plants due to gene-for-gene resistance. For instance, the cells located around the area where the fungus enters experience a process of programmed cell death (PCD) [68,69], cellular processes, metabolic processes [70] and catalytic activity, binding, and transporter activity [71–73] and are involved in plant immunity in order to impede the progression of the disease.

At the physical position of the six "QTL Hotspots", 1279 gene models were determined. Among these, 111 genes were identified as PCGs based on the study of GO enrichment, gene function, and existing literature. The actions of these 111 genes engage in activating the plant's immune response during a pathogen attack. Also, some of the genes activate secondary metabolite processes (lignin biosynthetic, flavonoid biosynthetic, and phenylpropanoid metabolic) as well as signal transduction. Of the 111 predicted PCGs, 23 are noted to provide resistance against *C. ilicicola* invasion in soybeans based on their gene expression data obtained from RNA-seq. Among the 23 PCGs, seven genes, namely, *Glyma.01G112300*, *Glyma.01G126600*, *Glyma.01G127100*, *Glyma.01G127200*, *Glyma.01G127700*, *Glyma.13G076200*, and *Glyma.18G287000*, are reported to offer resistance to crops upon pathogen infection either by activating defense response, signal transduction, or leucine-rich repeat responses/processes. For instance, in soybean, a gene annotated as defense response governs tolerance to *Phytophthora* root rot disease [74], while leucine-rich repeat has the highest number of documented disease resistance genes [75]. Similarly, *Glyma.07G133900* and *Glyma.07G134100* encode multicopper oxidase. In wheat, ascorbate oxidase and skewed5-similar proteins resulting from multicopper oxidase partake in the plant's immunity, and their silencing promotes the plant's immunity to *Verticillium wilt* [76]. The soybean *GmHSP40* results in cell death and enhances its immunity to mosaic virus, as its silencing results in making soybean more sensitive to the virus [77]. Similar roles are confirmed in pepper by overexpressing *CaHSP70a* [78]. Hence, *Glyma.18G289100* and *Glyma.18G289600* encoding the Hsp70 protein could possibly function to regulate RCR resistance.

We anticipated that the 23 predicted candidate genes may have a potential role in regulating RCR resistance in soybeans. However, additional testing and verification are necessary to demonstrate their specific involvement in soybean resistance to RCR, as well as their possible application in disease breeding programs. Also, the identified loci and QTL hotspot will aid in validating *C. ilicicola* resistance in soybeans through systematic breeding. Additionally, these genomic areas can serve as targets for enhancing our understanding of the RCR mechanism and improving soybean resistance to *C. ilicicola*. The identified harbor QTLs demonstrating promise will be employed for precise mapping and molecular cloning of crucial loci in the future. These regions can then be utilized to improve soybean plant immunity to *C. ilicicola* invasion.

5. Conclusions

The study examined the genetic architecture of soybean resistance to *C. ilicicola* invasion. A total of 29 QTLs were identified via linkage mapping using two different RIL populations. All the detected QTLs were grouped into six significant "QTL hotspots" and represent the major and consistent genomic areas that control the inheritance of soybean resistance to RCR. Among them, QTLs *qER-1-1mn* and *qSR-7-2zm6* accounted for 25.24% and 14.87% of the overall phenotypic variation, respectively. A common QTL *qDS-13-1zm6* and *qDS-13-1mn* were obtained from both populations colocalized in the same physical interval. We spotted 23 genes likely to regulate RCR resistance based on the six genomic areas known as "QTL hotspots". The identified genes serve as a ground for future research in developing genetic resources for enhancing soybean resistance. Similarly, additional functional studies are required for the validation and cloning of functional genes of the proposed candidate genes to determine their exact functions in regulating RCR resistance. These findings establish a crucial basis for generating fungal-resistant QTL for RCR in soybeans through cloning.

Supplementary Materials: The following supporting information can be downloaded at: https://www.mdpi.com/article/10.3390/agronomy14081693/s1, Table S1: Distribution of bin and SLAF markers mapped on soybean chromosomes/linkage groups in ZM6 and MN Populations. Table S2: Combined analysis of variance (ANOVA) for RCR evaluation trait (ER, SR, and DS) in ZM6 & MN RIL Population across two time screening (1st and 2nd time screening). Table S3: Model genes within hotspot A, hotspot B, hotspot C, hotspot D and hotspot E regions in both RIL Populations (ZM6 & MN). Table S4: Predicted candidate genes within hotspot A, hotspot B, hotspot C, hotspot D and hotspot E regions in both RIL Populations (ZM6 & MN) based on known functional annotation. Table S5: DEGs following infection by *C. ilicicola* Kobayashi et al., 2022 [54]. Figure S1: AgriGO annotation Palatino Linotypefor the QTL Hotspot A and B. (A) QTL Hotspot A, (B) QTL Hotspot B. Figure S2: AgriGO annotation information for the QTL Hotspot C and D. (C) QTL Hotspot C, (D) QTL Hotspot D. Figure S3: AgriGO annotation information for the QTL Hotspot E and F. (E) QTL Hotspot E, and (F) QTL Hotspot F. Figure S4: Heat map displaying the expression patterns of 23 selected genes in various soybean tissues from six QTL hotspots.

Author Contributions: A.A.-B., T.Z. and J.F. conceived and conducted the experiments; A.A., J.L., N.G. and C.C. performed part of the data collection; C.Z. and S.J. performed part of the data analysis work. All authors have read and agreed to the published version of the manuscript.

Funding: This work was supported by the National Natural Science Foundation of China (Grant Nos. 31871646, 32171965), the Core Technology Development for Breeding Program of Jiangsu Province (JBGS-2021-014), the Program of Collaborative Innovation Center for Modern Crop Production co-sponsored by Province and Ministry (CIC-MCP).

Data Availability Statement: The original contributions presented in the study are included in the article/Supplementary Materials, further inquiries can be directed to the corresponding author.

Conflicts of Interest: The authors declare no conflicts of interest.

References

1. Henchion, M.; Hayes, M.; Mullen, A.M.; Fenelon, M.; Tiwari, B. Future protein supply and demand: Strategies and factors influencing a sustainable equilibrium. *Foods* **2017**, *6*, 53. [CrossRef] [PubMed]
2. Cattelan, A.J.; Dall'Agnol, A. The rapid soybean growth in Brazil. *OCL* **2018**, *25*, D102. [CrossRef]
3. Ayalew, B.; Bekele, A.; Mazengia, Y. Analysis of cost and return of soybean production under small holder farmers in Pawe District, North Western Ethiopia. *J. Nat. Sci. Res.* **2018**, *8*, 28–34.
4. Liao, Z.; Pei, S.; Bai, Z.; Lai, Z.; Wen, L.; Zhang, F.; Li, Z.; Fan, J. Economic Evaluation and Risk Premium Estimation of Rainfed Soybean under Various Planting Practices in a Semi-Humid Drought-Prone Region of Northwest China. *Agronomy* **2023**, *13*, 2840. [CrossRef]
5. Sabagh, A.E.; Hossain, A.; Islam, M.S.; Barutçular, C.; Ratnasekera, D.; Kumar, N.; Meena, R.S.; Gharib, H.S.; Saneoka, H.; da Silva, J.A.T. Sustainable soybean production and abiotic stress management in saline environments: A critical review. *Aust. J. Crop Sci.* **2019**, *13*, 228–236. [CrossRef]
6. Miransari, M. *Environmental Stresses in Soybean Production: Soybean Production Volume 2*; Academic Press: Cambridge, MA, USA, 2016; Volume 2.
7. Lin, F.; Chhapekar, S.S.; Vieira, C.C.; Da Silva, M.P.; Rojas, A.; Lee, D.; Liu, N.; Pardo, E.M.; Lee, Y.-C.; Dong, Z. Breeding for disease resistance in soybean: A global perspective. *Theor. Appl. Genet.* **2022**, *135*, 3773–3872. [CrossRef]

8. Gai, Y.; Deng, Q.; Chen, X.; Guan, M.; Xiao, X.; Xu, D.; Deng, M.; Pan, R. Phylogenetic diversity of *Calonectria ilicicola* causing *Cylindrocladium* black rot of peanut and red crown rot of soybean in southern China. *J. Gen. Plant Pathol.* **2017**, *83*, 273–282. [CrossRef]
9. Fei, N.; Qi, Y.; Meng, T.; Fu, J.; Yan, X. First report of root rot caused by *Calonectria ilicicola* on blueberry in Yunnan Province, China. *Plant Dis.* **2018**, *102*, 1036. [CrossRef]
10. Zhang, Q.; Zhou, D.; Jiang, W.; Zhu, H.; Deng, S.; Wei, L. First report of soft rot of ginger caused by *Calonectria ilicicola* in Guangxi Province, China. *Plant Dis.* **2020**, *104*, 993. [CrossRef]
11. Roy, K.; McLean, K.; Lawrence, G.; Patel, M.; Moore, W. First report of red crown rot on soybeans in Mississippi. *Plant Dis.* **1989**, *73*, 273. [CrossRef]
12. Ochi, S.; Mimuro, G.; Kishi, S. Effect of red crown rot of soybean on occurrence of wrinkled seeds. *J. Gen. Plant Pathol.* **2022**, *88*, 232–238. [CrossRef]
13. Yamamoto, S.; Nomoto, S.; Hashimoto, N.; Maki, M.; Hongo, C.; Shiraiwa, T.; Homma, K. Monitoring spatial and time-series variations in red crown rot damage of soybean in farmer fields based on UAV remote sensing. *Plant Prod. Sci.* **2023**, *26*, 36–47. [CrossRef]
14. Kleczewski, N.; Plewa, D.; Kangas, C.; Phillippi, E.; Kleczewski, V. First report of red crown rot of soybeans caused by *Calonectria ilicicola* (anamorph: *Cylindrocladium parasiticum*) in Illinois. *Plant Dis.* **2019**, *103*, 1777. [CrossRef]
15. Akamatsu, H.; Fujii, N.; Saito, T.; Sayama, A.; Matsuda, H.; Kato, M.; Kowada, R.; Yasuta, Y.; Igarashi, Y.; Komori, H. Factors affecting red crown rot caused by *Calonectria ilicicola* in soybean cultivation. *J. Gen. Plant Pathol.* **2020**, *86*, 363–375. [CrossRef]
16. Win, K.T.; Kobayashi, M.; Tanaka, F.; Takeuchi, K.; Oo, A.Z.; Jiang, C.-J. Identification of *Pseudomonas* strains for the biological control of soybean red crown root rot. *Sci. Rep.* **2022**, *12*, 14510. [CrossRef] [PubMed]
17. Berner, D.; Berggren, G.; Pace, M.; White, E.; Gershey, J.; Freedman, J. Red crown rot: Now a major disease of soybeans. *La Agric.* **1986**, *29*, 4–5.
18. Lombard, L.; Crous, P.W.; Wingfield, B.D.; Wingfield, M.J. Phylogeny and systematics of the genus *Calonectria*. *Stud. Mycol.* **2010**, *66*, 31–69. [CrossRef] [PubMed]
19. Wang, H.; Wu, J.; Fang, L.; Xie, Y.; Wang, L. First Report of *Calonectria ilicicola* Causing Fruit Rot on Postharvest Prunus persica in Zhejiang Province, China. *Plant Dis.* **2023**, *107*, 3313. [CrossRef] [PubMed]
20. Liu, H.; Shen, Y.; Chang, H.; Tseng, M.; Lin, Y. First report of soybean red crown rot caused by *Calonectria ilicicola* in Taiwan. *Plant Dis.* **2020**, *104*, 979. [CrossRef]
21. Zhang, Y.; Chen, C.; Chen, C.; Chen, J.; Xiang, M.; Wanasinghe, D.N.; Hsiang, T.; Hyde, K.D.; Manawasinghe, I.S. Identification and characterization of *Calonectria* species associated with plant diseases in Southern China. *J. Fungi* **2022**, *8*, 719. [CrossRef]
22. Tazawa, J. Occurrence and control of red crown rot of soybean. *Plant Prot.* **2013**, *67*, 46–49.
23. Ochi, S. Studies on the epidemiology and control of red crown rot of soybean. *J. Gen. Plant Pathol.* **2017**, *83*, 427–428. [CrossRef]
24. Guan, M.; Pan, R.; Gao, X.; Xu, D.; Deng, Q.; Deng, M. First report of red crown rot caused by *Cylindrocladium parasiticum* on soybean in Guangdong, Southern China. *Plant Dis.* **2010**, *94*, 485. [CrossRef] [PubMed]
25. Jiang, C.-J.; Sugano, S.; Ochi, S.; Kaga, A.; Ishimoto, M. Evaluation of *Glycine max* and *Glycine soja* for resistance to *Calonectria ilicicola*. *Agronomy* **2020**, *10*, 887. [CrossRef]
26. Shafique, H.A.; Sultana, V.; Ehteshamul-Haque, S.; Athar, M. Management of soil-borne diseases of organic vegetables. *J. Plant Prot. Res.* **2016**, *56*, 221–230. [CrossRef]
27. Jiang, H.; Lv, S.; Zhou, C.; Qu, S.; Liu, F.; Sun, H.; Zhao, X.; Han, Y. Identification of QTL, QTL-by-environment interactions, and their candidate genes for resistance HG Type 0 and HG Type 1.2. 3.5. 7 in soybean using 3VmrMLM. *Front. Plant Sci.* **2023**, *14*, 1177345. [CrossRef]
28. Kantartzi, S.K.; McAllister, K.; Lee, Y.-C. Targeting and mapping resistance to *Cercospora sojina* in two elite soybean (*Glycine max* L.) populations. *J. Plant Breed. Crop Sci.* **2023**, *15*, 118–128.
29. Ferreira, E.G.C.; Marcelino-Guimarães, F.C. Mapping Major Disease Diseases Resistance Resistance Genes in Soybean by Genome-Wide Association Studies. *Methods Mol. Biol.* **2022**, *2481*, 313–340. [PubMed]
30. Xiong, H.; Chen, Y.; Pan, Y.-B.; Wang, J.; Lu, W.; Shi, A. A genome-wide association study and genomic prediction for *Phakopsora pachyrhizi* resistance in soybean. *Front. Plant Sci.* **2023**, *14*, 1179357. [CrossRef]
31. Sang, Y.; Liu, X.; Yuan, C.; Yao, T.; Li, Y.; Wang, D.; Zhao, H.; Wang, Y. Genome-wide association study on resistance of cultivated soybean to *Fusarium oxysporum* root rot in Northeast China. *BMC Plant Biol.* **2023**, *23*, 625. [CrossRef]
32. Zhao, F.; Cheng, W.; Wang, Y.; Gao, X.; Huang, D.; Kong, J.; Antwi-Boasiako, A.; Zheng, L.; Yan, W.; Chang, F. Identification of novel genomic regions for bacterial leaf pustule (BLP) resistance in soybean (*Glycine max* L.) via integrating linkage mapping and association analysis. *Int. J. Mol. Sci.* **2022**, *23*, 2113. [CrossRef]
33. Fournier, T.; Abou Saada, O.; Hou, J.; Peter, J.; Caudal, E.; Schacherer, J. Extensive impact of low-frequency variants on the phenotypic landscape at population-scale. *eLife* **2019**, *8*, e49258. [CrossRef] [PubMed]
34. Chanchu, T.; Yimram, T.; Chankaew, S.; Kaga, A.; Somta, P. Mapping QTLs Controlling Soybean Rust Disease Resistance in Chiang Mai 5, an Induced Mutant Cultivar. *Genes* **2022**, *14*, 19. [CrossRef]
35. Yamanaka, N.; Aoyagi, L.N.; Hossain, M.M.; Aoyagi, M.B.; Muraki, Y. Genetic Mapping of Seven Kinds of Locus for Resistance to Asian Soybean Rust. *Plants* **2023**, *12*, 2263. [CrossRef]

36. Kim, K.H.; Park, J.-H.; Kim, M.Y.; Heu, S.; Lee, S.-H. Genetic mapping of novel symptom in response to soybean bacterial leaf pustule in PI 96188. *J. Crop Sci. Biotechnol.* **2011**, *14*, 119–123. [CrossRef]
37. Li, D.; Sun, M.; Han, Y.; Teng, W.; Li, W. Identification of QTL underlying soluble pigment content in soybean stems related to resistance to soybean white mold (*Sclerotinia sclerotiorum*). *Euphytica* **2010**, *172*, 49–57. [CrossRef]
38. Huynh, T.; Bastien, M.; Iquira, E.; Turcotte, P.; Belzile, F. Identification of QTLs associated with partial resistance to white mold in soybean using field-based inoculation. *Crop Sci.* **2010**, *50*, 969–979. [CrossRef]
39. Han, Y.; Teng, W.; Yu, K.; Poysa, V.; Anderson, T.; Qiu, L.; Lightfoot, D.A.; Li, W. Mapping QTL tolerance to *Phytophthora* root rot in soybean using microsatellite and RAPD/SCAR derived markers. *Euphytica* **2008**, *162*, 231–239. [CrossRef]
40. Zhang, Z.; Hao, J.; Yuan, J.; Song, Q.; Hyten, D.L.; Cregan, P.B.; Zhang, G.; Gu, C.; Li, M.; Wang, D. *Phytophthora* root rot resistance in soybean E00003. *Crop Sci.* **2014**, *54*, 492–499. [CrossRef]
41. Niu, J.; Guo, N.; Sun, J.; Li, L.; Cao, Y.; Li, S.; Huang, J.; Zhao, J.; Zhao, T.; Xing, H. Fine mapping of a resistance gene RpsHN that controls *Phytophthora sojae* using recombinant inbred lines and secondary populations. *Front. Plant Sci.* **2017**, *8*, 538. [CrossRef]
42. McAllister, K.R.; Lee, Y.-C.; Kantartzi, S.K. QTL mapping for resistance to *Cercospora sojina* in Essex Forrest soybean (*Glycine max* L.) lines. *J. Plant Breed. Crop Sci.* **2021**, *13*, 14–22.
43. Pham, A.-T.; Harris, D.K.; Buck, J.; Hoskins, A.; Serrano, J.; Abdel-Haleem, H.; Cregan, P.; Song, Q.; Boerma, H.R.; Li, Z. Fine mapping and characterization of candidate genes that control resistance to *Cercospora sojina* K. Hara in two soybean germplasm accessions. *PLoS ONE* **2015**, *10*, e0126753. [CrossRef] [PubMed]
44. Wei, T.; Simko, V. R package "Corrplot": Visualization of a Correlation Matrix (Version 0.84). 2017. Available online: https://github.com/taiyun/corrplot (accessed on 7 January 2023).
45. Zhang, X.; Hina, A.; Song, S.; Kong, J.; Bhat, J.A.; Zhao, T. Whole-genome mapping identified novel "QTL hotspots regions" for seed storability in soybean (*Glycine max* L.). *BMC Genom.* **2019**, *20*, 499. [CrossRef] [PubMed]
46. Huang, X.; Feng, Q.; Qian, Q.; Zhao, Q.; Wang, L.; Wang, A.; Guan, J.; Fan, D.; Weng, Q.; Huang, T. High-throughput genotyping by whole-genome resequencing. *Genome Res.* **2009**, *19*, 1068–1076. [CrossRef] [PubMed]
47. Cao, Y.; Li, S.; Chen, G.; Wang, Y.; Bhat, J.A.; Karikari, B.; Kong, J.; Gai, J.; Zhao, T. Deciphering the genetic architecture of plant height in soybean using two RIL populations sharing a common M8206 parent. *Plants* **2019**, *8*, 373. [CrossRef] [PubMed]
48. Wang, S. *Windows QTL Cartographer 2.5*; Department of Statistics, North Carolina State University: Raleigh, NC, USA, 2007. Available online: https://brcwebportal.cos.ncsu.edu/qtlcart/WQTLCart.htm (accessed on 1 January 2019).
49. Song, Q.; Marek, L.; Shoemaker, R.; Lark, K.; Concibido, V.; Delannay, X.; Specht, J.E.; Cregan, P. A new integrated genetic linkage map of the soybean. *Theor. Appl. Genet.* **2004**, *109*, 122–128. [CrossRef] [PubMed]
50. Churchill, G.A.; Doerge, R.W. Empirical threshold values for quantitative trait mapping. *Genetics* **1994**, *138*, 963–971. [CrossRef] [PubMed]
51. Voorrips, R. MapChart: Software for the graphical presentation of linkage maps and QTLs. *J. Hered.* **2002**, *93*, 77–78. [CrossRef] [PubMed]
52. McCouch, S.R.; Chen, X.; Panaud, O.; Temnykh, S.; Xu, Y.; Cho, Y.G.; Huang, N.; Ishii, T.; Blair, M. Microsatellite marker development, mapping and applications in rice genetics and breeding. In *Oryza: From Molecule to Plant*; Sasaki, T., Moore, G., Eds.; Springer: Dordrecht, The Netherlands, 1997. [CrossRef]
53. Tian, T.; Liu, Y.; Yan, H.; You, Q.; Yi, X.; Du, Z.; Xu, W.; Su, Z. agriGO v2. 0: A GO analysis toolkit for the agricultural community, 2017 update. *Nucleic Acids Res.* **2017**, *45*, W122–W129. [CrossRef]
54. Kobayashi, M.; Win, K.T.; Jiang, C.-J. Soybean hypocotyls prevent *Calonectria ilicicola* invasion by multi-layered defenses. *Front. Plant Sci.* **2022**, *12*, 813578. [CrossRef]
55. Chen, C.; Chen, H.; He, Y.; Xia, R. TBtools, a toolkit for biologists integrating various biological data handling tools with a user-friendly interface. *BioRxiv* **2018**, *1*, 289660. [CrossRef]
56. Hung, H.Y.; Browne, C.; Guill, K.; Coles, N.; Eller, M.; Garcia, A.; Lepak, N.; Melia-Hancock, S.; Oropeza-Rosas, M.; Salvo, S.; et al. The relationship between parental genetic or phenotypic divergence and progeny variation in the maize nested association mapping population. *Heredity* **2012**, *108*, 490–499. [CrossRef] [PubMed]
57. Gelli, M.; Konda, A.R.; Liu, K.; Zhang, C.; Clemente, T.E.; Holding, D.R.; Dweikat, I.M. Validation of QTL mapping and transcriptome profiling for identification of candidate genes associated with nitrogen stress tolerance in sorghum. *BMC Plant Biol.* **2017**, *17*, 123. [CrossRef] [PubMed]
58. Win, K.T.; Jiang, C.-J. A fresh weight-based method for evaluating soybean resistance to red crown rot. *Breed. Sci.* **2021**, *71*, 384–389. [CrossRef] [PubMed]
59. Nakajima, T.; Sakai, S.; Gomi, T.; Kikuchi, A. Development of methods for assessing resistance to black root rot caused by *Calonectria crotalariae* in soybean [Glycine max] and screening for resistant germplasm. *Bull. Tohoku Natl. Agric. Exp. Stn.* **1994**, *88*, 39–56.
60. Antwi-Boasiako, A.; Jia, S.; Liu, J.; Guo, N.; Chen, C.; Karikari, B.; Feng, J.; Zhao, T. Identification and Genetic Dissection of Resistance to Red Crown Rot Disease in a Diverse Soybean Germplasm Population. *Plants* **2024**, *13*, 940. [CrossRef] [PubMed]
61. Cao, Y.; Li, S.; Wang, Z.; Chang, F.; Kong, J.; Gai, J.; Zhao, T. Identification of Major Quantitative Trait Loci for Seed Oil Content in Soybeans by Combining Linkage and Genome-Wide Association Mapping. *Front. Plant Sci.* **2017**, *8*, 1222. [CrossRef] [PubMed]
62. Li, Z.; Xu, Y. Bulk segregation analysis in the NGS era: A review of its teenage years. *Plant J.* **2022**, *109*, 1355–1374. [CrossRef] [PubMed]

63. Mackay, I.; Caligari, P. Efficiencies of F2 and backcross generations for bulked segregant analysis using dominant markers. *Crop Sci.* **2000**, *40*, 626–630. [CrossRef]
64. Jansen, R.; Van Ooijen, J.; Stam, P.; Lister, C.; Dean, C. Genotype-by-environment interaction in genetic mapping of multiple quantitative trait loci. *Theor. Appl. Genet.* **1995**, *91*, 33–37. [CrossRef]
65. Han, Y.; Zhang, Y.; Wu, D.; Zhao, X.; Teng, W.; Li, D.; Li, W. Identification of novel quantitative trait loci associated with tolerance to *Phytophthora* root rot in the soybean cultivar Hefeng 25 using two recombinant inbred line populations. *Can. J. Plant Sci.* **2017**, *97*, 827–834. [CrossRef]
66. Antwi-Boasiako, A.; Zheng, L.; Begum, N.; Amoah, S.; Zhao, T. Progress towards germplasm evaluation and genetic improvement for resistance to *Sclerotinia* white mold in soybean. *Euphytica* **2021**, *217*, 178. [CrossRef]
67. Huang, X.; Han, B. Natural variations and genome-wide association studies in crop plants. *Annu. Rev. Plant Biol.* **2014**, *65*, 531–551. [CrossRef] [PubMed]
68. Ding, L.-N.; Li, Y.-T.; Wu, Y.-Z.; Li, T.; Geng, R.; Cao, J.; Zhang, W.; Tan, X.-L. Plant disease resistance-related signaling pathways: Recent progress and future prospects. *Int. J. Mol. Sci.* **2022**, *23*, 16200. [CrossRef] [PubMed]
69. Ngou, B.P.M.; Ding, P.; Jones, J.D. Thirty years of resistance: Zig-zag through the plant immune system. *Plant Cell* **2022**, *34*, 1447–1478. [CrossRef] [PubMed]
70. Sertsuvalkul, N.; DeMell, A.; Dinesh-Kumar, S.P. The complex roles of autophagy in plant immunity. *FEBS Lett.* **2022**, *596*, 2163–2171. [CrossRef] [PubMed]
71. Devanna, B.N.; Jaswal, R.; Singh, P.K.; Kapoor, R.; Jain, P.; Kumar, G.; Sharma, Y.; Samantaray, S.; Sharma, T.R. Role of transporters in plant disease resistance. *Physiol. Plant.* **2021**, *171*, 849–867. [CrossRef] [PubMed]
72. Strobel, G.A. A mechanism of disease resistance in plants. *Sci. Am.* **1975**, *232*, 80–89. [CrossRef]
73. Xi, Y.; Cesari, S.; Kroj, T. Insight into the structure and molecular mode of action of plant paired NLR immune receptors. *Essays Biochem.* **2022**, *66*, 513–526.
74. Zhou, L.; Deng, S.; Xuan, H.; Fan, X.; Sun, R.; Zhao, J.; Wang, H.; Guo, N.; Xing, H. A novel TIR-NBS-LRR gene regulates immune response to *Phytophthora* root rot in soybean. *Crop J.* **2022**, *10*, 1644–1653. [CrossRef]
75. Ameline-Torregrosa, C.; Wang, B.-B.; O'Bleness, M.S.; Deshpande, S.; Zhu, H.; Roe, B.; Young, N.D.; Cannon, S.B. Identification and characterization of nucleotide-binding site-leucine-rich repeat genes in the model plant *Medicago truncatula*. *Plant Physiol.* **2008**, *146*, 5–21. [CrossRef] [PubMed]
76. Wang, F.; Lu, T.; Zhu, L.; Cao, A.; Xie, S.; Chen, X.; Shen, H.; Xie, Q.; Li, R.; Zhu, J. Multicopper oxidases GbAO and GbSKS are involved in the *Verticillium dahliae* resistance in *Gossypium barbadense*. *J. Plant Physiol.* **2023**, *280*, 153887. [CrossRef] [PubMed]
77. Liu, J.Z.; Whitham, S.A. Overexpression of a soybean nuclear localized type–III DnaJ domain-containing HSP40 reveals its roles in cell death and disease resistance. *Plant J.* **2013**, *74*, 110–121. [CrossRef] [PubMed]
78. Kim, N.H.; Hwang, B.K. Pepper heat shock protein 70a interacts with the type III effector AvrBsT and triggers plant cell death and immunity. *Plant Physiol.* **2015**, *167*, 307–322. [CrossRef] [PubMed]

Disclaimer/Publisher's Note: The statements, opinions and data contained in all publications are solely those of the individual author(s) and contributor(s) and not of MDPI and/or the editor(s). MDPI and/or the editor(s) disclaim responsibility for any injury to people or property resulting from any ideas, methods, instructions or products referred to in the content.

agronomy

MDPI

Article

Uncovering *Fusarium* Species Associated with Fusarium Wilt in Chickpeas (*Cicer arietinum* L.) and the Identification of Significant Marker–Trait Associations for Resistance in the International Center for Agricultural Research in the Dry Areas' Chickpea Collection Using SSR Markers

Sojida M. Murodova [1], Tohir A. Bozorov [1,*], Ilkham S. Aytenov [1], Bekhruz O. Ochilov [1], Dilafruz E. Qulmamatova [1], Ilkhom B. Salakhutdinov [2], Marufbek Z. Isokulov [1], Gavkhar O. Khalillaeva [1], Laylo A. Azimova [1] and Sodir K. Meliev [1]

1 Laboratory of Molecular and Biochemical Genetics, Institute of Genetics and Plants Experimental Biology, Uzbek Academy of Sciences, Yukori-Yuz, Kibray 111226, Tashkent Region, Uzbekistan; murodovasojida1433@gmail.com (S.M.M.); ilhamaytenov@gmail.com (I.S.A.); behruz.ochilov1995@mail.ru (B.O.O.); dilafruz10.10@mail.ru (D.E.Q.); isoqulovmarufbek@gmail.com (M.Z.I.); xalillaevagavhar@gmail.com (G.O.K.); laylobio@gmail.com (L.A.A.); meliev.sodir@mail.ru (S.K.M.)

2 Laboratory of Plant Resistance Genomics, Center of Genomics and Bioinformatics, Uzbek Academy of Sciences, University Street 2, Kibray 111226, Tashkent Region, Uzbekistan; ilkhom.salakhutdinov@genomics.uz

* Correspondence: tohirbozorov@yahoo.com

Abstract: Enhancing plants' resistance against FW is crucial for ensuring a sustainable global chickpea production. The present study focuses on the identification of fungal pathogens and the assessment of ninety-six chickpea samples for Fusarium wilt from the International Center for Agricultural Research in the Dry Areas (ICARDA)'s collection. Eight fungal isolates were recovered from the symptomatic chickpeas. Polyphasic identification was conducted by comparing the internal transcribed spacer region (ITS), the *elongation factor 1-α* (*tef1-α*), and *beta-tubulin* (*tub2*). Among them, *Neocosmospora solani*, *N. nelsonii*, *N. falciformis*, *N. brevis*, *Fusarium brachygibbosum*, and *F. gossypinum* were identified. An analysis of the genetic diversity of chickpeas, using 69 polymorphic simple sequence repeat (SSR) markers, revealed a total of 191 alleles across all markers, with, on average, each SSR marker detecting approximately 2.8 alleles. A STRUCTURE analysis delineated lines into two distinct subgroups (K = 2). Association mapping, using the generalized linear model (GLM) and mixed linear model (MLM) approaches, identified six and five marker–trait associations (MTAs) for FW resistance, respectively. Notably, these TA42, TA125 (A) and TA125 (B), TA37, and TAASH MTAs, commonly found in both models, emerge as potential candidates for the targeted enhancement of FW resistance in chickpeas. To our knowledge, this study represents an inaugural report on the association mapping of genomic loci governing FW resistance in chickpeas from the ICARDA's accessions.

Keywords: association mapping; chickpeas; *Cicer arietinum* L.; Fusarium wilt; SSR markers

1. Introduction

The chickpea (*Cicer arietinum* L.), of the Fabaceae family, holds significant agricultural importance globally, and ranks as the second most important legume crop, following the common bean (*Phaseolus vulgaris* L.) and the dry pea (*Pisum sativum* L.) [1]. It is an important source of dietary protein for humans and provides crucial feed for livestock. Its contribution is vital to the sustainable cultivation of cereals, particularly within wheat-based cropping systems, as it enhances the soil's fertility through nitrogen fixation. Globally, chickpea production has reached 14.2 million metric tons, and is mostly concentrated in Asian regions (83% of total production) (FAO, 2019) [2].

The chickpea possesses a relatively compact genome size of 750 Mbp and exhibits minimal genetic polymorphisms [3]. To study its diversity or perform a linkage analysis for this crop, a substantial quantity of polymorphic markers is necessary [4]. Previous investigations of chickpeas, employing RAPD and RFLP markers, uncovered minimal polymorphic variations. Simple sequence repeat (SSR) markers have become extensively employed in the genetic characterization and diversity analyses of crop plants [5].

Numerous diseases pose substantial challenges to global chickpea production, where Fusarium wilt and root rot are caused by different *Fusarium* spp. [6]. *Fusarium oxysporum* f. sp. *ciceris* (Foc) is recognized as one of the most critical. Managing Fusarium wilt has predominantly involved the breeding of resistant cultivars as a component of integrated management strategies. However, the extensive pathogenic variability within populations of *F. oxysporum* f. sp. *ciceris* poses challenges to the long-term sustainability of resistant cultivars [7]. Ascochyta blight (*Didymella rabiei*), Fusarium wilt, cyst nematodes (*Heterodera* spp.), and insect pests are the primary biotic factors contributing to significant yield gaps and reduced planted areas. Fusarium wilt, in particular, stands out as a crucial soil-borne disease, affecting chickpea production in the Mediterranean region, in South Asia, and in East Africa [8–10]. The efficient identification of pathogenic races of *F. oxysporum* f. sp. *ciceris* is essential, given a reliance on resistant cultivars for managing Fusarium wilt disease. The traditional non-molecular methods for determining an organism's taxonomic level of 'formae speciales' (special forms) of organisms are resource-intensive, in terms of both time and materials. To overcome these obstacles, a PCR-based molecular assay has been devised [11].

Many countries report annual yield losses ranging from 10 to 15%; under optimal environmental conditions favourable for pathogens, this can potentially result in total crop failure [12–14]. Nevertheless, significant attention has been directed towards developing chickpea genotypes with increased levels of resistance against Fusarium wilt, in both national and international chickpea breeding programs [15]. In chickpea breeding, there is a strong emphasis on identifying and developing genotypes that exhibit a broad resistance to multiple races and populations of Foc. Recognizing the diversity of races within Foc is crucial for developing effective breeding programs aimed at enhancing resistance in chickpeas. It should be noted that the plant's resistance may diminish when exposed to a combination of *Fusarium* and other soil-borne pathogens [16]. Various races and genetically diverse populations of Foc have been documented across India, Turkey, Tunisia, Spain, Sudan, Iran, and Ethiopia [17]. The use of chickpea differential lines and molecular markers has enabled researchers to identify eight distinct races (0, 1A, 1B/C, 2, 3, 4, 5, and 6) of Foc. Furthermore, based on their above-ground symptoms, the following two distinct pathotypes were identified: yellowing and wilting [5,9]. The wilting symptom is mainly characterized by severe chlorosis and flaccidity, accompanied by vascular discoloration leading to plant death, and is induced by the races 1A, 2, 3, 4, 5, and 6. Conversely, the races 0 and 1B/C induce the yellowing symptom [5,15]. Races 0 and 1B/C demonstrate lower levels of virulence compared to other known races responsible for wilting [8,11]. California (USA) and Spain have documented the races 0, 1A, 1B/C, 5, and 6, while Syria, Tunisia, and Turkey have reported the races 0 and 1B/C, and Israel has observed the races 0, 1A, and 6. Morocco has recorded the races 1A and 6, and Lebanon has confirmed the presence of the race 0 [18,19].

Molecular-marker technology is essential for examining the genetic diversity of crops [20]. Various markers, including restriction fragment length polymorphisms (RFLPs), random amplified polymorphic DNA (RAPD), amplified fragment length polymorphisms (AFLPs), simple sequence repeats (SSRs), and single nucleotide polymorphisms (SNPs) are employed in genetic diversity studies [21]. SSRs, in particular, are favoured for their high polymorphism, reliable reproducibility, co-dominance, and straightforward application [22]. Numerous studies have highlighted the extensive use of SSRs in chickpea genetics, genomics, and breeding [21]. SSRs are applied in areas such as cultivar identification, allele mining, genetic mapping, association studies, genetic diversity analyses, population structure investigations, and phylogenetic relationship establishment [23]. In this research, we conducted experiments on 96 chickpea samples with Fusarium wilt from the International Center for Agricultural Research in the Dry Areas (ICARDA)'s chickpea collections, including one local variety, which were analysed using SSR markers to determine the marker–trait associations (MTAs) for Fusarium wilt resistance. The results of our study will also provide genetic resources for breeding new chickpea varieties and help categorize accessions based on their resistance capabilities.

2. Materials and Methods

2.1. Plant Materials and Fungal Pathogens

In this study, we conducted experiments on 96 chickpea samples from the ICARDA's CIFWN (Chickpea International Fusarium Wilt Nursery 2020) and CIENMED (Chickpea International Elite Nursery for Mediterranean Environments 2021) collections, including one local variety (Supplementary Table S1). The traits were studied over a period of two years (2022–2023) in their respective field conditions.

Fungi pathogens were isolated from the infected chickpea samples. Briefly, the infected plant stems were ground using a home blender in sterile conditions. The ground tissues were placed on a potato dextrose agar (potato starch 4 gm L^{-1}, dextrose 20 gm L^{-1}, and agar 15 gm L^{-1}, pH 5.6) (Cat#P8931-250G, Solarbio, Beijing, China) and put in a thermostatic incubator at 25 °C, until the appearance of fungal colonies were observed. The DNA of the fungus isolates was extracted following the method described by Liu et al. [24]; the internal transcribed spacer region (ITS) of 18S RNA, *elongation factor 1-α* (*tef1-α*), and *beta-tubulin* (*tub2*) gene fragments were sequenced, analysed, and molecularly identified, as described by Liu et al. [25].

2.2. *Fusarium* Infection

The twenty-days-old chickpea samples were artificially infected and their disease severity levels were studied. Each plant sample was studied with 10 plants in two replicates. The fungal inoculation was performed following the method described by Armstrong-Cho et al. [26]. Briefly, 30 mL of a freshly prepared conidia suspension (1×100 conidia $mL^{-1)}$ was applied to the base of each seedling growing in a 2 L pot, a method known as drenching. In the non-inoculated control group, water was applied to the base of the seedlings, instead. This inoculation was repeated three times. The plants were cultivated in a glasshouse environment at a temperature of 25 ± 1 °C, a 16 h light and 8 h dark photoperiod, and a relative humidity ranging between 50% and 55%. The disease severity in the infected plants was assessed under laboratory conditions, according to the method described by Chumakov et al. [27], and the FW resistance was assessed in a manner similar to the methodology described by Jha et al. (2018) [28]. The genotypes were categorized based on their disease–incidence percentages; those with less than 10.0% were deemed resistant, while those ranging between 10.1% and 20.0% were classified as moderately tolerant. The genotypes exhibiting incidences between 20.1% and 50.0% were labelled as moderately susceptible, and those exceeding 50.0% were categorized as susceptible.

2.3. *Polymerase Chain Reaction (PCR) Analysis*

The CTAB method described by Saghai-Maroof et al. [29] was employed to extract the genomic DNA from the 3-weeks-old chickpea seedlings. The PCR protocol, a reaction mixture of 10 μL volume, comprised 6.6 μL distilled water, 1.00 μL template DNA (25 ng), 0.2 μL forward and 0.2 μL reverse primers (5 μM each), 1.00 μL 10 × PCR buffer (10 mM Tris-HCl, 50 mM KCl, pH 8.3), 0.2 μL dNTP mix (0.04 mM each of dATP, dGTP, dCTP, and dTTP), and 0.2 μL Taq polymerase, was prepared. The polymerization of this reaction mixture was conducted using a TC1000-G Thermocycler (DLAB, Beijing, China) employing a touchdown PCR profile for amplification, which included an initial denaturation at 94 °C for 10 min, followed by 35 cycles at 94 °C for 15 s (denaturation), an appropriate temperature of primers for 30 s (annealing), and 72 °C for 1 min (elongation), with a final extension at 72 °C for 10 min. The amplified fragments were separated on a 2% agarose gel using a 1 × TBE running buffer, and the images were analysed using gel analyser software (23.1.1 version).

2.4. Genetic Diversity and Population Structure Analysis

The genetic diversity parameters, including the number of alleles per locus (Na), gene diversity (He), and polymorphism information content (PIC), were calculated using an online marker efficiency calculator (https://irscope.shinyapps.io/iMEC/, accessed on 1 April 2024). A neighbour-joining tree with 1000 bootstrap values was conducted using the software TASSEL v. 5.0 to visualize the genetic relationships among the genotypes. A population structure (Q) and subpopulation (K) determination was performed using a model-based analysis with the software STRUCTURE v. 2.3.4 [30]. Ten independent runs employing the admixture model were conducted with 100,000 Markov Chain Monte Carlo iterations for each K value ranging from 2 to 10, with a burn-in length of 280. Additionally, the best K value was determined using the ΔK method, proposed by Evanno et al. [31], through the processing of STRUCTURE results with STRUCTURE HARVESTER (available at https://github.com/dentearl/structureHarvester, accessed on 1 April 2024).

2.5. Association Mapping for the Determination of Marker–Trait Associations (MTA)

The wilt disease scores and genotypic data were analysed to identify the significant MTAs using both the generalized linear model (GLM) and mixed linear model (MLM) approaches in TASSEL v. 5.0 [32,33]. To detect the MTAs, thresholds of $p < 0.05 < 0.01$ were applied.

3. Results

3.1. The Identification of Pathogenic Fungi Isolates

The fungi isolates from the diseased chickpeas were collected. Based on the colonial morphologies of these isolates, we selected those that closely resembled *Fusarium* patterns for molecular identification. The identification of the selected fungal isolates using a sequence of internal transcribed spacer (ITS) regions demonstrated that all selected isolates belonged to the *Fusarium* species. Among them, eight *Fusarium* isolates were chosen to further identify their species level (Figure 1).

The DNA from the eight selected *Fusarium* species was extracted, and the ITS, tef1-α, and tub2 regions were used as DNA barcodes. For each barcode, a parsimonious tree (MP) was built. The results indicated that isolates 1 and 4, based on three barcodes, were identified as *F. solani* and *F. brachygibbosum*, respectively, in all three DNA barcodes. The other isolates represented different species of *Fusarium* (Figure 1B). However, the polyphasic identification (https://www.fusarium.org/, accessed on 1 April 2024) of isolates using the three barcodes revealed that six of the isolates were distinct *Fusarium* species, while isolates 3, 5, and 6 were identified as *F. falciformis* (Table 1).

Further, to select the highly virulent fungal species, we examined their virulency with respect to the plants showing the greatest wilting symptoms in order to inoculate the chickpea samples. The results showed that isolate number 4 was identified as *F. brachygibbosum* (which demonstrated a strong wilting symptom in the three independently repeated experiments (Supplementary Figure S1)).

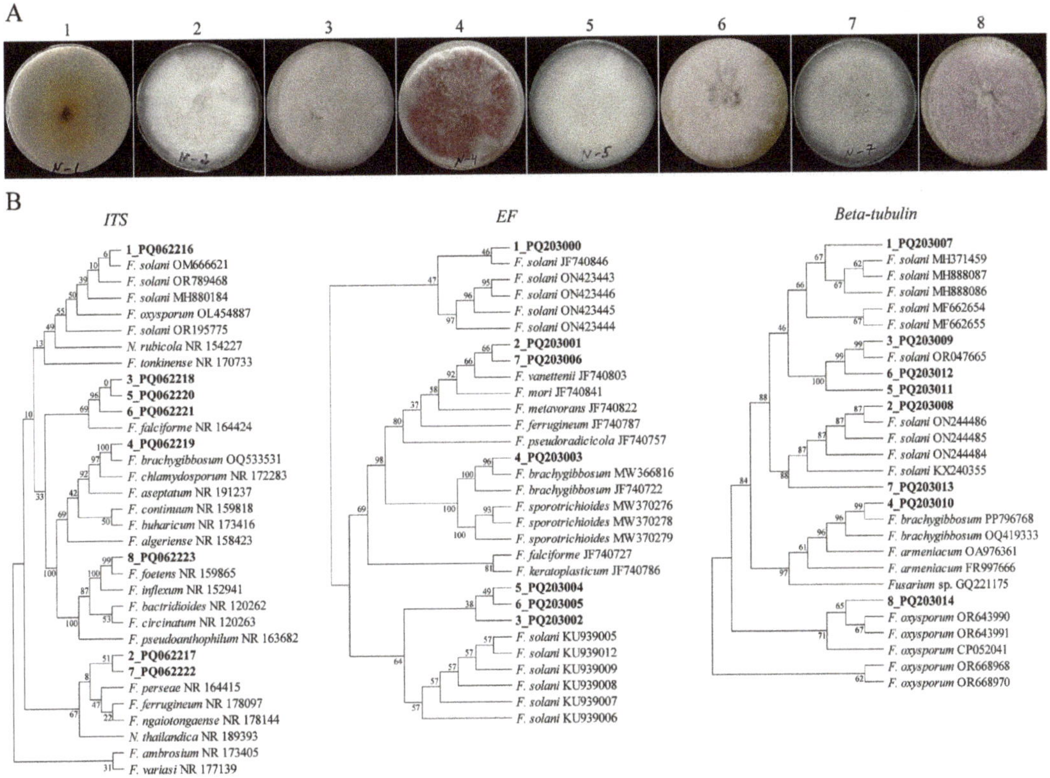

Figure 1. The morphological and molecular data of the *Fusarium* isolates. (**A**) A colonial morphology of *Fusarium* species. The numbers indicate the number of isolates. (**B**) The phylograms of the fungal isolates were constructed using a MP analysis based on ITS, tef1-α, and beta-tubulin barcodes. The evolutionary history was inferred using the Maximum Parsimony method. The percentages of replicate trees in which the associated taxa are clustered together in the bootstrap test (500 replicates) are shown next to the branches. The MP tree was obtained using the Subtree-Pruning-Regrafting (SPR) algorithm with the search level 1, in which the initial trees were obtained by the random addition of sequences (10 replicates). The evolutionary analyses were conducted in MEGA11.

Table 1. The polyphasic identification of fungal isolates using three DNA barcodes.

Isolate	Predicted Fungi	Strain	Similarity, %
1	*Neocosmospora solani* (previously the *F. solani*)	NRRL 46643	99.71
2	*Neocosmospora nelsonii* (previously the *F. caricae*)	CBS 309.75	99.58
3	*Neocosmospora falciformis* (previously the *F. falciforme*)	NRRL 32798	99.8
4	*Fusarium brachygibbosum*	NRRL 34033	99.4
5	*Neocosmospora falciformis* (previously the *F. falciforme*)	NRRL 32798	99.8
6	*Neocosmospora falciformis* (previously the *F. falciforme*)	NRRL 32339	99.9
7	*Neocosmospora brevis* (previously the *F. brevis*)	F93	99.69
8	*Fusarium gossypinum*	CBS 116613	100

3.2. Genetic Diversity for FW Resistance

An analysis of the FW response in 96 chickpea genotypes revealed a wide range of genetic variability. Figure 1 illustrates the frequency distribution of disease incidence among the tested chickpea genotypes throughout the year. In combining the disease scoring data from the year, a total of 68 genotypes, or 70.8%, were classified as resistant (R); 15, or 15.6%, as moderately resistant (MR); 8, or 8.3%, as moderately susceptible, and 5, or 5.2%, as susceptible (S) (Figure 2; Table 2).

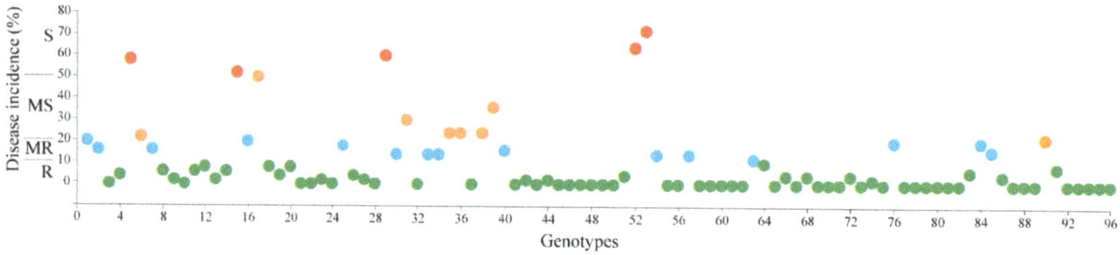

Figure 2. The distribution of the *Fusarium* response of the 96 (CIFWN and CIENMED) chickpea genotypes. R: resistant (green), MR: moderately resistant (blue), MS: moderately susceptible (orange), S: susceptible (red).

Table 2. Details of the 96 chickpea genotypes used for the study.

№	Genotype	Type	FW Response	№	Genotype	Type	FW Response *
1	17102	Germplasm line	MR	49	17174	Germplasm line	R
2	17103	Germplasm line	MR	50	17201	Variety	R
3	17104	Variety	R	51	17206	Germplasm line	R
4	17106	Germplasm line	R	52	17212	Germplasm line	S
5	17108	Germplasm line	S	53	17222	Variety	S
6	17109	Germplasm line	MS	54	17223	Germplasm line	MR
7	17110	Variety	MR	55	17225	Variety	R
8	17111	Germplasm line	R	56	17236	Germplasm line	R
9	17112	Germplasm line	R	57	17244	Germplasm line	MR
10	17114	Germplasm line	R	58	17265	Germplasm line	R
11	17115	Germplasm line	R	59	17269	Germplasm line	R
12	17116	Variety	R	60	17270	Variety	R
13	17117	Germplasm line	R	61	M1	Germplasm line	R
14	17118	Germplasm line	R	62	M2	Germplasm line	R
15	17120	Germplasm line	S	63	M3	Germplasm line	MR
16	17121	Germplasm line	MR	64	M4	Germplasm line	R
17	17123	Germplasm line	MS	65	M5	Germplasm line	R
18	17124	Germplasm line	R	66	M6	Germplasm line	R
19	17125	Variety	R	67	M7	Germplasm line	R
20	17126	Germplasm line	R	68	M8	Germplasm line	R
21	17127	Germplasm line	R	69	M9	Germplasm line	R
22	17130	Germplasm line	R	70	M10	Germplasm line	R
23	17132	Germplasm line	R	71	M11	Variety	R
24	17135	Germplasm line	R	72	M12	Germplasm line	R
25	17136	Germplasm line	MR	73	M13	Germplasm line	R
26	17139	Germplasm line	R	74	M14	Germplasm line	R
27	17141	Germplasm line	R	75	M15	Germplasm line	R
28	17142	Germplasm line	R	76	M16	Germplasm line	MR
29	17143	Variety	S	77	M17	Germplasm line	R
30	17144	Germplasm line	MR	78	M18	Variety	R
31	17145	Germplasm line	MS	79	M19	Germplasm line	R
32	17147	Germplasm line	R	80	M20	Germplasm line	R
33	17148	Germplasm line	MR	81	M21	Germplasm line	R
34	17150	Germplasm line	MR	82	M22	Germplasm line	R
35	17151	Germplasm line	MS	83	M23	Germplasm line	R
36	17153	Germplasm line	MS	84	M24	Germplasm line	MR
37	17154	Germplasm line	R	85	M25	Variety	MR

Table 2. *Cont.*

№	Genotype	Type	FW Response	№	Genotype	Type	FW Response *
38	17156	Germplasm line	MS	86	M26	Germplasm line	R
39	17157	Germplasm line	MS	87	M27	Germplasm line	R
40	17158	Variety	MR	88	M28	Germplasm line	R
41	17159	Germplasm line	R	89	M29	Germplasm line	R
42	17161	Variety	R	90	M30	Germplasm line	MS
43	17162	Germplasm line	R	91	M31	Germplasm line	R
44	17163	Germplasm line	R	92	M32	Germplasm line	R
45	17165	Germplasm line	R	93	M33	Germplasm line	R
46	17166	Germplasm line	R	94	M34	Germplasm line	R
47	17169	Germplasm line	R	95	M35	Variety	R
48	17172	Germplasm line	R	96	M36	Germplasm line	R

* R: resistant, MR: moderately resistant, MS: moderately susceptible, S: susceptible, FW: Fusarium wilt.

3.3. A SSR-Based Molecular Diversity Analysis

A total of 180 SSR markers were screened from the genotypes used in this study (Supplementary Table S2). Among them, 69 markers demonstrated a varied degree of polymorphism, 60 markers showed monomorphic fragments, and the rest of the markers did not produce any fragments, as observed from PCR products on 2.5% agarose gel. These SSRs, previously reported by various research groups [34,35], cover all eight linkage groups (LGs) of chickpeas. An assessment of 96 chickpea genotypes using 69 polymorphic SSRs resulted in the detection of 191 alleles with an average of 2.8 alleles per marker (Supplementary Table S3). The number of observed alleles ranged from two to five, while the PIC values varied between 0.1 and 0.37. Additionally, the gene diversity spanned from 0.1 to 0.5, with an average value of 0.4.

3.4. A Structure Analysis and Cluster Analysis

Using the Bayesian approach with the STRUCTURE software, the population structure of the 96 chickpea genotypes was examined. The population structure study was conducted using a pre-determined number of clusters (K) from 1 to 10 with the optimal number of clusters determined based on the maximum value of delta K (an ad hoc quantity). The highest measure of delta K was attained at K = 2 [31], indicating the existence of two subpopulations in the entire collection (Figure 3A). Based on their Q-values, genotypes with a shared ancestry ≥80% diverged into two major clusters, cluster QI (11 genotypes), and cluster QII (68 genotypes). The remaining 17 genotypes (17.7%) were admixtures with a shared ancestry of <80% (Figure 3B).

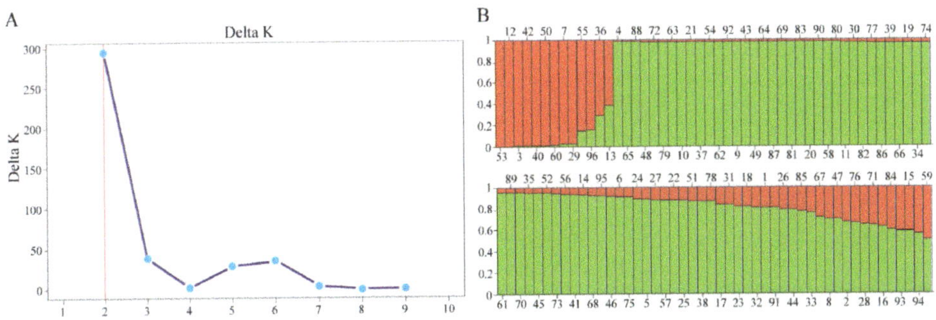

Figure 3. An analysis of the population structure of the chickpea genotypes. (**A**) An estimation of the hypothetical subpopulations based on the ΔK values. (**B**) A population structure analysis based on the Q-values. The maximum measure of Evanno's delta K, determined by the STRUCTURE HARVESTER, was K = 2, which indicates that the entire population can be grouped into 2 subpopulations (red and green).

A Neighbour-joining clustering analysis was performed with a Bayesian-based population structure. The pairwise, the genetic distances among the 96 genotypes ranged from 0.087912 to 0.478022. The entire set of 96 genotypes was grouped into two clusters, based on unweighted neighbour-joining clustering. Cluster 1 contained 62 genotypes, while cluster 2 had 34 genotypes (Figure 3). The delta K-based clusters corresponded to the NJ tree clusters that were assigned according to the population structures QI (red) and QII (green) (Figure 4).

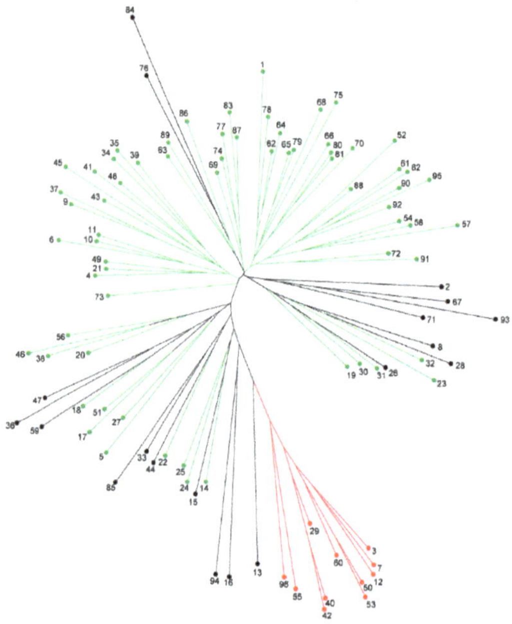

Figure 4. The cluster and population structure analyses. A neighbour-joining tree of the 96 chickpea genotypes for FW resistance phenotyping. Green and red colors indicate Cluster I and II, respectively, while the black color indicates the admixtures identified by STRUCTU RE analysis.

3.5. *Identifying the Associations Related to FW Resistance*

We utilized both the general linear model (GLM) and mixed linear model (MLM) approaches to identify significances in the association mappings for the Fusarium wilt (FW) incidence data and the simple sequence repeat (SSR) dataset. The GLM analysis identified an associative group showing a significance level higher than $p < 0.05$ for FW-resistant traits. Figure 5 presents the Manhattan plots depicting the FW resistance based on both the GLM and MLM analyses, with a threshold p-value set at 1.8.

Both approaches identified the following five common marker–trait associations (MTAs) for FW resistance: TA42, TA125 (A), TA125 (B), TA37, and TAASH. TR2, however, was detected as significant only in the GLM analysis. Table 3 provides detailed data on these findings. The GLM analysis revealed significant associations with the R2 values ranging from 0.06927 to 0.091, while the MLM analysis demonstrated significant associations with the R2 values ranging from 0.0596 to 0.0912 for two MTAs. Interestingly, both the GLM and MLM analyses detected additional significant MTAs for the TA125 SSR marker. These SSR markers displayed the highest associations, and a p-value of 0.05 was particularly noteworthy for its significance. This indicates the robustness of the associations detected across different analytical approaches.

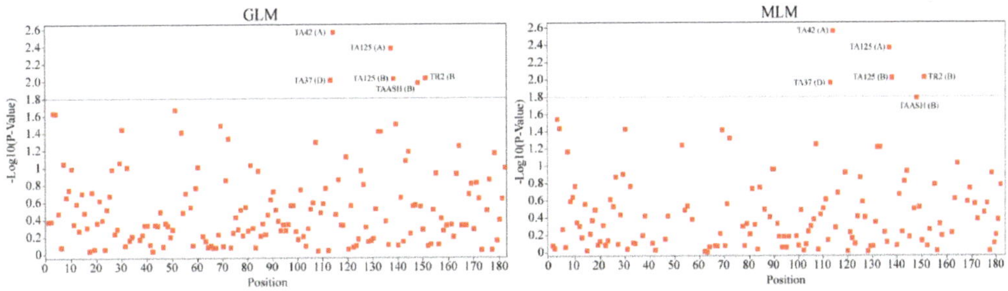

Figure 5. A mapping of the genomic loci linked with Fusarium wilt resistance in chickpeas (GLM and MLM analyses for association mapping).

Table 3. The significant MTAs for FW resistance in the chickpeas obtained in the year 2023 from the GLM and MLM approaches to association mapping.

MTA	Locus	*p*-Value	R^2	LG	Reference *
		GLM mean in year 2023			
TA42	A	0.0028	0.0911	LG5	[36]
TA125	A	0.0043	0.0834	LG1	[36,37]
TR2	B	0.0096	0.0692	LG6	[35]
TA125	B	0.0098	0.0688	LG1	[36,37]
TA37	D	0.01	0.0684	LG2	[35]
TAASH	B	0.0108	0.067	LG1	[37]
		MLM mean in year 2023			
TA42	A	0.0028	0.0912	LG5	[36]
TA125	A	0.0043	0.0834	LG1	[36,37]
TA125	B	0.0098	0.0688	LG1	[36,37]
TA37	D	0.0111	0.0666	LG2	[35]
TAASH	B	0.0165	0.0596	LG1	[37]

* The LG indicates a linkage group based on the literature.

4. Discussion

The purpose of this study was to identify the SSR (simple sequence repeat) markers that are linked to Fusarium wilt (FW) traits in chickpeas. SSR markers are abundant throughout the genome, valued for their high reproducibility and cost-effectiveness, and are applicable to a broad range of species. Their widespread occurrence allows researchers to pinpoint markers that are closely associated with specific traits, which is beneficial in marker-assisted breeding. In our research, we examined 96 chickpea genotypes using 69 SSR markers exhibiting polymorphism. This analysis revealed a total of 191 alleles, with an average of 2.8 alleles per marker. These results are in line with previous studies of chickpeas [28,38].

Diseases caused by *Fusarium* species, causing root rot and Fusarium wilt, have a considerable impact on chickpea yields worldwide. These *Fusarium* species are notable pathogens that can cause damping-off in chickpea seedlings and affect mature plants, as well. [39]. Among these, *Fusarium oxysporum* f. sp. *ciceri* is particularly significant, and is the primary cause of Fusarium wilt in chickpeas across various production regions globally [40,41]. However, there remains a notable gap in the accurate identification of the *Fusarium* pathogens responsible for root rot in chickpeas. This study addresses this issue by focusing on both morphological and molecular methods for identifying *Fusarium* species. Previous research highlights the challenges involved in accurately identifying these pathogens and emphasizes the necessity of employing a thorough approach that combines both morphological observation and genetic analysis [42].

Extensive research has been conducted on various genetic markers in the identification of *Fusarium* species, including the *EF-1α* gene, the ribosomal DNA internal transcribed spacer (ITS) region, and the *β-tubulin* gene. Among these markers, the partial sequencing of the *EF-1α* gene has been particularly recognized for its precision and effectiveness in identifying *Fusarium* species, compared to the other markers [43]. The current study utilized ITS, *tef1-α*, and *tub2* as DNA barcodes to identify *Fusarium* species. Notably, a BLAST analysis of these individual markers identified *Fusarium* species that were similar but distinct (Figure 1B). However, employing a polyphasic approach, which is well-suited for developing taxonomic hypotheses [44], revealed six distinct *Fusarium* species in the Fusarioid-ID database. Our findings align with those of Moparthi et al. [6], who identified several *Fusarium* species from wilted chickpea plants. Among the identified species, *F. brachygibbosum* is known to be a fungal pathogen responsible for causing wilting in a variety of plants [45,46]. In the current study, *F. brachygibbosum* was selected to infect the chickpea samples for the association mapping study due to its ability to induce severe wilting symptoms.

The use of association mapping to identify breeding-relevant traits has gained significant attention for speeding up crop improvement. Notable examples of MTA identification in chickpeas include the mapping of traits such as drought tolerance [47], heat stress resistance [28], seed weight and seed protein content [48], and grain zinc content [49]. Previous studies have elucidated the genetic basis of FW resistance in chickpeas [17,50,51], identifying MTAs through biparental quantitative trait locus (QTL) mapping.

Overall, the potential of association genetics to identify the genomic regions responsible for FW resistance has been partially explored. Subsequent scientific research on chickpeas, focused on their resistance to FW, particularly in the Foc-2 race, was conducted. This involved initial efforts in constructing an association map [38]. In the present study, five markers showed a significant association with the FW response, using both the GLM and MLM methods. Both analyses identified common markers associated with wilt resistance. Earlier studies of chickpeas have reported the presence of QTLs associated with FW resistance in LG01 [52]. Notably, over the course of year, four SSR markers, TA37, TA42, TA125, and TAASH, exhibited consistent significance in the association mappings for FW resistance in both the GLM and MLM analyses. Several studies reported that chickpea resistance to three Foc races (1, 2 and 3) in pot culture experiments was identified as flanking and tightly linked with DNA markers associated with resistance genes. According to molecular analyses, the TA125 marker is the most prominent for FW resistance, producing the highest number of alleles (9) [53]. Moreover, the TA125 SSR marker is associated with drought tolerance in chickpeas [54], indicating that the associated locus may respond to both abiotic and biotic stresses. The closest marker, TA37, was identified at a distance of 0.2 centimorgan (cM) from Foc-1 in LG2. Additionally, TR2 was identified 3.0 cM from the Foc-1 locus, flanking it. These markers, TA37 and TR2, can be utilized in marker-assisted selection, the introgression of resistance genes (R-genes) into economically important cultivars, and, ultimately, to clone the candidate gene for FW resistance [55]. The genomic regions associated with a resistance to FW have been identified and are flanked by the markers TA37 and CaM1402-CaM1101 on the chromosome Ca6 [56]. TA37 is reported to be in the cluster containing genes conferring FW resistance in the linkage group 2 [57]. Moreover, it has successfully amplified an allele of 290 bp in resistant genotypes and shows great potential in the distinction between resistant and susceptible genotypes. It can be reliably used in marker-assisted selection for enhancing wilt resistance in chickpeas. The marker TR2 was significantly associated with a FW response in the GLM analysis. Interestingly, this marker also linked to botrytis grey mould (BGM) resistance, making it useful for developing chickpea genotypes that are tolerant to BGM [58]. A QTL analysis in a similar study revealed the presence of QTLs for both early and late wilting in LG2 [59]. The QTL mapping of FW resistance placed the TAASH marker in LG5 and the TA125 marker in LG1 [50,60]. LG2 in chickpeas notably contains genes/QTLs resistant to the FW races 1 (Foc1), 3 (Foc3), 4 (Foc4), and 5 (Foc5), as reported by various studies [17,50,51,61,62]. Through composite interval mapping, it was determined that there are two main loci in LG1 and LG2. These two loci are considered the primary loci of FW resistance, explaining up to 76.66% of the phenotypic variations in FW resistance [63]. The QTLs in LG4 and LG6 were reported to be associated with Ascochyta blight resistance [64].

Therefore, SSR markers hold promise as indicators to select the desired level of resistance against FW in chickpeas. To enhance the efficiency of identifying chickpea lines resistant to Fusarium wilt disease within the existing germplasm, molecular markers can be employed for screening. This approach will aid in gene pyramiding and molecular breeding [65]. Previous researchers have identified the genetic linkage of resistance genes by utilizing various RAPD and SSR markers for the different races of Foc (Foc 1, 2, 3, 4, and 5) in inbred chickpea lines derived from both resistant and susceptible parent combinations [66]. In our study, we used SSR markers to identify the markers linked to resistance against the *F. brachygibbosum* pathogen in chickpeas. However, further investigation is needed to uncover the races of this pathogen. This will aid in the selection of chickpea lines resistant to specific races. It is worth mentioning that the SSR markers associated with FW resistance in chickpeas might also be linked to other quantitative and qualitative traits. This enhances the reliability of these marker associations. Therefore, in our future research, we will also conduct association mapping studies on other traits in chickpeas.

5. Conclusions

We conducted an identification of *Fusarium* isolates and an investigation into the disease response of 96 chickpea genotypes for Fusarium wilt (FW), using an association mapping approach. This study represents the first MTA analysis for FW in chickpeas, utilizing an FW-responsive chickpea collection and genome-wide SSRs. Significant associations with FW resistance were consistently identified by both the GLM and MLM analyses for the following four SSR markers: TA37, TA42, TA125, and TAASH. Nevertheless, a more detailed mapping is required to accurately pinpoint the responsible gene(s) within the candidate genomic regions for future studies and breeding initiatives aimed at improving FW resistance in chickpeas.

Supplementary Materials: The following supporting information can be downloaded at: https://www.mdpi.com/article/10.3390/agronomy14091943/s1, Figure S1: Infection chickpea with isolate 4 (*Fusarium brachygibbosum*); Table S1: Information of lines used for this study; Table S2: SSR markers used in this study; Table S3: Number of alleles, gene diversity, and polymorphism recorded in 96 chickpea genotypes.

Author Contributions: S.M.M. and T.A.B. wrote the manuscript and prepared all the tables and figures. D.E.Q. provided the chickpea seeds. S.M.M., I.S.A., L.A.A. and B.O.O. isolated and purified. DNA., S.M.M., G.O.K. and M.Z.I. contributed to fungal isolation. S.M.M., I.S.A. and T.A.B. performed fungal identification. S.M.M., S.K.M. and G.O.K. conducted the field work. I.B.S. was responsible for data analysis. T.A.B. supervised the experiment. T.A.B. and D.E.Q. were responsible for funding acquisition. All authors have read and agreed to the published version of the manuscript.

Funding: This research was funded by Ministry of Innovative Development of Uzbekistan (Grant № IL-402104268 and Crant № PZ-20200929166).

Data Availability Statement: The authors declare that the experimental data published in this paper are made accessible upon request for interested readers. All gene sequences of the genes can be found under the following accession numbers: PQ203000, PQ203001, PQ203002, PQ203003, PQ203004, PQ203005, PQ203006, PQ203007, PQ203008, PQ203009, PQ203010, PQ203011, PQ203012, PQ203013, PQ203014, PQ062216, PQ062218, PQ062220, PQ062221, PQ062219, PQ062223, PQ062217, and PQ062222.

Acknowledgments: We appreciate the support provided by Sirojbek Isokulov, from the Institute of Genetics and Experimental Biology of Plants, Uzbek Academy of Sciences, for his assistance with timely chemical ordering.

Conflicts of Interest: The authors declare that they have no conflicts of interest.

References

1. Merga, B.; Haji, J.; Yildiz, F. Economic importance of chickpea: Production, value, and world trade. *Cogent Food Agric.* **2019**, *5*, 1615718. [CrossRef]
2. FAOSTAT. Available online: https://www.fao.org/faostat/en/#home (accessed on 2 June 2024).
3. Varshney, R.K.; Mir, R.R.; Bhatia, S.; Thudi, M.; Hu, Y.; Azam, S.; Zhang, Y.; Jaganathan, D.; You, F.M.; Gao, J.; et al. Integrated physical, genetic and genome map of chickpea (*Cicer arietinum* L.). *Funct. Integr. Genom.* **2014**, *14*, 59–73. [CrossRef]
4. Varshney, R.K.; Graner, A.; Sorrells, M.E. Genic microsatellite markers in plants: Features and applications. *Trends Biotechnol.* **2005**, *23*, 48–55. [CrossRef]
5. Udupa, S.M.; Robertson, L.D.; Weigand, F.; Baum, M.; Hahl, G. Allelic variation at (TAA)n microsatellite loci in a world collection of chickpea (*Gicer arietinum* L.) germplasm. *Mol. Gen. Genet.* **1999**, *261*, 354–363. [CrossRef]
6. Moparthi, S.; Perez-Hernandez, O.; Burrows, M.E.; Bradshaw, M.J.; Bugingo, C.; Brelsford, M.; McPhee, K. Identification of *Fusarium* spp. associated with chickpea root rot in Montana. *Agriculture* **2024**, *14*, 974. [CrossRef]
7. Cunnington, J.; Lindbeck, K.; Jones, R.H. National diagnostic protocol for the detection of Fusarium wilt of Chickpea (*Fusarium oxysporum* f. sp. *ciceris*). In *Plant Health Australia*; SPHD: Canberra, Australia, 2007.
8. Trapero-Casas, A.; Jimenez-Díaz, R.M. Fungal wilt and root rot diseases of chickpea in southern Spain. *Phytopathol. Mediterr.* **1985**, *75*, 1146–1151. [CrossRef]
9. Jalali, B.L.; Chand, H. *Diseases of International Importance, Diseases of Cereals and Pulses*; Singh, U.S., Chaube, H.S., Kumar, J., Mukhopadhyay, A.N., Eds.; Prentice Hall: Englewood Cliff, NJ, USA, 1992; Volume 1, pp. 429–444.
10. Jiménez-Díaz, R.M.; Castillo, P.; Jiménez-Gasco, M.; Landa, B.B.; Navas-Cortés, J.A. Fusarium wilt of chickpeas: Biology, ecology and management. *Crop Prot.* **2015**, *73*, 16–27. [CrossRef]
11. Del Mar Jimenez-Gasco, M.; Jimenez-Diaz, R.M. Development of a specific polymerase chain reaction-based assay for the identification of *Fusarium oxysporum* f. sp. *ciceris* and its pathogenic races 0, 1A, 5, and 6. *Phytopathology* **2003**, *93*, 200–209. [CrossRef]
12. Halila, M.H.; Strange, R.N. Identification of thecausal agent of wilt of chickpea in Tunisia as *Fusarium oxysporum* f. sp. *ciceris* race 0. *Phytopathol. Mediterr.* **1996**, *35*, 67–74.
13. Navas-Cortes, J.A.; Hau, B.; Jimenez-Diaz, R.M. Yield loss in chickpeas in relation to development of fusarium wilt epidemics. *Phytopathology* **2000**, *90*, 1269–1278. [CrossRef]
14. Arvayo-Ortiz, R.M.; Esqueda, M.; Acedo-Felix, E.; Sanchez, A.; Gutierrez, A. Morphological variability and races of *Fusarium oxysporum* f.sp. *ciceris* associated with chickpea (*Cicer arietinum*) crops. *Am. J. Agric. Biol. Sci.* **2011**, *6*, 114–121. [CrossRef]
15. Infantino, A.; Kharrat, M.; Riccioni, L.; Coyne, C.J.; McPhee, K.E.; Grünwald, N.J. Screening techniques and sources of resistance to root diseases in cool season food legumes. *Euphytica* **2006**, *147*, 201–221. [CrossRef]
16. Tesso, T.T.; Ochanda, N.; Little, C.R.; Claflin, L.T.; Tuinstra, M.R. Analysis of host plant resistance to multiple *Fusarium* species associated with stalk rot disease in sorghum [*Sorghum bicolor* (L.) Moench]. *Field Crops Res.* **2010**, *118*, 177–182. [CrossRef]
17. Caballo, C.; Castro, P.; Gil, J.; Millan, T.; Rubio, J.; Die, J.V. Candidate genes expression profiling during wilting in chickpea caused by *Fusarium oxysporum* f. sp. *ciceris* race 5. *PLoS ONE* **2019**, *14*, e0224212. [CrossRef]
18. Jimenez-Diaz, R.M.; Trapero-Casas, A.; de La Colina, J.C. Races of *Fusarium oxysporum* f. sp. *ciceri* infecting chickpeas in southern Spain. In *Vascular Wilt Diseases of Plants: Basic Studies and Control*; Springer: Berlin/Heidelberg, Germany, 1989; pp. 515–520.
19. Gurjar, G.; Barve, M.; Giri, A.; Gupta, V. Identification of Indian pathogenic races of *Fusarium oxysporum* f. sp. *ciceris* with gene specific, *ITS* and random markers. *Mycologia* **2009**, *101*, 484–495. [CrossRef]
20. Kesawat, M.S.; Das Kumar, B. Molecular markers: It's application in crop improvement. *J. Crop Sci. Biotechnol.* **2009**, *12*, 169–181. [CrossRef]
21. Kumar, R.; Sharma, V.K.; Rangari, S.K.; Jha, U.C.; Sahu, A.; Paul, P.J.; Gupta, S.; Gangurde, S.S.; Kudapa, H.; Mir, R.R.; et al. High confidence QTLs and key genes identified using Meta-QTL analysis for enhancing heat tolerance in chickpea (Cicer arietinum L.). *Front. Plant Sci.* **2023**, *14*, 1274759. [CrossRef]
22. Zane, L.; Bargelloni, L.; Patarnello, T. Strategies for microsatellite isolation: A review. *Mol. Ecol.* **2002**, *11*, 1–16. [CrossRef] [PubMed]
23. Morgante, M.; Salamini, F. SSR (Simple Sequence Repeat) markers in crop improvement. In *Genetic Diversity in Plants*; Springer: Berlin/Heidelberg, Germany, 2003.
24. Liu, D.; Coloe, S.; Baird, R.; Pederson, J. Rapid mini-preparation of fungal DNA for PCR. *J. Clin. Microbiol.* **2000**, *38*, 471. [CrossRef]
25. Liu, X.; Li, X.; Bozorov, T.A.; Ma, R.; Ma, J.; Zhang, Y.; Yang, H.; Li, L.; Zhang, D. Characterization and pathogenicity of six Cytospora strains causing stem canker of wild apple in the Tianshan Forest, China. *For. Pathol.* **2020**, *50*, e12587. [CrossRef]
26. Armstrong-Cho, C.; Sivachandra Kumar, N.T.; Kaur, R.; Banniza, S. The chickpea root rot complex in Saskatchewan, Canada-detection of emerging pathogens and their relative pathogenicity. *Front. Plant Sci.* **2023**, *14*, 1117788. [CrossRef] [PubMed]
27. Chumakov, A.E.; Minkevich, I.I.; Vlasov, Y.; Gavrilova, E.A. *The Main Methods of Phytopathological Research*; Kolos: Moscow, Russia, 1974.
28. Jha, U.C.; Jha, R.; Bohra, A.; Parida, S.K.; Kole, P.C.; Thakro, V.; Singh, D.; Singh, N.P. Population structure and association analysis of heat stress relevant traits in chickpea (*Cicer arietinum* L.). *3 Biotech* **2018**, *8*, 43. [CrossRef]

29. Saghai-Maroof, M.A.; Soliman, K.M.; Jorgensen, R.A.; Allard, R.W. Ribosomal DNA spacer-length polymorphisms in barley: Mendelian inheritance, chromosomal location, and population dynamics. *Proc. Natl. Acad. Sci. USA* **1984**, *81*, 8014–8018. [CrossRef]
30. Pritchard, J.K.; Stephens, M.; Donnelly, P. Inference of population structure using multilocus genotype data. *Genetics* **2000**, *155*, 945–959. [CrossRef]
31. Evanno, G.; Regnaut, S.; Goudet, J. Detecting the number of clusters of individuals using the software STRUCTURE: A simulation study. *Mol. Ecol.* **2005**, *14*, 2611–2620. [CrossRef] [PubMed]
32. Bradbury, P.J.; Zhang, Z.; Kroon, D.E.; Casstevens, T.M.; Ramdoss, Y.; Buckler, E.S. TASSEL: Software for association mapping of complex traits in diverse samples. *Bioinformatics* **2007**, *23*, 2633–2635. [CrossRef]
33. Zhang, Z.; Ersoz, E.; Lai, C.Q.; Todhunter, R.J.; Tiwari, H.K.; Gore, M.A.; Bradbury, P.J.; Yu, J.; Arnett, D.K.; Ordovas, J.M.; et al. Mixed linear model approach adapted for genome-wide association studies. *Nat. Genet.* **2010**, *42*, 355–360. [CrossRef]
34. Gujaria, N.; Kumar, A.; Dauthal, P.; Dubey, A.; Hiremath, P.; Bhanu Prakash, A.; Farmer, A.; Bhide, M.; Shah, T.; Gaur, P.M.; et al. Development and use of genic molecular markers (GMMs) for construction of a transcript map of chickpea (*Cicer arietinum* L.). *Theor. Appl. Genet.* **2011**, *122*, 1577–1589. [CrossRef] [PubMed]
35. Choudhary, S.; Gaur, R.; Gupta, S. EST-derived genic molecular markers: Development and utilization for generating an advanced transcript map of chickpea. *Theor. Appl. Genet.* **2012**, *124*, 1449–1462. [CrossRef]
36. Gaur, R.; Sethy, N.K.; Choudhary, S.; Shokeen, B.; Gupta, V.; Bhatia, S. Advancing the STMS genomic resources for defining new locations on the intraspecific genetic linkage map of chickpea (*Cicer arietinum* L.). *BMC Genom.* **2011**, *12*, 117. [CrossRef]
37. Winter, P.; Pfaff, T.; Udupa, S.M.; Huttel, B.; Sharma, P.C.; Sahi, S.; Arreguin-Espinoza, R.; Weigand, F.; Muehlbauer, F.J.; Kahl, G. Characterization and mapping of sequence-tagged microsatellite sites in the chickpea (*Cicer arietinum* L.) genome. *Mol. Gen. Genet.* **1999**, *262*, 90–101. [CrossRef] [PubMed]
38. Jha, U.C.; Jha, R.; Bohra, A.; Manjunatha, L.; Saabale, P.R.; Parida, S.K.; Singh, N.P. Association mapping of genomic loci linked with Fusarium wilt resistance (Foc2) in chickpea. *Plant Genet. Resour. Charact. Util.* **2021**, *19*, 195–202. [CrossRef]
39. Jendoubi, W.; Bouhadida, M.; Boukteb, A.; Béji, M.; Kharrat, M. Fusarium Wilt Affecting Chickpea Crop. *Agriculture* **2017**, *7*, 23. [CrossRef]
40. Choudhary, A.K.; Kumar, S.; Patil, B.S.; Sharma, M.; Kemal, S.; Ontagodi, T.P.; Datta, S.; Patil, P.; Chaturvedi, S.K.; Sultana, R.; et al. Narrowing yield gaps through genetic improvement for Fusarium wilt resistance in three pulse crops of the semi-arid tropics. *SABRAO J. Breed. Genet.* **2013**, *45*, 341–370.
41. Lakmes, A.; Jhar, A.; Sadanandom, A.; Brennan, A.C.; Kahriman, A. Inheritance of resistance to chickpea Fusarium wilt disease (*Fusarium oxysporum* f. sp. *ciceris* Race 2) in a wide-cross *Cicer arietinum* x *Cicer reticulatum* mapping family. *Genes* **2024**, *15*, 819. [CrossRef] [PubMed]
42. O'Donnell, K.; Whitaker, B.K.; Laraba, I.; Proctor, R.H.; Brown, D.W.; Broders, K.; Kim, H.S.; McCormick, S.P.; Busman, M.; Aoki, T.; et al. DNA sequence-based identification of *Fusarium*: A Work in Progress. *Plant Dis.* **2022**, *106*, 1597–1609. [CrossRef] [PubMed]
43. Knutsen, A.K.; Torp, M.; Holst-Jensen, A. Phylogenetic analyses of the *Fusarium poae*, *Fusarium sporotrichioides* and *Fusarium langsethiae* species complex based on partial sequences of the translation elongation factor-1 alpha gene. *Int. J. Food Microbiol.* **2004**, *95*, 287–295. [CrossRef]
44. Gannibal, P.B. Polyphasic Approach to Fungal Taxonomy. *Biol. Bull. Rev.* **2022**, *12*, 18–28. [CrossRef]
45. Namsi, A.; Rabaoui, A.; Masiello, M.; Moretti, A.; Othmani, A.; Gargouri, S.; Gdoura, R.; Werbrouck, S.P.O. First report of leaf wilt caused by *Fusarium proliferatum* and *F. brachygibbosum* on aate palm (*Phoenix dactylifera*) in Tunisia. *Plant Dis.* **2020**, *105*, 1217. [CrossRef]
46. Qiu, R.; Li, J.; Zheng, W.; Su, X.; Xing, G.; Li, S.; Zhang, Z.; Li, C.; Wang, J.; Chen, Y.; et al. First report of root rot of tobacco caused by *Fusarium brachygibbosum* in China. *Plant Dis.* **2021**, *105*, 4170. [CrossRef]
47. Thudi, M.; Upadhyaya, H.D.; Rathore, A.; Gaur, P.M.; Krishnamurthy, L.; Roorkiwal, M.; Nayak, S.N.; Chaturvedi, S.K.; Basu, P.S.; Gangarao, N.V.; et al. Genetic dissection of drought and heat tolerance in chickpea through genome-wide and candidate gene-based association mapping approaches. *PLoS ONE* **2014**, *9*, e96758. [CrossRef]
48. Upadhyaya, H.D.; Bajaj, D.; Das, S.; Kumar, V.; Gowda, C.L.; Sharma, S.; Tyagi, A.K.; Parida, S.K. Genetic dissection of seed-iron and zinc concentrations in chickpea. *Sci. Rep.* **2016**, *6*, 24050. [CrossRef] [PubMed]
49. Upadhyaya, H.D.; Bajaj, D.; Narnoliya, L.; Das, S.; Kumar, V.; Gowda, C.L.; Sharma, S.; Tyagi, A.K.; Parida, S.K. Genome-Wide Scans for Delineation of Candidate Genes Regulating Seed-Protein Content in Chickpea. *Front. Plant Sci.* **2016**, *7*, 302. [CrossRef]
50. Winter, P.; Benko-Iseppon, A.M.; Hüttel, B.; Ratnaparkhe, M.; Tullu, A.; Sonnante, G.; Pfaff, T.; Tekeoglu, M.; Santra, D.; Sant, V.J.; et al. A linkage map of the chickpea (*Cicer arietinum* L.) genome based on recombinant inbred lines from a C. arietinum × C. reticulatum cross: Localization of resistance genes for fusarium wilt races 4 and 5. *Theor. Appl. Genet.* **2000**, *101*, 1155–1163. [CrossRef]
51. Sharma, K.D.; Winter, P.; Kahl, G.; Muehlbauer, F.J. Molecular mapping of *Fusarium oxysporum* f. sp. *ciceris* race 3 resistance gene in chickpea. *Theor. Appl. Genet.* **2004**, *108*, 1243–1248. [CrossRef] [PubMed]
52. Jingade, P.; Ravikumar, R.L. Development of molecular map and identification of QTLs linked to Fusarium wilt resistance in chickpea. *J. Genet.* **2015**, *94*, 723–729. [CrossRef]

53. Soi, S.; Chauhan, U.S.; Yadav, R.; Kumar, J.; Yadav, S.S.; Yadav, H.; Kumar, R. STMS based diversity analysis in chickpea (*Cicer arietinum* L.) for Fusarium wilt. *New Agric.* **2014**, *25*, 243–250.
54. Maqbool, M.A.; Aslam, M.; Ali, H.; Shah, T.M. Evaluation of advanced chickpea (*Cicer arietinum* L.) accessions based on drought tolerance indices and SSR markers against different water treatments. *Pak. J. Bot.* **2016**, *48*, 1421–1429.
55. Barman, P.; Handique, A.K.; Tanti, B. Tagging STMS markers to Fusarium wilt race-1 resistance in chickpea (*Cicer arietinum* L.). *Indian J. Biotechnol.* **2014**, *13*, 370–375.
56. Raghu, R.; Ravikumar, R.L. Development of novel microsatellite markers using genome sequence information in chickpea (*Cicer arietinum* L.). *Mysore J. Agric. Sci.* **2016**, *50*, 395–399.
57. Millan, T.; Clarke, H.J.; Siddique, K.H.M.; Buhariwalla, H.K.; Gaur, P.M.; Kumar, J.; Juan, G.; Kahl, G.; Winter, P. Chickpea molecular breeding: New tools and concepts. *Euphytica* **2006**, *147*, 81–103. [CrossRef]
58. Sachdeva, S.; Dawar, S.; Rani, U.; Patil, B.S.; Soren, K.R.; Singh, S.; Sanwal, S.K.; Chauhan, S.K.; Bharadwaj, C. Identification of SSR markers linked to Botrytis grey mould resistance in chickpea (*Cicer arietinum*). *Phytopathol. Mediterr.* **2019**, *58*, 283–292.
59. Patil, B.S.; Ravikumar, R.L.; Bhat, J.S.; Soregaon, C.D. Molecular mapping of qtls for resistance to early and late Fusarium wilt in chickpea. *Czech J. Genet. Plant Breed.* **2014**, *50*, 171–176. [CrossRef]
60. Sabbavarapu, M.M.; Sharma, M.; Chamarthi, S.K.; Swapna, N.; Rathore, A.; Thudi, M.; Gaur, P.M.; Pande, S.; Singh, S.; Kaur, L.; et al. Molecular mapping of QTLs for resistance to Fusarium wilt (race 1) and Ascochyta blight in chickpea (*Cicer arietinum* L.). *Euphytica* **2013**, *193*, 121–133. [CrossRef]
61. Ratnaparkhe, M.B.; Santra, D.K.; Tullu, A.; Muehlbauer, F.J. Inheritance of inter-simple-sequence-repeat polymorphisms and linkage with a fusarium wilt resistance gene in chickpea. *Theor. Appl. Genet.* **1998**, *96*, 348–353. [CrossRef]
62. Tullu, A.; Muehlbauer, F.J.; Simon, C.J.; Mayer, M.S.; Kumar, J.; Kaiser, W.J.; Kraft, J.M. Inheritance and linkage of a gene for resistance to race 4 of fusarium wilt and RAPD markers in chickpea. *Euphytica* **1998**, *102*, 227–232. [CrossRef]
63. Lal, D.; Ravikumar, R.L.; Jingade, P.; Subramanya, S. Validation of molecular markers linked to Fusarium wilt resistance (Foc 1) in recombinant inbred lines of chickpea (*Cicer arietinum*). *Plant Breed.* **2022**, *141*, 429–438. [CrossRef]
64. Tar'an, B.; Warkentin, T.D.; Tullu, A.; Vandenberg, A. Genetic mapping of ascochyta blight resistance in chickpea (Cicer arietinum L.) using a simple sequence repeat linkage map. *Genome* **2007**, *50*, 26–34. [CrossRef]
65. Soregoan, C.D.; Ravi Kumar, R.L. Marker assisted characterization of wilt resistance in productive chickpea genotypes. *Electron. J. Plant Breed.* **2010**, *1*, 1159–1163.
66. Sharma, K.D.; Muehlbauer, F.J. Fusarium wilt of chickpea: Physiological specialization, genetics of resistance and resistance gene tagging. *Euphytica* **2007**, *157*, 1–14. [CrossRef]

Disclaimer/Publisher's Note: The statements, opinions and data contained in all publications are solely those of the individual author(s) and contributor(s) and not of MDPI and/or the editor(s). MDPI and/or the editor(s) disclaim responsibility for any injury to people or property resulting from any ideas, methods, instructions or products referred to in the content.

agronomy

MDPI

Article

Transcriptomics and Physiological Analyses of Soybean Stay-Green Syndrome

Dagang Wang [1,†], Yanan Wang [1,†], Ruidong Sun [1], Yong Yang [1], Wei Zhao [2], Guoyi Yu [3], Yueying Wang [4], Feng Wang [5], Lin Zhou [5] and Zhiping Huang [1,*]

1 Crop Research Institute, Anhui Academy of Agricultural Sciences, Hefei 230031, China; wangdagang@aaas.org.cn (D.W.); 15856979430@163.com (Y.W.); 2018201070@njau.edu.cn (R.S.); soyivy1988@163.com (Y.Y.)
2 Institute of Plant Protection and Agro-Products Safety, Anhui Academy of Agricultural Sciences, Hefei 230031, China; bioplay@sina.com
3 Anhui Longkang Farm of Land-Reclamation, Bengbu 233426, China; yuguoyi8523@163.com
4 Soybean Research Institute, Suzhou Academy of Agricultural Sciences, Suzhou 234000, China; wyysznky@163.com
5 Agricultural Technology Promotion Center of Jieshou, Jieshou 230065, China; jswf488644@163.com (F.W.); jsymzl@163.com (L.Z.)
* Correspondence: huangzhiping@aaas.org.cn
† These authors contributed equally to this work.

Abstract: Stay-green syndrome (SGS) is an important factor that causes soybean (*Glycine max*) yield reduction. Despite progress being made, the regulatory mechanism remains largely unclear. Therefore, in this study, an SGS-sensitive soybean variety, "HD0702", was employed to investigate the underlying mechanism. Transcriptomic analyses were performed in a tissue-specific manner to investigate differentially expressed genes (DEGs) in soybeans impacted by SGS and in those without SGS. A total of 1858 DEGs were identified in the pods, and 2814 DEGs were identified in the leaves. Further investigation revealed that SGS mainly affected the expression levels of key genes involved in the regulation of photosynthesis, starch and sucrose metabolism, and plant hormone signal transduction. To support this finding, the chlorophyll content of the pods was to be found increased by 320% for chlorophyll *a* and 260% for chlorophyll *b*. In leaves, soluble sugar levels significantly increased, whereas phytohormones IAA and ABA decreased in SGS pods. DEGs were classified using gene ontology (GO) terms, and photosynthesis-related genes *α-glucosidase*, *β-mannosidase*, *β-amylase 5* (*GmBAM5*), and *starch synthase 2* (*GmSS2*) were up-regulated. This study demonstrates a molecular and physiological basis for SGS that merits further investigation to allow for SGS management.

Keywords: soybean; stay-green syndrome; RNA-seq; regulation pathway

1. Introduction

Soybean (*Glycine max* (L.) Merr.) is a crop with important economic properties for China; the seeds are abundant in protein content and oil-suitable for both human consumption and livestock feed. However, soybean production is often hampered by factors such as the occurrence of stay-green syndrome (SGS), resulting in a severe yield reduction in soybean crops [1–5]. Soybean SGS, a condition that leads to delayed leaf senescence (stay-green), occurs mainly during the reproductive period, causing flat pods and yield reductions of up to 75%, even no production in some cases [2,6].

The primary causes of SGS remain unclear, but soybean virus diseases, insect pests, unbalanced nutrients, and climate issues have all been implicated [1,4,7,8]. It is also believed that SGS is likely influenced by extreme temperatures during the reproductive growth period of soybean, genetic background, field management, continuous cropping, alternate cropping, and sowing dates [9–12]. A novel geminivirus, designated soybean stay-green associated geminivirus (SoSGV), has been identified as a significant pathogen of soybean SGS. A large amount of SoSGV can accumulate in the seed coat after SoSGV infection, where it may prevent transportation photoassimilates from vegetative tissues (including the seed coat) to the cotyledon and embryo [1], and the common brown leafhopper (*Orosius orientalis*) is the transmission vector of *SoSGV* [13]. Previous transcriptome profiling indicated that some *FLOWERING LOCUS T* (*FT*) genes are suppressed by *Riptortus pedestris* infection. The overexpression of *GmFT2a*, a floral inducer, attenuates SGS through the mediation of soybean defense responses and photosynthesis [7]. Liang treated soybeans with dry heat, and several plants exhibited stay-green symptoms, indicating that high temperature is capable of causing soybean SGS [9].

During the development of the soybean, both the leaves and the pods serve as photosynthetic organs [14]. Soybeans possess several advantageous characteristics, including the photosynthetic ability of their leaves and pods, seeds that receive photoassimilates, and vascular systems to transport photoassimilates. These attributes collectively contribute to the high yield of soybeans [15,16]. The development of SGS soybean seeds is inhibited, which results in an imbalance between the source and sink of soybean plants [1,3,4]. Leaves affected by SGS are capable of performing photosynthesis as usual; however, the inability to transport nutrients from the leaves to the grains is a significant factor contributing to the abnormal development of seeds [16,17]. The process of leaf senescence results in a reduction in the synthesis of photosynthetic products and an acceleration in the degradation of proteins [18]. Phytohormones play a pivotal role in numerous biological processes, including plant senescence [19]. The combined effect of various phytohormones may exert a positive or negative influence on leaf senescence [20]. In *Arabidopsis*, jasmonate (JA) interacts with auxin, ethylene, and gibberellin to regulate leaf senescence [21]. The stay-green phenotype appears to be antithetical to the senescence phenotype, with the former leaves remaining green and not falling off after pod maturation. The formation of the SGS phenotype may also be associated with the influence of phytohormones, although this hypothesis has not been widely investigated.

Through the profiling of genome-wide gene expression, we used RNA-seq to uncover genes involved in SGS and investigate their expression patterns. From DEG analysis, we determined that SGS impacts multiple biological processes, including photosynthesis and hormonal signaling. To characterize the transcriptomic profiles, we also investigated physiological alterations, such as photosynthesis-associated traits and hormone levels, which enhanced our understanding of molecular mechanisms for SGS. The results allowed us an insight into the molecular mechanism as well as the identification of an array of genes that may be targeted for further characterization.

2. Materials and Methods

2.1. Experimental Materials

The SGS-sensitive soybean variety "HD0702" was planted at Longkang Farm, an experimental site of the Crop Research Institute of Anhui Academy of Agricultural Sciences. The HD702 was planted in two field plots, and the plants exhibited SGS and non-SGS phenotypes. The SoSGV infection of soybean is characterized by the typical symptoms of virus infection, including leaf shrinkage, abnormal filling of soybean pods, and plant dwarfing. Samples with or without SGS were sampled independently. The collected materials included Z1 (SGS pods) and Z2 (SGS leaves), as well as control samples Z3 (normal developing pods) and Z4 (normal developing leaves). The four groups of samples from SGS and normal plants were each collected in three biological replicates. The collected samples were flash-frozen using liquid nitrogen and stored in a −80 °C freezer. Transcriptome sequencing and analysis were performed at Biomarker Technologies (Beijing, China). Three biological replicates per sample were utilized.

2.2. RNA Extraction and Real-Time Quantitative PCR (qPCR)

Total RNA was extracted using a MiniBEST Plant RNA Extraction Kit (TaKaRa, Osaka, Japan). The purity, concentration, and integrity of the RNA samples were accurately determined by 1% agarose gel electrophoresis and Nanodrop 2000 (Thermo Scientific, Waltham, MA, USA) measurement.

Twenty-five differentially expressed genes were randomly selected for real-time quantitative PCR analysis, with the soybean Tubulin gene (No.AY907703) used as an internal reference gene. cDNA was synthesized using the PrimeScript™ RT Master Mix (Perfect Real Time) kit (TaKaRa, Japan). Primer 5.0 software was used for primer design for qPCR (Table S1). qPCR reactions were prepared using the HieffTM qPCR SYBR® Green Master Mix Kit (Yeasen, Shanghai, China) and conducted on a Bio-Rad iCycler iQ5 instrument (Bio-Rad, Hercules, CA, USA) according to the manufacturer's instructions. The reaction conditions were pre-denaturation at 95 °C for 5 min, followed by 40 cycles of 95 °C for 10 s and 60 °C for 30 s. Each reaction was performed in triplicate using three biological replicates, and the relative gene expression was calculated using the $2^{-\Delta\Delta CT}$ method [22].

2.3. Library Construction and Transcriptome Sequencing

Isolated mRNA was enriched using magnetic beads alongside oligo (dT) primers, which was randomly interrupted by adding a Fragmentation Buffer. With the use of random hexamers, the obtained mRNA was then used for first-strand cDNA synthesis. The cDNA was then mixed with d buffer, dNTPs, RNase H, and DNA polymerase I to synthesize the second cDNA strand. AMPure XP beads were used to purify the cDNA. The purified double-stranded cDNA underwent end-repair, A tailing, and sequencing adapter ligation. Fragment size selection was conducted using AMPure XP beads. PCR enrichment was performed to obtain the complete cDNA library. Once the library was constructed, qPCR was used to quantify the effective concentration of the library (effective library concentration >2 $nmol \cdot L^{-1}$) and verify the library quality. After determining the library quality, multiple libraries were pooled and sequenced on the Illumina platform following previously published protocols.

2.4. Transcriptome Analysis

The Illumina high-throughput sequencing platform performs paired-end sequencing on cDNA libraries. Based on Sequencing by Synthesis (SBS) technology, it produces many high-quality raw reads. The quality of each sample file was determined using FastQC (v0.11.9). Reads with adapters and/or low-quality reads (a ratio of N greater than 10%; quality value of $Q \leq 10$) were removed. After quality control steps, high-quality clean reads were obtained.

The clean reads were aligned to the soybean reference genome (Wm82.a2.v1) using Hisat2 software (v2.2.1). Thereafter, htseq-count software (v0.13.5) was used to quantify the transcript abundance. Differential expression analysis was conducted using DESeq2 (v1.32.0), and the screening criteria were as follows: absolute value of $\log_2$ Fold Change ≥ 1.0 and p-value < 0.05. Gene ontology (GO) enrichment and Kyoto Encyclopedia of Genes and Genomes (KEGG) biochemical pathway enrichment analyses for the identified DEGs were performed using the 'clusterProfiler' (v4.0.5) package in R.

2.5. Determination of Chlorophyll Level

Samples were obtained and weighed. Briefly, to determine chlorophyll levels, 100 mL of the extract (absolute ethanol–acetone = 1:1) was added to a flask containing leaf samples, followed by incubation in the dark at room temperature for 5 h. The resulting filtrate was used to quantify the chlorophyll content. The measurement of the absorbance value of each sample was carried out at 645 nm or 663 nm using a UV–visible spectrophotometer (Shanghai, China) [23].

2.6. Determination of Phytohormones

Extraction was conducted on fresh leaves (0.2 g) homogenized with 4 mL of 50% acetonitrile aqueous solution at 4 °C. The mixture was shaken for 3 min allowed to settle for 4 h. Samples were then centrifuged at 12,000 rpm at room temperature for 10 min, and the supernatant was passed through a 0.22 μm organic membrane. Thm Measurement of phytohormones was performed using a High-Performance Liquid Chromatography (HPLC) system, with external standard solutions of indole-3-acetic acid (IAA), abscisic acid (ABA), jasmonic acid (JA), and salicylic acid (SA) used to integrate peaks. Phytohormone contents were represented as $\mu g \cdot kg^{-1}$ [24].

2.7. Detection of Soluble Sugar Content

Soluble sugar contents were measured from the leaf tissues of the SGS and non-SGS soybeans. A 0.1 g sample of fresh leaves homogenized with 80% ethanol was prepared for this extraction. Samples were then centrifuged at 8000 rpm for 20 min, and the supernatant was purified for the determination of glucose, galactose, mannose, and rhamnose levels. An HPLC system was used to determine the soluble sugar content, and external standard solutions of glucose, galactose, mannose, and rhamnose were used to integrate peaks [25].

2.8. Morphological Trait Measurement

Soybean plants were grown in a soybean field at the Experimental Station of the Crop Research Institute, Anhui Academy of Agricultural Sciences, Bengbu (32°92′ N, 117°38′ E), Anhui province. The experiment was carried out in a randomized complete-block design containing three replicates. Each tested soybean was planted in an experimental plot of approximately 24 m^2 (600 plants for a single plot). The plant-to-plant spacing was 10 cm and row-to-row spacing was 40 cm. Fifty plants from each replicate were randomly selected for agronomic trait measurement including plant height, number of fruit branches per plant, and number of pods per plant. Plant height was determined as the height of the main stem at the podding stage. All available soybean pods from a single plant were collected for measurements to determine the number of pods per plant. All pods from a single plot were collected and dried at 37 °C, and 100 randomly selected pods were used for bean weight measurement. All pods from a single plot were collected and treated as described above to measure soybean yield.

2.9. Statistical Analysis

The statistical analysis was conducted using SPSS 27.0 software. Statistical analysis was performed to determine the significance of differences between the treatment and control groups using Student's t-test. The experiment was performed in triplicate using three independent biological replicates. Data are means ± SD (n = 3 biologically independent samples). Different letters indicate a significant difference between means ($p \leq 0.05$) according to Tukey's HSD test.

3. Results

3.1. SGS Had a Negative Impact on Agronomic Traits of Soybean

Soybean stay-green syndrome results in serious yield loss mainly due to its impacts on two factors: green pods with abnormal seeds and/or long-term green plants after the harvesting season. The soybean cultivar "HD0702" is sensitive to SGS, and we found that plants growing in two different plots exhibited SGS and non-SGS phenotypes. The phenotypic comparison of "HD0702" plants with and without SGS during the developmental period indicated that the pods of SGS soybeans do not swell, and some of their developed seeds are much smaller than normally developed ones (Figure 1A,B). Moreover, the leaves and pods remained green in SGS soybean plants compared to light brown leaves and pods for non-SGS plants due to programed senescence (Figure 1C). These results suggest that the syndrome impacts plant maturation and seed filling. Whether SGS is caused by environmental or physiological stress still remains to be determined.

Figure 1. Developing pods and leaves of HD0702 with and without SGS. Developing pods and seeds of normal (non-SGS HD0702) (**A**) and stay-green (**B**) soybeans; (**C**) Phenotype of SGS and normal control plants. Left, an SGS plant; right, a normal plant. Scale bar = 1 cm in (**A**,**B**); scale bar = 10 cm in (**C**).

In addition to remaining green, the SGS plants also exhibited poor field performance, with a significant reduction in plant height and soybean yield compared to control plants. The plant height (PH) of SGS plants was significantly reduced by about 10 cm ($p < 0.01$) compared to control plants (Table 1). The SGS plants produced 29–35 pods per plant, 20% lower ($p < 0.01$) than non-SGS plants (producing 37–43). Moreover, compared to normal plants, the number of effective pods (EP) per SGS plant decreased significantly, by approximately 23% ($p < 0.01$) on average. In comparison with normal soybeans, the number of grains per plant (GNpP) and the grain weight per plant (GWpP) were found to be significantly reduced in the SGS plants ($p < 0.01$) (Table 1). The agronomic characteristics related to soybean yields indicated that the 100-grain weight (HGW) and yield per unit area (YpUA) were decreased in SGS plants by 50% ($p < 0.01$) and 38% ($p < 0.01$), respectively, compared to non-SGS plants (Table 1). These observations indicated that SGS dramatically impacts plant performance, including yield, and is likely a result of a failure to transport photosynthate for the promotion of seed development. Ultimately, this results in an overall soybean yield loss.

Table 1. Agronomic traits of SGS and non-SGS plants under field conditions.

	PH (cm)	PT (cm)	NoNoMs	TPN	EP	GNpP	GWpP (g)	HGW (g)	YpUA (kg)
Control	80.5 ± 1.2 *	18.5 ± 0.6 *	16.8 ± 0.4 *	40.2 ± 1.0 *	37.8 ± 0.9 *	79.6 ± 1.0 *	13.3 ± 0.2 *	18.0 ± 0.4 *	155.6 ± 2.0 *
SGS	69.4 ± 2.1	15.7 ± 0.3	14.0 ± 0.7	32.0 ± 1.0	14.4 ± 1.0	26.8 ± 1.2	2.4 ± 0.2	9.0 ± 0.4	95.8 ± 1.8
p-value	0.01	0.01	0.01	0.01	0.01	0.01	0.01	0.01	0.01

Note: plant height (PH), pod height (PT), number of the nods on main stem (NoNoMs), total pod numbers (TPN), effective pods (EP), grain numbers per plant (GNpP), grain weight per plant (GNpP), hundred-grain weight (HGW), and yield per unit area (YpUA). Significance of differences between soybean HD0702 SGS plants and non-SGS plants was determined using Student's *t*-tests. * $p \leq 0.05$.

3.2. Transcriptome Sequencing Revealed Underlying Molecular Mechanism

To determine SGS's molecular mechanism, we carried out transcriptome sequencing of the developing pods and leaves in both SGS and non-SGS HD0702. After filtration, we obtained an average of 50,338,134 clean reads per sample (93% reads with a QC greater than 30), with an average GC content of 45.04%. The filtered data (clean reads) were aligned to a soybean reference genome using Hisat2. The average alignment rate of each sample was 93.65%, with rates ranging from 88.01% to 96.16%. Our results had high-quality transcriptome reads and exceptional alignment with the reference genome, meaning that the results could be used in further analysis (Table S2).

DEGs were identified using a significant difference threshold of $p < 0.05$ with $|\log_2$ Fold Change$| \geq 1$. In SGS pods, compared to control pods, a total of 1858 DEGs were identified, with 959 genes up-regulated and 899 genes down-regulated (Figure 2A). When comparing leaves, a total of 2814 DEGs were identified, including 1725 up-regulated genes and 1089 down-regulated genes in SGS leaves (Figure 2A). Venn diagram analysis of the DEGs revealed 362 DEGs in common, while 1496 and 2452 genes were specific to pods and leaves, respectively (Figure 2B).

To validate our RNA-seq results, qPCR was performed on 15 randomly selected genes in the pods and leaves, respectively. A comparison of the qPCR results revealed a regression slope of RNA-seq vs. qPCR of $R^2 = 0.9139$ in the pods (Figure S1A), and $R^2 = 0.9187$ in the leaves (Figure S1B). These results suggest a strong positive correlation between RNA-seq and qPCR results, indicating the robustness of our RNA-seq data.

3.3. SGS Increased Photosynthesis-Related Processes

To fully appreciate the gene expression changes in SGS plants, we performed gene ontology (GO)-term enrichment analysis on the DEGs [26]. GO terms associated with photosynthesis were highly enriched in the pod DEGs (Table S6). Of note, 13 chlorophyll *a-b* binding protein genes (including *LHCA1*, *2*, *4*, and *LHCB2*, *5*, *6*) were all up-regulated in the SGS pods compared to control pods (Figure 3A, Table S6). However, photosynthesis-related processes were not significantly enriched in the leaf DEGs. Similarly, KEGG terms related to photosynthesis and photosynthesis-antenna proteins were also markedly enriched in the pod DEGs (Figure 2D, Table S9). A close assessment showed that the transcript levels of 46 photosynthesis-related DEGs were up-regulated in the SGS pods relative to control pods.

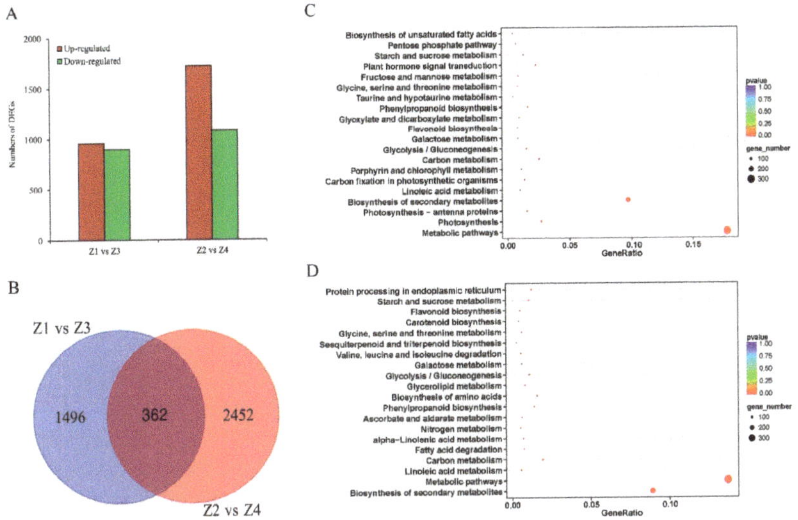

Figure 2. Differentially expressed genes (DEGs) and KEGG enrichment analysis of pods and leaves from SGS plants. (**A**) DEGs in Z1 (SGS pods) vs. Z3 (normal developing pods) and Z2 (SGS leaves) vs. Z4 (normal developing leaves); (**B**) Venn diagram analysis showed common DEGs across the pods and leaves of SGS plants; (**C**) KEGG enrichment analysis of the top 20 pathways in SGS pods; (**D**) KEGG enrichment analysis of the top 20 pathways in SGS leaves.

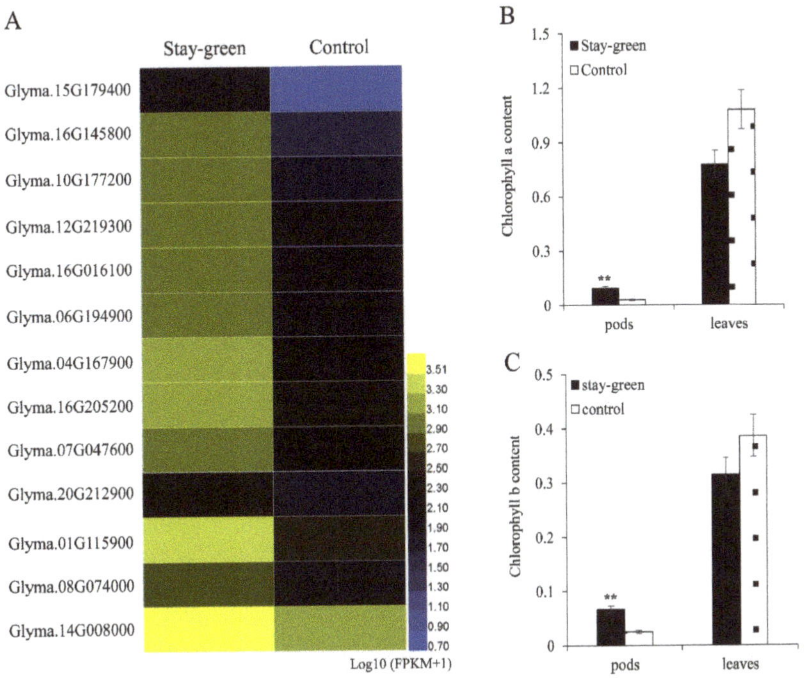

Figure 3. SGS-enhanced photosynthesis-related processes compared to control plants. (**A**) Expression of photosynthesis-related DEGs was significantly increased in SGS pods relative to control pods; (**B**) chlorophyll *a* content in SGS and non-SGS control plants; (**C**) chlorophyll *b* content in SGS and non-SGS control plants. Values represent means ± S.D. of three biological replicates. ** $p \leq 0.01$. Student's *t*-test.

To further clarify the impact of SGS on photosynthesis, we measured the chlorophyll *a* and *b* content in developing pods and leaves. Significant differences in chlorophyll pigment content were detected in pods of SGS plants and normal plants, with 0.095 mg/g and 0.030 mg/g for chlorophyll *a* and 0.066 mg/g and 0.025 mg/g for chlorophyll *b*, respectively. Therefore, higher chlorophyll *a* and *b* contents were found in SGS pods compared to non-SGS pods (Figure 3B,C). However, no significant differences were observed between SGS and normal leaves (Figure 3B,C). These results demonstrate that SGS impacted several photosynthesis-related process-related genes and enhanced photosynthesis-related processes in soybean pods.

3.4. SGS-Affected Carbohydrate Metabolism in Soybean

In the above analyses, we found that genes related to carbohydrate metabolism were significantly enriched in SGS plants (Table S11). In the pods, the expression of several genes with roles in glycolysis/gluconeogenesis, galactose metabolism, fructose and mannose metabolism, and starch and sucrose metabolism were differentially expressed in SGS plants compared to control plants (Figure 2C, Table S9). For example, the transcript levels of two *α-glucosidase* and five *β-mannosidase* genes were significantly up-regulated in SGS pods compared to control pods (Table S3). In contrast, the transcript levels of a *β-amylase* gene were down-regulated approximately five-fold in SGS pods (Table S3). With respect to the leaves, the KEGG pathways pertaining to glycolysis/gluconeogenesis, galactose metabolism, and starch and sucrose metabolism were significantly enriched (Figure 2C, Table S10). For example, *β-amylase 5* (*GmBAM5*) and *starch synthase 2* (*GmSS2*) were significantly up-regulated, while *starch synthase 1* (*GmSS1*) and *sucrose synthase 4* (*GmSuS4*) were down-regulated (Table S4).

We measured and compared the levels of four soluble sugars in both SGS and control plants. Our comparisons indicated the presence of over 1.5-fold and 3.5-fold increases in glucose, and 1.5-fold and 2.5-fold in galactose concentration in the pods and leaves, respectively, of SGS plants compared to control plants (Figure 4A,B). In our RNA-Seq results, SGS retained higher mannose and rhamnose levels than control plants (Figure 4C,D). These measurements confirmed the enhancement of many carbohydrate metabolic pathways, including sugar-associated metabolism, and found them to be associated with SGS.

3.5. Phytohormone Signal Transduction Was Related to Soybean SGS

Phytohormones are required for stress tolerance in plants [27]. Through KEGG analysis, we found that phytohormone signal transduction pathways were significantly enriched, and interestingly, they were significantly enriched in pods compared to leaves (Table S9). An analysis of identified DEGs reveals 39 involved in phytohormone signal transduction, of which 26 genes were down-regulated in SGS pods compared to control pods (Figure S2). For example, the transcript levels of twelve auxin-related genes (including *IAA4*, *GH3.1*, *3.2*, *3.3*, and *3.6*) were significantly down-regulated in SGS pods (Table 2). In contrast, the transcript level of a jasmonate signaling-responsive gene *jasmonate-zim-domain protein 12* (*JAZ12*) was up-regulated by approximately 4.8-fold in SGS pods compared to pods of control plants (Table 2). The expression of seven genes that respond to abscisic acid (ABA) were significantly down-regulated in the SGS pods, including *Glyma.14G162100*, *Glyma.11G018000*, *Glyma.19G194500*, *Glyma.13G229300*, *Glyma.10G071700*, *Glyma.15G083200*, and *Glyma.13G153200* (Table 2).

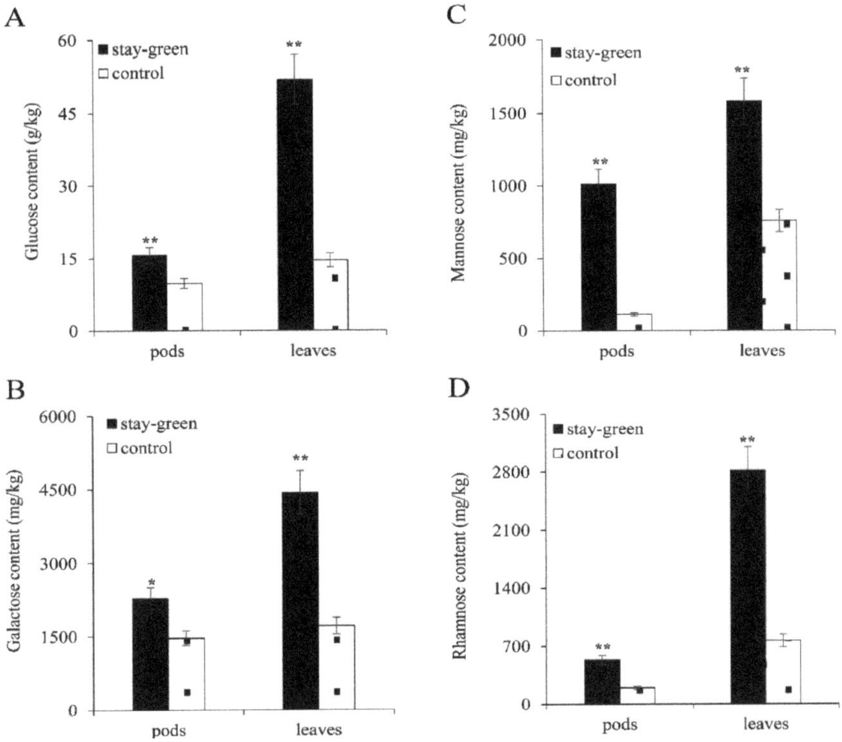

Figure 4. Soluble sugar content in SGS and control soybean pods and leaves. Total glucose (**A**), galactose (**B**), mannose (**C**), and rhamnose (**D**) concentrations in 10-week-old plant pods and leaves. Values represent means ± S.D. of three biological replicates. * $p \leq 0.05$, ** $p \leq 0.01$. Student's *t*-test.

Table 2. DEGs related to phytohormone signal transduction pathway.

	Gene ID	log2FC (Z1/Z3)	Description	Gene
IAA	*Glyma.08G207900*	5.670477896	auxin-regulated protein (Aux22)	*IAA19*
	Glyma.10G180100	4.628120766	auxin-induced protein ali50	*IAA14*
	Glyma.18G029000	3.685020814	auxin transporter-like protein 14	*AUX1*
	Glyma.03G063600	3.498404808	auxin transporter-like protein 3	*AUX1*
	Glyma.10G031900	3.222101295	auxin-induced protein AUX28-like	*IAA7*
	Glyma.12G035800	3.111470625	protein small auxin up-regulated RNA 10	
	Glyma.15G017500	2.213751664	auxin-responsive protein IAA27-like	*IAA27*
	Glyma.16G020700	−2.007459234	SAUR-like auxin-responsive family protein	
	Glyma.02G142600	−2.494455806	auxin-induced protein 22D-like	*IAA4*
	Glyma.20G210500	−2.56056018	AUX/IAA family protein	*IAA4*
	Glyma.11G051600	−2.783551656	probable indole-3-acetic acid-amido synthetase GH3.1	*GH3.1*
	Glyma.06G260800	−2.811652061	indole-3-acetic acid-amido synthetase GH3.6	*GH3.6*
	Glyma.10G031800	−2.822819592	auxin-induced protein 22D-like	*IAA4*
	Glyma.01G190600	−2.92913841	probable indole-3-acetic acid-amido synthetase GH3.1	*GH3.1*
	Glyma.19G161000	−3.653749063	auxin-induced protein 22E	*IAA4*
	Glyma.12G124100	−4.025510264	indole-3-acetic acid-induced protein ARG7	*ARG7*
	Glyma.12G124900	−5.114374267	auxin-induced protein 6B-like	*ARG7*
	Glyma.05G101300	−5.899642393	auxin-responsive protein GH3	*GH3.2*
	Glyma.17G165300	−6.18193335	indole-3-acetic acid-amido synthetase GH3.3	*GH3.3*

Table 2. *Cont.*

	Gene ID	log2FC (Z1/Z3)	Description	Gene
ABA	*Glyma.02G131700*	7.381977493	abscisic acid responsive element-binding factor 1	*ABF1*
	Glyma.14G162100	−2.599498056	highly ABA-induced PP2C gene 3	*HAI3*
	Glyma.11G018000	−4.252300358	highly ABA-induced PP2C gene 3	*HAI3*
	Glyma.19G194500	−4.405553285	Basic-leucine zipper (bZIP) transcription factor family protein	*ABI5,GIA1*
	Glyma.13G229300	−4.492990584	abscisic acid receptor PYL12	*PYL6*
	Glyma.10G071700	−4.618257272	protein ABSCISIC ACID-INSENSITIVE 5	*ABI5,GIA1*
	Glyma.15G083200	−5.099195179	abscisic acid receptor PYL12	*PYL6*
	Glyma.13G153200	−7.101027188	protein ABSCISIC ACID-INSENSITIVE 5	*bZIP67*
Ethylene	*Glyma.03G251700*	−2.443456546	ethylene sensor	*EIN4*
	Glyma.10G007000	−4.762101059	ethylene-responsive transcription factor 1B	*ERF1*
	Glyma.03G162700	−7.31908808	ethylene-responsive transcription factor 15	*ERF15*
JA	*Glyma.17G043700*	2.284024564	jasmonate-zim-domain protein 12	*JAZ12*

Based on the above results, we determined that the phytohormones may also be differentially expressed between SGS and non-SGS plants (Figure 5). Our measurements indicated that indoleacetic acid (IAA) and abscisic acid (ABA) were reduced in SGS pods relative to control pods (Figure 5A,B). However, the levels of IAA and ABA did not significantly differ between SGS leaves and non-SGS leaves (Figure 5A,B). In contrast, jasmonic acid (JA) and salicylic acid (SA) contents were increased in both pods and leaves of SGS plants relative to the control plants (Figure 5C,D). These results demonstrate that the changes in endogenous hormone levels such as IAA and ABA play a key role in the development of SGS.

3.6. The Accumulation of ROS Was Inhibited in SGS Soybean

In addition to the above observations, we found that the GO term related to oxidoreductase activity (Tables S6 and S7) was enriched for the DEGs of both pods and leaves. In SGS pods, 27 DEGs were related to oxidoreductase activity, and the number of up-regulated genes was approximately twice the number of down-regulated genes (Table S12). All three DEGs involved in hydrogen peroxide catabolic processes were up-regulated, including *APX3* (*Glyma.11G078400*), *APX4* (*Glyma.14G177200*), and *2CPB* (*Glyma.02G198400*) (Table S3). Additionally, four DEGs, *PER64* (*Glyma.01G163100*), *PER12* (*Glyma.10G222500*), *APX3* (*Glyma.11G078400*), and *APX4* (*Glyma.14G177200*), associated with peroxidase activity were up-regulated, and another four, *PER12* (*Glyma.20G169200*), *PER43* (*Glyma.08G181500*), *GPX6* (*Glyma.01G219400*), and *PER21* (*Glyma.10G050800*), were down-regulated in SGS pods (Table S3). In SGS leaves, 17 DEGs associated with oxidoreductase activity were up-regulated, and 13 oxidoreductase genes were down-regulated (Table S12).

In plants, glutathione (GSH) is an important antioxidant, and it can alleviate plant symptoms through the activation of ROS-scavenging gene expression, thereby reducing ROS accumulation [28]. In KEGG enrichment analysis, 15 DEGs in pods and 16 DEGs in leaves were identified to be involved in glutathione metabolism. Of these, six "glutathione metabolism" genes were up-regulated in SGS pods, and nine were down-regulated (Table S13). In leaves, 12 "glutathione metabolism" genes were up-regulated in SGS leaves, while 4 were down-regulated (Table S13).

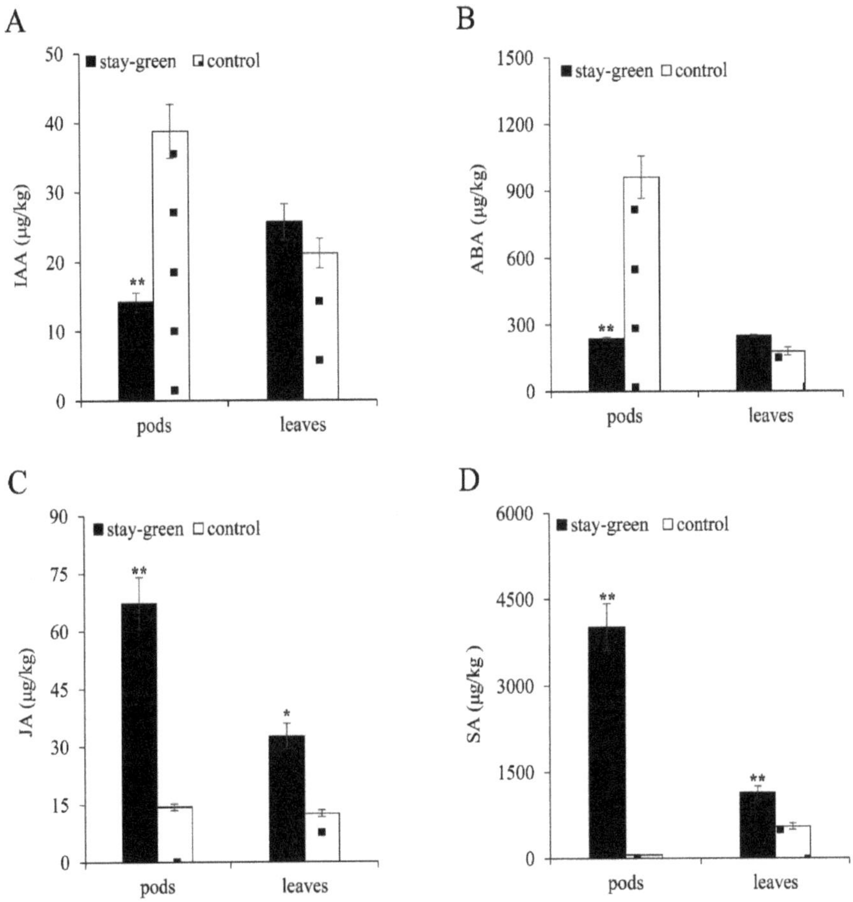

Figure 5. Phytohormone content of SGS and control plants. IAA (**A**), ABA (**B**), JA (**C**), and SA (**D**) contents of SGS and control plants. Values represent means ± S.D. of three replicates. IAA, indoleacetic acid; ABA, abscisic acid; JA, jasmonic acid; SA, salicylic acid. * $p \leq 0.05$, ** $p \leq 0.01$. Student's *t*-test.

4. Discussion

Soybeans are a critically important economic crop worldwide, which is critical for food, feed, and industrial raw materials. The Yellow-Huai-Hai region is one of the primary soybean-producing areas in China, and diseases and pests act to restrict soybean yield [29]. Recently, soybean SGS has become a widespread and serious problem in the Yellow-Huai-Hai region, resulting in large soybean yield losses [30]. To adapt to environments including those containing the pathogen, immune mechanisms have been evolved and activated in plants [31]. Physical and chemical defenses are two key mechanisms of defense against pathogens or diseases and include approaches such as synthetic methyl salicylate, flavonoids, and terpenes [32]. In this study, KEGG analysis of the 362 DEGs indicated that the sesquiterpenoid and triterpenoid biosynthesis pathways were enriched in the leaves and pods of SGS plants (Table S11), suggesting that SGS may be caused by disease, which is regulated by these pathways. The consequence of an increased chlorophyll content is a delayed leaf senescence, whereby pods and stems remain green at the late stage of the reproductive period. This results in an imbalance in growth, a disruption in the transport of nutrients, and a reduction in the yield of soybeans.

The accurate identification of SGS genotypes is difficult predominantly because of the continuous variation in characteristics, caused by complex genetic factors [33]. Several studies have suggested that the SGS phenotype is often caused by mutations in genes related to photosynthesis, chlorophyll metabolism, leaf senescence, and antioxidative enzyme activity [34,35]. In addition to the genetic factors, environmental influences may also play a causative role, including temperature [36], drought [37], diseases [38,39], and insect infestation [40]. The SGS leaves (Figure 3B,C) indicated that plants retained the ability to synthesize photosynthetic products alongside impacted nutrient delivery from maternal tissues to filial cotyledons. Therefore, we speculate that the accumulation of nutrients and the absence of signaling substances in pods may be crucial for determining whether the SGS phenotype affects the development of soybeans. It is well established that pods are an important organ for photosynthesis. The high expression of photosynthesis-related genes (Tables S6 and S9) and significantly increased chlorophyll *a* and *b* contents (Figure 3B,C) indicate that pod photosynthesis is likely constant or enhanced due to SGS.

Phytohormones play crucial roles in plant growth and development, regulating the plant growth rate, and coordinating and integrating various components to promote the normal growth and reproductive processes [24]. The auxin level is important during the grain-filling process [41]. In this study, IAA content was significantly decreased in the SGS pods (Figure 5A), which likely results in failed seed development and filling compared to non-SGS plants. ABA induces the transcription of *AtNAP*, with the transcription factor AtNAP inducing the expression of *SAG113*, encoding the protein phosphatase 2C to control the senescence phenotype [42]. RNA-seq data indicated that two protein phosphatase 2C genes were significantly down-regulated (Table S3). The ABA content in SGS plants is lower than in normal plants (Figure 5B), suggesting that ABA is important for senescence. The JA and SA contents in the leaves and pods of SGS plants are significantly higher than in normal plants (Figure 5C,D). As important signaling molecules within the plant defense system, JA and SA can likely respond to SGS. This is due to their reported role in defense against biological and abiotic stresses in plants [43]. How phytohormones are involved in SGS tolerance remains to be determined.

SGS results in abnormal soybean development. Moving forward, we can explore whether signaling substances inhibit the nutrient transport process to the seeds, or if the transport of substances from the leaves was prevented. Moreover, we will explore the process of protein biosynthesis in SGS soybeans, as proteins are the main components of soybeans.

5. Conclusions

In summary, SGS results in retained green in leaves and pods, even during the seed filling stage. We determined that photosynthesis, carbohydrate metabolism, and oxidative stress were significantly enhanced in SGS pods, which likely act to protect the plants. Additionally, the IAA and ABA concentrations were reduced in the SGS pods. In contrast, the concentrations of JA and SA were increased, indicating that phytohormones are important in the development of SGS. However, how these processes are involved in the regulation of SGS requires further investigation. Our study evaluated the possible biological processes essential for soybean SGS development, which provides insight into the molecular mechanism to allow for further studies.

Supplementary Materials: The following supporting information can be downloaded at https://www.mdpi.com/article/10.3390/agronomy15010082/s1. Table S1: Primers for qPCR amplification of differentially expressed genes; Table S2: Data assembly for transcriptome sequencing; Table S3: DEGs in SGS soybean pods; Table S4: DEGs in SGS soybean leaves; Table S5: DEGs both of SGS pods and leaves; Table S6: GO enrichment analysis of soybean pods; Table S7: GO enrichment analysis of soybean leaves; Table S8: GO enrichment analysis both of pods and leaves; Table S9: KEGG enrichment analysis of soybean pods; Table S10: KEGG enrichment analysis of soybean leaves; Table S11: KEGG analysis both of soybean pods and leaves; Table S12: GO enrichment of ROS-related DEGs; Table S13: KEGG analysis of glutathione metabolism pathway; Figure S1: Correlation between qPCR and RNA-seq analyses of 15 DEGs in the pods (A) and the leaves (B) of SGS plants; Figure S2: Heat map illustrating the expression of 39 DEGs involved in phytohormone signal transduction pathways.

Author Contributions: D.W.: investigation, formal analysis, and writing—original draft. Z.H.: conceptualization, project administration, funding acquisition, and writing—review and editing. Y.W. (Yanan Wang): data curation, formal analysis, and writing—original draft. R.S.: visualization. Y.Y.: data curation. W.Z.: investigation. G.Y.: data curation. Y.W. (Yueying Wang): investigation. F.W.: data curation. L.Z.: data curation. All authors have read and agreed to the published version of the manuscript.

Funding: This work was jointly supported by the Key Research and Development Program of Anhui Province (202004a06020034, 2023n06020007), the Program on Industrial Technology System of National Soybean (CARS-04-PS07), the Talent Program of Anhui Academy of Agricultural Sciences (QNYC-201909), and the Establish Science and Technology Enhancement Project of National Modern Agricultural Industrial Park (Anhui Guoyang).

Data Availability Statement: The original contributions presented in the study are included in the article, further inquiries can be directed to the corresponding author.

Conflicts of Interest: Author Guoyi Yu was employed by the company Anhui Longkang Farm of Land-Reclamation. The remaining authors declare that the research was conducted in the absence of any commercial or financial relationships that could be construed as a potential conflict of interest.

References

1. Cheng, R.; Mei, R.; Yan, R.; Chen, H.; Miao, D.; Cai, L.; Fan, J.; Li, G.; Xu, R.; Lu, W.; et al. A new distinct geminivirus causes soybean stay-green disease. *Mol. Plant* **2022**, *15*, 927–930. [CrossRef] [PubMed]
2. Li, K.; Zhang, X.; Guo, J.; Penn, H.; Wu, T.; Li, L.; Jiang, H.; Chang, L.; Wu, C.; Han, T. Feeding of *Riptortus pedestris* on soybean plants, the primary cause of soybean staygreen syndrome in the Huang-Huai-Hai river basin. *Crop J.* **2019**, *7*, 360–367. [CrossRef]
3. Wang, D.; Yu, G.; Chen, S.; Li, J.; Wu, Q.; Hu, G.; Huang, Z. Research progress on comprehensive hazard factors of soybean staygreen syndrome. *Soybean Sci.* **2021**, *40*, 708–714.
4. Wang, X.; Wang, M.; Wang, L.; Feng, H.; He, X.; Chang, S.; Wang, D.; Wang, L.; Yang, J.; An, G.; et al. Whole-plant microbiome profiling reveals a novel geminivirus associated with soybean stay-green disease. *Plant Biotechnol. J.* **2022**, *20*, 2159–2173. [CrossRef] [PubMed]
5. He, H.; Li, H.; Wang, Y.; Xu, Y.; Cui, X.; Zhou, X.; Li, F. Soybean stay-green associated geminivirus: A serious threat to soybean production in China. *Virology* **2025**, *602*, 110312. [CrossRef]
6. Guo, J.; Ma, W.; Lei, Q.; Yang, X.; Li, Y.X. Tentative analysis of "Zhengqing" phenomena of soybean in the Huanghuai valleys. *J. Henan Agric. Sci.* **2012**, *41*, 45–48.
7. Wei, Z.; Guo, W.; Jiang, S.; Yan, D.; Shi, Y.; Wu, B.; Xin, X.; Chen, L.; Cai, Y.; Zhang, H.; et al. Transcriptional profiling reveals a critical role of GmFT2a in soybean staygreen syndrome caused by the pest Riptortus pedestris. *New Phytol.* **2023**, *237*, 1876–1890. [CrossRef]
8. Yin, J.; Hu, Z.; Xu, S.; Hong, X.; Qiu, Y.; Cheng, X.; Wang, L.; Shen, W.; Zhi, H.; Li, K.; et al. Leafhopper transmits soybean stay-green associated virus to leguminous plants. *Phytopathol. Res.* **2023**, *5*, 17. [CrossRef]
9. Liang, J.M.; Mo, X.S.; Qian, C.A.O.; Chen, H.M.; Shang, B.Z.; Zhang, M.J.; Li, D.X. Empirical research of dry heatwave simulation causing soybean staygreen syndrome. *Chin. J. Oil Crop Sci.* **2023**, *45*, 175–182.
10. Xiao, J.; Wei, L.; Liu, B.; Duan, X.; Yang, H. Analysis on the cause of "greenness symptoms" phenomenon of summer soybean in southern. *Shanxi Agric. Sci.* **2020**, *48*, 1305–1308.

11. Luo, J.; Shi, S.; Liu, J.; Liu, G.; Chong, Y.; Jia, J.; Ma, D.; Zhao, D.; Yin, J. Analysis on the causes of soybean stay green syndrome. *Bull. Agric. Sci. Technol.* **2020**, *579*, 143–146.
12. Zhang, S.-j.; Hou, L.-x. Study on the Cause Mechanism for Pods without Peas in Soybean under Drought Stress. *Acta Agric. Boreali-Sin.* **2005**, *20*, 61–63.
13. Cheng, R.; Yan, R.; Mei, R.; Wang, Y.; Niu, W.; Ai, H.; Qiao, S.; Xu, M.; Yu, W.; Ye, W.; et al. Epidemiological evaluation and identification of the insect vector of soybean stay-green associated virus. *Phytopathol. Res.* **2023**, *5*, 20. [CrossRef]
14. Qin, S.; Yan, X.; Li, M.; Sun, X.; Lu, X.; Li, N.; Wang, S. Analysis of relationships among source, sink and flux of soybean. *Hunan Agric. Sci.* **2010**, *248*, 54–56.
15. Mo, X.; Liang, J.; Li, D.; Liang, F.; Zhang, M.; Liu, M.; Xiao, J.; Zhang, P. Exploring the causes and precautions of soybean staygreen syndrome in Huang-Huai-Hai regions. *Soybean Sci.* **2019**, *38*, 770–778.
16. Zhang, J. Analyzing of the "Staygreen Syndrome" Phenomenon in Soybean from the Perspective of Source-Sink-Flow. Master's Thesis, Zhengzhou University, Zhengzhou, China, 2021.
17. Yang, C.; Zhang, G.; Li, K.; Wang, X.; Shu, H.; Guo, J.; Liu, R. Biomass and nutrient accumulation and distribution characteristics of soybean "Zhengqing" plants. *Jiangsu Agric. Sci.* **2021**, *49*, 66–72.
18. Ergo, V.V.; Veas, R.E.; Vega, C.R.C.; Lascano, R.; Carrera, C.S. Leaf photosynthesis and senescence in heated and droughted field-grown soybean with contrasting seed protein concentration. *Plant Physiol. Biochem.* **2021**, *166*, 437–447. [CrossRef]
19. Chen, K.; Li, G.J.; Bressan, R.A.; Song, C.P.; Zhu, J.K.; Zhao, Y. Abscisic acid dynamics, signaling, and functions in plants. *J. Integr. Plant Biol.* **2020**, *62*, 25–54. [CrossRef]
20. He, D.; Li, Z.; Zhao, L.; Chi, X.; Zheng, G.; Liu, P.; Zhang, W.; Cao, S. Relationship between endogenous hormones, gene expression and senescence in soybean male sterile lines. *Soybean Sci.* **2020**, *39*, 205–211.
21. Hu, Y.; Jiang, Y.; Han, X.; Wang, H.; Pan, J.; Yu, D. Jasmonate regulates leaf senescence and tolerance to cold stress: Crosstalk with other phytohormones. *J. Exp. Bot.* **2017**, *68*, 1361–1369. [CrossRef]
22. Livak, K.J.; Schmittgen, T.D. Analysis of relative gene expression data using real-time quantitative PCR and the 2(-Delta Delta C(T)) Method. *Methods* **2001**, *25*, 402–408. [CrossRef] [PubMed]
23. Kong, L.; Song, Q.; Wei, H.; Wang, Y.; Lin, M.; Sun, K.; Zhang, Y.; Yang, J.; Li, C.; Luo, K. The AP2/ERF transcription factor PtoERF15 confers drought tolerance via JA-mediated signaling in Populus. *New Phytol.* **2023**, *240*, 1848–1867. [CrossRef] [PubMed]
24. Zhang, Y.; Li, Y.; Hassan, M.J.; Li, Z.; Peng, Y. Indole-3-acetic acid improves drought tolerance of white clover via activating auxin, abscisic acid and jasmonic acid related genes and inhibiting senescence genes. *BMC Plant Biol.* **2020**, *20*, 150. [CrossRef]
25. Kim, Y.J.; Choi, J.; Lee, G.; Lee, K.G. Analysis of furan and monosaccharides in various coffee beans. *J. Food Sci. Technol.* **2021**, *58*, 862–869. [CrossRef]
26. Chen, L.; Zhang, Y.H.; Wang, S.; Zhang, Y.; Huang, T.; Cai, Y.D. Prediction and analysis of essential genes using the enrichments of gene ontology and KEGG pathways. *PLoS ONE* **2017**, *12*, e0184129. [CrossRef]
27. Waadt, R.; Seller, C.A.; Hsu, P.K.; Takahashi, Y.; Munemasa, S.; Schroeder, J.I. Plant hormone regulation of abiotic stress responses. *Nat. Rev. Mol. Cell Biol.* **2022**, *23*, 680–694. [CrossRef]
28. Zhu, F.; Zhang, Q.P.; Che, Y.P.; Zhu, P.X.; Zhang, Q.Q.; Ji, Z.L. Glutathione contributes to resistance responses to TMV through a differential modulation of salicylic acid and reactive oxygen species. *Mol. Plant Pathol.* **2021**, *22*, 1668–1687. [CrossRef]
29. Savary, S.; Willocquet, L.; Pethybridge, S.J.; Esker, P.; McRoberts, N.; Nelson, A. The global burden of pathogens and pests on major food crops. *Nat. Ecol. Evol.* **2019**, *3*, 430–439. [CrossRef]
30. Zhang, X.; Wang, M.; Wu, T.; Wu, C.; Jiang, B.; Guo, C.; Han, T. Physiological and molecular studies of staygreen caused by pod removal and seed injury in soybean. *Crop J.* **2016**, *4*, 435–443. [CrossRef]
31. Underwood, W.; Somerville, S.C. Focal accumulation of defences at sites of fungal pathogen attack. *J. Exp. Bot.* **2008**, *59*, 3501–3508. [CrossRef]
32. Ahuja, I.; Kissen, R.; Bones, A.M. Phytoalexins in defense against pathogens. *Trends Plant Sci.* **2012**, *17*, 73–90. [CrossRef] [PubMed]
33. Yoo, S.C.; Cho, S.H.; Zhang, H.; Paik, H.C.; Lee, C.H.; Li, J.; Yoo, J.H.; Lee, B.W.; Koh, H.J.; Seo, H.S.; et al. Quantitative trait loci associated with functional stay-green SNU-SG1 in rice. *Mol Cells* **2007**, *24*, 83–94. [CrossRef] [PubMed]
34. Li, Q.; Yang, H.; Guo, J.; Huang, Q.; Zhong, S.; Tan, F.; Ren, T.; Li, Z.; Chen, C.; Luo, P. Comparative transcriptome analysis revealed differential gene expression involved in wheat leaf senescence between stay-green and non-stay-green cultivars. *Front. Plant Sci.* **2022**, *13*, 971927. [CrossRef]
35. Zhang, J.; Li, H.; Xu, B.; Li, J.; Huang, B. Exogenous Melatonin Suppresses Dark-Induced Leaf Senescence by Activating the Superoxide Dismutase-Catalase Antioxidant Pathway and Down-Regulating Chlorophyll Degradation in Excised Leaves of Perennial Ryegrass (*Lolium perenne* L.). *Front. Plant Sci.* **2016**, *7*, 1500. [CrossRef]

36. Li, Z.; Tang, M.; Hassan, M.J.; Zhang, Y.; Han, L.; Peng, Y. Adaptability to High Temperature and Stay-Green Genotypes Associated with Variations in Antioxidant, Chlorophyll Metabolism, and γ-Aminobutyric Acid Accumulation in Creeping Bentgrass Species. *Front. Plant Sci.* **2021**, *12*, 750728. [CrossRef]
37. Borrell, A.K.; Hammer, G.L.; Douglas, A.C.L. Does maintaining green leaf area in sorghum improve yield under drought? I. Leaf growth and senescence. *Crop. Sci.* **2000**, *40*, 1026–1037. [CrossRef]
38. Joshi, A.K.; Kumari, M.; Singh, V.P.; Reddy, C.M.; Kumar, S.; Rane, J.; Chand, R. Stay green trait: Variation, inheritance and its association with spot blotch resistance in spring wheat (*Triticumaestivum* L.). *Euphytica* **2007**, *153*, 59–71. [CrossRef]
39. Li, X.; Liu, T.; Chen, W.; Zhong, S.; Zhang, H.; Tang, Z.; Chang, Z.; Wang, L.; Zhang, M.; Li, L.; et al. Wheat WCBP1 encodes a putative copper-binding protein involved in stripe rust resistance and inhibition of leaf senescence. *BMC Plant Biol.* **2015**, *15*, 239. [CrossRef]
40. Li, Q.; Zhong, S.; Sun, S.; Fatima, S.A.; Zhang, M.; Chen, W.; Huang, Q.; Tang, S.; Luo, P. Differential effect of whole-ear shading after heading on the physiology, biochemistry and yield index of stay-green and non-stay-green wheat genotypes. *PLoS ONE* **2017**, *12*, e0171589. [CrossRef]
41. Teng, Z.; Yu, H.; Wang, G.; Meng, S.; Liu, B.; Yi, Y.; Chen, Y.; Zheng, Q.; Liu, L.; Yang, J.; et al. Synergistic interaction between ABA and IAA due to moderate soil drying promotes grain filling of inferior spikelets in rice. *Plant J.* **2022**, *109*, 1457–1472. [CrossRef]
42. Zhang, K.; Gan, S.S. An abscisic acid-AtNAP transcription factor-SAG113 protein phosphatase 2C regulatory chain for controlling dehydration in senescing Arabidopsis leaves. *Plant Physiol.* **2012**, *158*, 961–969. [CrossRef] [PubMed]
43. Zhao, B.; Liu, Q.; Wang, B.; Yuan, F. Roles of Phytohormones and Their Signaling Pathways in Leaf Development and Stress Responses. *J. Agric. Food Chem.* **2021**, *69*, 3566–3584. [CrossRef]

Disclaimer/Publisher's Note: The statements, opinions and data contained in all publications are solely those of the individual author(s) and contributor(s) and not of MDPI and/or the editor(s). MDPI and/or the editor(s) disclaim responsibility for any injury to people or property resulting from any ideas, methods, instructions or products referred to in the content.

agronomy

MDPI

Article

Identification, Characterization, and Chemical Management of *Fusarium asiaticum* Causing Soybean Root Rot in Northeast China

Jinxin Liu [1], Wanqiu Cui [1], Qingyi Zhao [1], Zhipeng Ren [2], Lin Li [3], Yonggang Li [1,*], Lei Sun [4] and Junjie Ding [5,*]

1 College of Plant Protection, Northeast Agricultural University, Harbin 150030, China; mary610123@126.com (J.L.); cuiwanqiu0904@163.com (W.C.); qingyizhao202501@126.com (Q.Z.)
2 Heilongjiang Academy of Agricultural Sciences, Harbin 150030, China; renzhipeng3@163.com
3 Heilongjiang Ecology Institute, Harbin 150030, China; lilin_1002@163.com
4 Heilongjiang Acacemy of Black Soil Conservation & Utilization, Harbin 150030, China; tufeisuosunlei@163.com
5 Jiamusi Branch of Heilongjiang Academy of Agricultural Sciences, Jiamusi 154000, China
* Correspondence: neaulyg@126.com (Y.L.); me999@126.com (J.D.)

Abstract: Soybean root rot, a soil-borne fungal disease, is caused by multiple pathogens that seriously affect soybean production. During spring 2021, 92 pathogenic fungal strains were isolated from soybean plants with root rot in Hailun City, Heilongjiang Province, China. Through morphological and molecular identification, these strains were identified as *Fusarium oxysporum* (39.1%), *F. asiaticum* (30.4%), *F. graminearum* (13.0%), *Pythium macrosporum* (8.7%), and *Rhizoctonia solani* (8.7%). Among them, *F. oxysporum* was the dominant species, and *F. asiaticum*, not previously reported as a soybean root rot pathogen in Northeast China. Approximately 50% of the *F. asiaticum* isolates were moderately pathogenic. In addition, *F. asiaticum* had a wide host range, infecting black soybean, French bean, white hyacinth bean, mung bean, and adzuki bean but not corn, peanut, rice, and oat roots. Regarding field management, fludioxonil and pyraclostrobin had the best control effects of 73.8% and 69.4%, with EC_{50} values of 0.0029–0.0071 $\mu g \cdot mL^{-1}$ and 0.0045–0.0076 $\mu g \cdot mL^{-1}$, respectively. The study reported that *F. asiaticum* is a pathogen causing soybean root rot in northeast China. The application of chemical fungicides and non-host crop rotation can effectively control the disease caused by *F. asiaticum*.

Keywords: soybean; root rot; *Fusarium asiaticum*; host range; identification; fungicide efficacy

1. Introduction

Soybeans [*Glycine max* (Linn.) Merr.] hold the highest economic value among food and oil crops worldwide. They are abundant in protein, oil, vitamins, and various mineral nutrients. Moreover, they can be easily adapted to different environments and are widely cultivated globally for human consumption, animal feed, and biodiesel production [1]. With the adjustment of the supply side structure in China, soybean planting area in this country is increasing every year [2]. Soybean planting area will increase by 667,000 hectares to 45,700,000 hectares from 2021 to 2022 in Heilongjiang Province, accounting for nearly 50% of China's total soybean planting area. Located in one of the three major black soil zones in the world, Heilongjiang has fertile soil, suitable climatic conditions, and a high-quality ecological environment, providing a suitable production environment for soybeans. However, soybean root rot has become a major obstacle to soybean production [3].

https://doi.org/10.3390/agronomy15020388

Pathogenic fungi can infect root cells, causing serious damage to soybean roots. Infected plants often exhibit growth retardation due to the weakened ability of their roots to absorb water and nutrients [4]. In severe cases, it can lead to plant death, significantly reducing the yield and quality of soybeans [5]. The economic loss caused by the impact of various pathogenic fungi on soybeans has increased in recent years [6].

The community structure of fungal pathogens that cause soybean root rot is complex. At present, 64 fungal pathogens causing soybean root rot have been reported internationally, including species from the genus *Fusarium* such as *Fusarium pseudograminearum, F. proliferatum, F. sporotrichioides, F. fujikuroi, F. graminearum, F. armeniacum, F. commune, F. tricinctum, and F. asiaticum* [7–15]; species from the genus *Pythium,* including *P. oopapillum, P. macrosporum, P. aphanidermatum,* and *P. deliense* [4,16–18]; *Rhizoctonia solani; Helicobasidium mompa; Thielaviopsis basicola; Stachybotrys chartarum; Sclerotium rolfsii; Mycoleptodiscus terrestris*; and *Phymatotrichopsis omnivora* [19–25]. Thirty species have been reported domestically, including *F. oxysporum* var. rendolens, *F. oxysporum, F. graminearum, F. chlamydosporum, F. merismoidescorda, F. episphaeria, F. camptoceras,* '*Candidatus* Pythium huanghuaiense', *Phytophthora sojae, Phytophthora sansomeana, Rhizoctonia solani, Phomopsis longicolla,* and *Pratylenchus coffeae* [26–36]. Sixteen species have been reported in Heilongjiang Province, e.g., Fusarium *graminearum, Phytophthora sojae* [28,37]. In addition, *Rhizoctonia solani, Phomopsis longicolla* have also been reported. The above results indicate that the species diversity of fungal pathogens that cause soybean root rot in the Heilongjiang Province, the main soybean production area, is potentially very complex. However, *F. asiaticum,* as a pathogenic fungus of soybean root rot, has not been reported in Heilongjiang province.

Soybean root rot caused by *Fusarium* spp. is a major disease, mainly transmitted in the soil [38,39]. Fusarium root rot of soybeans can endanger any stage of soybean development, resulting in water-soaked decay after sowing and before germination, which affects the germination rate of seeds after infection. Seedling infection leads to the decay of the root epidermis, browning of vascular bundles, withered yellow leaves, and plant death in serious cases, along with shriveled grains and serious economic costs [40].

Reducing the impact of diseases during soybean cultivation is crucial for increasing yield [41,42]. Currently, the most cost-effective way to control soybean root rot is to cultivate disease-resistant soybean varieties [43]. However, there are few specially bred Fusarium-resistant cultivars [44]. Therefore, fungicide treatment is one of the most effective disease management strategies for controlling soybean root rot [45]. The fungicides commonly used to control Fusarium root rot are pyraclostrobin, fludioxonil, and azoxystrobin [46,47]. However, *F. asiaticum* has unique genetic variations that may make it resistant to commonly used fungicides, and existing control methods for other *Fusarium* species may no longer be effective [48]. Therefore, it is necessary to screen fungicides and determine the sensitivity of target pathogens causing soybean root rot in the field.

In 2020 and 2021, soybean root rot occurred in Hailun city, Heilongjiang Province, with a diseased seedling rate of 20–30% in general plots and over 60% in severely infected plots. The objectives of this study were to identify the pathogenic microorganisms causing soybean root rot, analyze their pathogenicity, and determine the sensitivity and efficacy of common fungicides against these pathogens, providing a basis for formulating control strategies.

2. Materials and Methods

2.1. Pathogen Isolation and Assessment of Their Pathogenicity

Field investigation of soybean root rot at the seedling stage was conducted at 3 sites (5 hectares per field) in an important soybean planting area in Hailun city. The local soil, mainly black sandy clay with 4–5% organic matter, had a disease incidence of 10–20% in the surveyed fields (about 5 ha each). Soybean plants (n = 182) with root rot symptoms were collected using five-spot sampling (Figure 1). The roots were thoroughly rinsed under running tap water for 10 min to remove soil and debris. Pathogens were isolated from symptomatic root tissues following a published method and cultured on potato dextrose agar (PDA) at 28 °C in the dark [49]. After three days, hyphal tips were transferred to isolate and purify the fungal cultures. One diseased tissue sample from each infected seedling was selected and isolated. The number of isolated and purified pathogens was recorded and the percentage isolated to each species was calculated.

Figure 1. Soybean plants with root rot symptoms in the field.

The pathogenicity of isolated and purified strains was evaluated according to the Koch hypothesis [50]. A method of inoculating soybeans with fungus by embedding roots of sorghum grain [51]. All strains were re-isolated from diseased soybean seedlings and observed. The specific method was as follows: A total of 1/3 of the volume of the sterile soil was inserted into a flowerpot (diameter of 15 cm), and 20 g of sorghum grains that are already overgrown with pathogenic fungi were evenly sprinkled. Then, 12 soybean seeds (the variety used was HeNong 511) were evenly placed on the surface, covered with a layer of culture soil (1 cm), and after the seedlings emerged, 10 were kept in each flowerpot [2]. Seedlings cultured without isolation strains were used as the control group. Each treatment had three replicates, and the experiment was conducted twice. After 20 days, the incidence of root rot was investigated, and the infected plants were re-isolated and morphologically identified according to Koch's postulates; the disease index was calculated according to the classification standard of root rot by Wang et al. [52]. Ten highly pathogenic strains were selected for subsequent experiments.

Disease severity was visually scored on a scale of 0–7 based on the growth status of soybean seedlings: 0 = no symptoms; 1 = the taproot was basically unchanged or slightly browned, the fibrous root was not long, the growth point was browned, and the plant growth was normal; 3 = the taproot turned black but continued to grow through the infection point, the fibrous root tip turned black, and the plant grew normally; 5 = the taproot was seriously blackened and could not continue growing through the infection point, the fibrous root was obviously reduced or absent, the aboveground growth was poor, and the plant growth was short; and 7 = root rot, failure of normal growth or emergence, partial cotyledon rot, or plant death [52].

The percentage of disease index (PDI) for soybean root rot was calculated using the following formula: PDI = $\sum$(the number of diseased plants at each level × the corresponding relative ratings)/(the total number of surveyed plants × the highest disease level rating) × 100.

The pathogenicity of the isolates was classified based on the average disease index of the two repeated experiments. Isolates with a disease index less than 50 were classified as having weak pathogenicity (designated as W), those with a disease index between 50 (inclusive) and 60 (exclusive) were classified as having moderate pathogenicity (designated as M), and those with a disease index of 60 or greater were classified as having high pathogenicity (designated as H).

2.2. Identification of the Pathogen

Isolates responsible for soybean root rot were identified through their morphological characteristics and molecular methods [53]. For molecular identification, genomic DNA was extracted from the mycelia of representative isolates using a Tiangen Genome Extraction Kit (Tiangen Biotech, Beijing, China). The internal transcribed spacer region (ITS), translation elongation factor 1-α (*Tef1*), and β-tubulin (*Tub2*) genes were amplified and sequenced using primers ITS1/ITS4, EF1-728F/EF1-986R, and Bt2a/Bt2b, respectively [54–56]. Subsequently, the obtained sequences were submitted to the GenBank database (Table A1). Polymerase chain reaction (PCR) was carried out in a 50 μL reaction system containing 10 μM of each primer, 2 × Taq Master Mix, and 10 ng of template DNA. The PCR conditions were as follows: initial denaturation at 94 °C for 5 min, followed by 35 cycles, each cycle including 1 min denaturation at 94 °C, 1 min annealing at 52 °C, 1.5 min extension at 72 °C, and finally a 10 min final extension at 72 °C. The PCR products were purified and sequenced by GENEWIZ (Azenta Life Sciences, Suzhou, China).

Phylogenetic trees of representative isolates were constructed using PhyloSuite v1.2.2 (http://phylosuite.jushengwu.com/, accessed on 2 June 2023) following the MrBayes method and were further edited in FigTree v1.4.4 (http://tree.bio.ed.ac.uk/software/figtree/, accessed on 2 June 2023) [57,58].

2.3. Biological Characteristics of Fusarium asiaticum

To determine the optimum pH and temperature for the isolates, the mycelium growth rate method was employed [59,60]. Ten isolates were cultured on PDA at different pH levels (5.0, 6.0, 7.0, 8.0, 9.0, 10.0 and 11.0) and temperatures (10 °C, 20 °C, 25 °C, 28 °C, 30 °C, and 35 °C). A 0.5 cm diameter mycelial plug from a 96 h PDA-grown isolate was transferred to different treated PDA plates and incubated under the corresponding conditions. Each treatment had three replicates, and the experiment was conducted twice. The colony diameters of the isolates were measured after 72 h.

2.4. Host Range Determination of Fusarium asiaticum

The *F. asiaticum* isolates causing soybean root rot were inoculated on crops commonly grown in Heilongjiang, which were black soybean (*Glycine max* Linn.), French bean (*Phaseolus vulgaris* Linn.), white hyacinth bean (*Dolicho Lablab* Linn.), mung bean (*P. radiatus* Linn.), corn (*Zea mays* Linn.), peanut (*Arachis hypogaea* Linn.), rice (*Oryza sativa* Linn.), adzuki bean (*Vigna umbellata* Thunb.), and oat (*Avena sativa* Linn.) seedlings. The inoculation method was consistent with that used to determine the pathogenicity of the above isolates. Each treatment was replicated thrice. Approximately 20 days after inoculation, the pathogenicity of the isolate in each crop was investigated and evaluated. The pathogen was re-isolated and identified from the inoculated seedlings to complete Koch's postulates. All experiments were performed twice.

2.5. Sensitivity of Fusarium asiaticum to Fungicides

The mycelial growth rate method was employed to evaluate the sensitivity of the isolates to the following fungicides that are commonly utilized for controlling *Fusarium* root rot [46,47]: pyraclostrobin [25% flowable concentrate (Jiangsu Tuoqiu Agrochemicals Co., Ltd., Yancheng, China)], prochloraz (450 g·L^{-1} EW) [Shanghai Hulian Biopharmaceutical (Xiayi) Co., Ltd., Shanghai, China], fludioxonil (25% FSB) [Syngenta (Nantong) Crop Protection Co., Ltd., Nantong, China], and a mixture of 11.7% propiconazole + 7% azoxystrobin [18.7% suspoemulsion (Syngenta Nantong Crop Protection Co., Ltd., Nantong, China)] [61].

The fungicides were dissolved in 1000 mL of sterile distilled water. Stock solutions of the four fungicides were then added to PDA at concentrations of 0 (control), 0.001, 0.002,

0.005, 0.01, and 0.02 μg·mL^{-1}. The PDA plates were incubated at 26 °C for five days. Subsequently, a 0.7 cm diameter mycelial plug of the isolate was placed at the center of each fungicide-amended PDA plate and incubated in the dark at 26 °C for seven days. Each treatment was replicated three times, and the entire experiment was conducted twice. After the incubation period, the colony diameters were measured. The effective concentration resulting in 50% mycelial growth reduction (EC_{50}) of the four fungicides was calculated according to the method described by Lehner et al. [62]. Data from the two replicate experiments were pooled, and EC_{50} values were calculated. The inhibitory effect was expressed as a percentage relative to the control, using the formula: 1 − [(diameter of treated colony − 0.5)/(diameter of control colony − 0.5) × 100] [63].

2.6. *Efficacy of Pyraclostrobine and Fludioxonil Against Soybean Root Rot Caused by Fusarium asiaticum*

Pot experiments were conducted in 20 cm diameter plastic pots in a greenhouse at the experimental station of Northeast Agricultural University, Harbin, China. The greenhouse conditions were set at 25 ± 3 °C with a 12 h/12 h (light/dark) photoperiod. The specific treatments applied were as follows: (1) pyraclostrobine at effective doses of 62.5, 125, and 250 μg·mL^{-1}; (2) fludioxonil at effective doses of 62.5, 125, and 250 μg·mL^{-1}; (3) controls treated with sterile water. The sorghum seed inoculation method was once again applied for pathogen inoculation to determine pathogenicity. Ten soybean seedlings were kept in each bowl, with three replicates for each treatment. The experiment was carried out by soaking seeds, first washing the seeds with sterile water, and then soaking with different concentrations of fungicide liquids configured for 20–30 min. Control seeds are soaked in equal amounts of sterile distilled water. After 20 days, the disease severity was determined using the same method as the pathogenicity determination of the isolates. The disease index was calculated as described above. Seedling height, mass, and root length were measured using a graduated ruler (1 mm) and balance (1 mg). The control efficacy was calculated using the following formula:

The control efficacy was calculated using the formula: Control efficacy = (Disease index of the control group − Disease index of the fungicide − treatment group)/Disease index of the control group × 100%.

2.7. *Data Analysis*

Analysis of variance (ANOVA) was conducted using SPSS Statistics 17.0 (IBM/SPSS, Armonk, NY, USA). The treatment means were compared and separated by applying the least significant difference (LSD) test at a significance level of $p = 0.05$. The EC_{50} values were estimated with the assistance of GraphPad Prism 8 (GraphPad Software Inc., USA).

3. Results

3.1. *Disease Symptoms and Identification of Causal Organisms*

In May 2021, diseased soybean seedlings were detected in Hailun City, Heilongjiang Province, China (47.46093° N, 126.9682° E). Lesion plaques were evident at the base of the stem and were initially reddish-brown and then gradually enlarged, followed by blackening of the cortex, decay, and necrosis. The above-ground parts of infected plants were dwarfed by healthy plants, with the green leaves fading. According to the Koch postulates, 92 pathogenic isolates were isolated from 182 symptomatic seedlings. Based on morphological and molecular identification, these isolates were classified into five species (Table 1, Figure 2): *F. oxysporum* (39.1%), *Fusarium asiaticum* (30.4%), *F. graminearum* (13.0%), *Pythium macrosporum* (8.7%), and *Rhizoctonia solani* (8.7%). In addition to *F. asiaticum*, other pathogenic fungi that cause soybean root rot have been reported in China. Therefore, further systematic identification of *F. asiaticum* was performed in this study. Twenty-eight isolates from infected soybean roots (Figure 3B,C) were white in color, flocculent, luxuriant, and dense with a rose red pigment (Figure 3A). The average growth rate of mycelium was 20.7 mm·d^{-1} on PDA at 28 °C. The macroconidia were thick, with curved apical and basal cells, usually having 4–6 septa, and measuring 44.9 to 44.2 × 3.4 to 5.4 μm on carnation leaf agar. The apical cells were beak-shaped and slightly curved, and the podocydia were not obvious (Figure 3D–F). The chlamydospores were globose to subglobose. Based on these characteristics, the isolates were identified as *F. asiaticum* [64–66].

Table 1. Frequency of pathogens isolated from soybean root rot samples in Hailun, Heilongjiang province, China.

Pathogens	Number of Isolates	Frequency (%)
Fusarium oxysporum	36	39.1
F. asiaticum	28	30.4
F. graminearum	12	13.0
Pythium spp.	8	8.7
Rhizoctonia solani	8	8.7

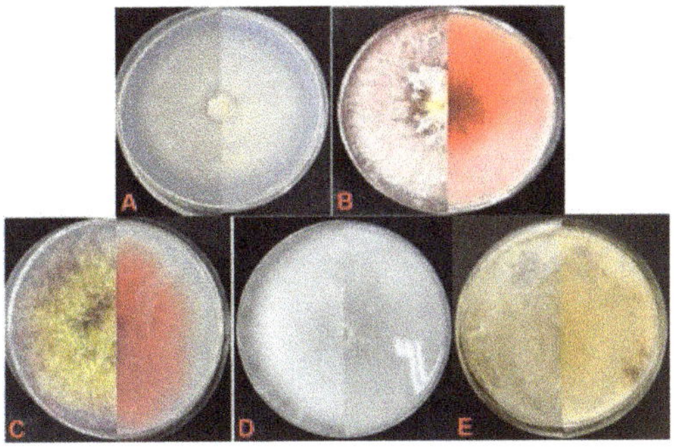

Figure 2. Colony of pathogenic fungi causing soybean root rot on PDA. (**A**) *Fusarium oxysporum*; (**B**) *F. asiaticum*; (**C**) *F. graminearum*; (**D**) *Pythium* spp.; (**E**) *Rhizoctonia solani*.

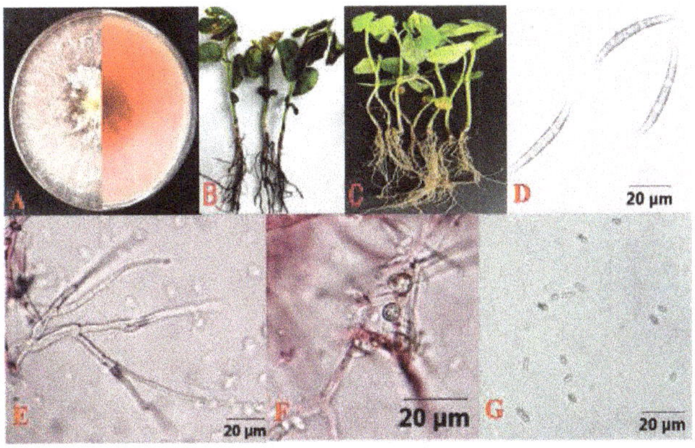

Figure 3. *Fusarium asiaticum* causing soybean root rot. (**A**) Colony of *F. asiaticum* isolate HL35 on PDA; (**B**) Field symptoms of soybean root rot caused by *F. asiaticum*; (**C**) Indoor potting symptoms of soybean root rot caused by *F. asiaticum*; (**D**) Macroconidia; (**E**) Conidiophores; (**F**) Chlamydospore; (**G**) Microconidia.

Genomic DNA from four single-conidium cultures (HL9, HL15, HL25, and HL35) was extracted and amplified using fungal universal primers ITS, *Tef1*, and *Tub2*. The obtained sequences were deposited in the GenBank (accession numbers are shown in Table A1). BLAST analysis revealed that the ITS1/4, EF1-728F/986R, and Bt2a/Bt2b sequence amplicons of HL9, HL15, HL25, and HL35 shared high similarity with those of *F. asiaticum* strain MTLYB02 (OM100564.1), strain RTH17 (LC500693.1), and strain HBTS484 (KM062027.1), respectively. In addition, the phylogenetic analysis showed that isolates HL9, HL15, HL25, and HL35 belonged to the same evolutionary branch as *F. asiaticum*, with high similarity (Figure 4). The combination of molecular and morphological methods confirmed the twenty-eight isolates were *F. asiaticum*.

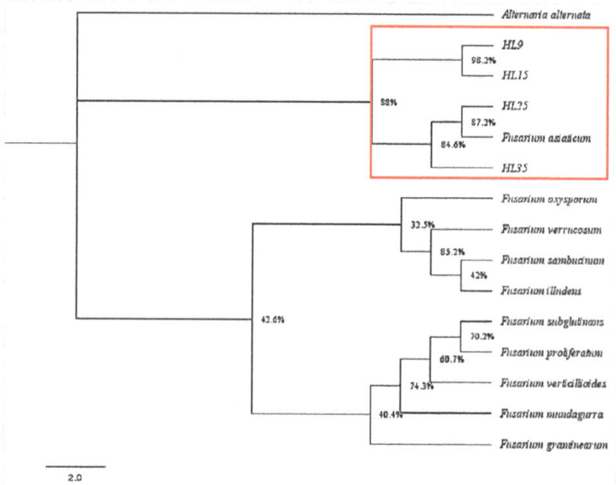

Figure 4. A phylogenetic tree of *Fusarium asiaticum* isolates HL9, HL15, HL25, and HL35, along with members of *Fusarium* spp., was constructed based on Bayesian inference. The analysis was performed on the combined dataset of internal transcribed spacer region (ITS), translation elongation factor 1-α (*Tef1*), and β-tubulin (*Tub2*) gene sequences. The tree-sampling frequency was set at 1000 generations. Branches with Bayesian posterior probabilities of 0.997 were considered significantly supported. *F. asiaticum* was designated as the outgroup.

3.2. Pathogenicity of *Fusarium asiaticum* on Soybean Roots

Differences in pathogenicity were detected among the 28 isolates of *F. asiaticum*. Among these were four highly pathogenic isolates, 10 moderately pathogenic isolates, and 14 weakly pathogenic isolates (Table 2). The four highly pathogenic isolates and six moderately pathogenic isolates were selected for subsequent tests.

Table 2. Disease index and pathogenicity of representative *Fusarium asiaticum* isolates isolated from soybean root rot samples in Hailun, Heilongjiang province, China.

No.	Isolates	Disease Index	Pathogenicity [1]	No.	Isolates	Disease Index	Pathogenicity [1]
1	HL7	51.4	M	15	HL40	27.1	W
2	HL9	27.1	W	16	HL42	47.1	W
3	HL12	54.3	M	17	HL43	58.6	M
4	HL15	62.9	H	18	HL45	57.1	M

Table 2. *Cont.*

No.	Isolates	Disease Index	Pathogenicity [1]	No.	Isolates	Disease Index	Pathogenicity [1]
5	HL18	30.0	W	19	HL48	31.4	W
6	HL19	54.3	M	20	HL51	48.6	W
7	HL20	30.0	W	21	HL55	65.6	H
8	HL24	21.4	W	22	HL56	54.3	M
9	HL25	57.1	M	23	HL58	47.1	W
10	HL28	35.7	W	24	HL59	51.4	M
11	HL32	54.3	M	25	HL62	31.4	W
12	HL35	71.4	H	26	HL66	45.7	W
13	HL37	57.1	M	27	HL73	62.9	H
14	HL38	32.8	W	28	HL81	42.6	W

[1] Isolates with a disease index of less than 50 were classified as having weak pathogenicity (designated as W), those with a disease index between 50 (inclusive) and 60 (exclusive) were classified as having moderate pathogenicity (designated as M), and those with a disease index of 60 or greater were classified as having high pathogenicity (designated as H).

3.3. Biological Characteristics of *Fusarium asiaticum*

The *F. asiaticum* isolates grew in the pH range 5.0–11.0, but mycelial growth varied significantly at different pH values ($p < 0.05$), with an optimal pH of 8.0 (Figure 5A). The ten isolates could grow in the temperatures of 10–30 °C and did not grow at 5 °C and 37 °C, with an optimal temperature of 20 °C (Figure 5B).

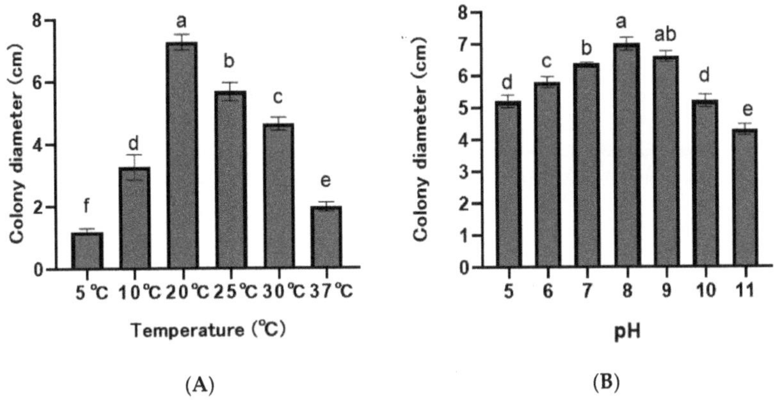

Figure 5. Colony diameters of *Fusarium asiaticum* isolates at different pHs and temperatures. (**A**) Temperature. (**B**) pH. According to the least significant difference test ($p = 0.05$), the different letters above the bar indicate significant differences for each isolate.

3.4. Host Range Determination of *Fusarium asiaticum*

The pathogenicity tests of ten *F. asiaticum* isolates on different crops showed that they were pathogenic to the roots of black soybean, French bean, white hyacinth bean, mung bean, and adzuki bean roots, but not peanut, corn, adzuki bean, and oat roots (Figure 6). No symptoms were observed in the control seedlings of each crop that were treated with sterile water. *Fusarium asiaticum* isolates inoculated on seedlings of different crops were successfully re-isolated from the diseased parts of inoculated black soybean, French bean, white hyacinth bean, mung bean, and adzuki bean roots but could not be isolated from the inoculated parts of corn, peanut, rice, and oat roots.

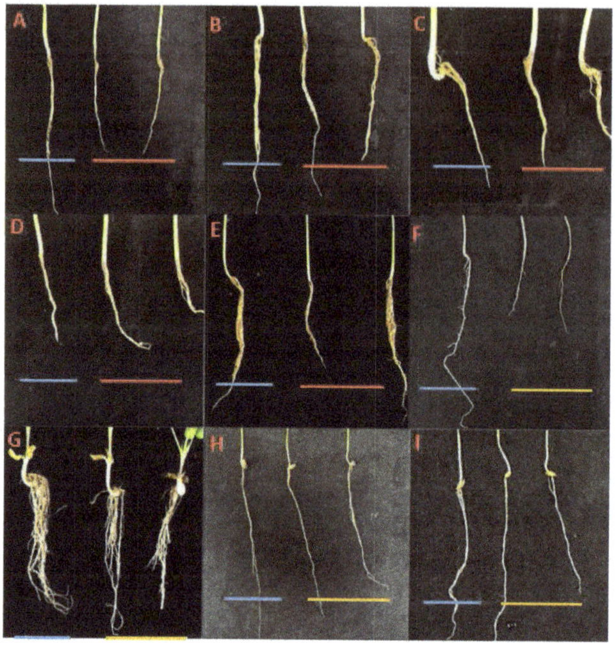

Figure 6. Host range of *Fusarium asiaticum* determined. (**A**) black soybean (*Glycine max* L.) roots. (**B**) French bean (*Phaseolus vulgaris* L.) roots. (**C**) white hyacinth bean (*Dolicho Lablab* L.) roots. (**D**) mung bean (*P. radiatus* L.) roots. (**E**) adzuki bean (*Vigna umbellata* T.) roots. (**F**) rice (*Oryza sativa* L.) roots. (**G**) peanut (*Arachis hypogaea* L.) roots. (**H**) corn (*Zea mays* L.) roots. (**I**) oats (*Avena sativa* L.) roots. In each picture, the blue line represents the control plant; the red line represents the plants inoculated with *F. asiaticum*, whose roots showed significant underdevelopment and even decay compared to the control; the yellow line represents the plants inoculated with *F. asiaticum* but showed no significant difference in root morphology compared to the control.

3.5. Sensitivity to Fungicides

The ten tested *F. asiaticum* isolates showed consistent sensitivity to pyraclostrobine, prochloraz, fludioxonil, and propiconazole·azoxystrobin. Fludioxoni had the strongest inhibitory effect on *F. asiaticum*, with an EC_{50} value of 0.0029–0.0071 μg·mL^{-1}, followed by pyraclostrobine and prochloraz, with EC_{50} values of 0.0045–0.0076 and 0.0059–0.0126 μg·mL^{-1}, respectively. In addition, propiconazole·azoxystrobin had the weakest inhibitory effect, with an EC_{50} value of 0.0101–0.0187 μg·mL^{-1} (Table 3, Figure 7).

Table 3. Sensitivity of the ten tested *Fusarium asiaticum* isolates to frequently used fungicides for the control of soybean root rot in northeast China.

Fungicides	EC_{50} (μg·mL^{-1})	Fungal Sensitivity to Fungicides [1]
Fludioxonil	0.0029–0.0071	S
Pyraclostrobine	0.0045–0.0076	MR
Prochloraz	0.0059–0.0126	MR
Propiconazole·azoxystrobin	0.0101–0.0187	R

[1] S (sensitive): EC_{50} < 0.0050 μg·mL^{-1}; MR (moderately resistant): EC_{50} > 0.0050–0.01 μg·mL^{-1}; R (resistant): EC_{50} > 0.01 μg·mL^{-1}.

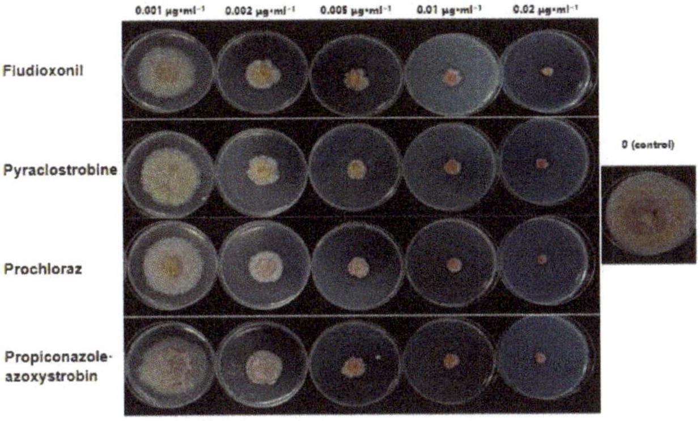

Figure 7. Sensitivity of *Fusarium asiaticum* isolate HL35 to fungicides.

3.6. Efficacy of Fungicides on Soybean Root Rot Caused by Fusarium asiaticum

Based on the result of the sensitivity of *F. asiaticum* to fungicides, fludioxonil and pyraclostrobin were selected as field control agents. In the tested doses, the higher the effective dose of fludioxonil and pyraclostrobin, the better the pot control effects (Table 4). As shown in Table 4, fludioxonil at 250 $\mu g \cdot mL^{-1}$ markedly reduced the severity of soybean root rot caused by *F. asiaticum* and had the best control efficacy of 73.8%. Pyraclostrobin at 250 $\mu g \cdot mL^{-1}$ also had good control efficacy of 69.4%. Overall, fludioxonil at 250 $\mu g \cdot mL^{-1}$ was the most effective dose for controlling soybean root rot. Moreover, the average plant height, root length, and fresh weight of all treated plants were significantly greater than those of the control group ($p < 0.05$).

Table 4. Control effect of fludioxonil and pyraclostrobin on soybean root rot through pot experiments in a greenhouse.

Fungicide	Effective Dose ($\mu g \cdot mL^{-1}$)	Control Efficacy (%) [1]	Plant Height (cm) [1]	Root Length (cm) [1]	Fresh Weight (g) [1]
fludioxonil	250	73.8 ± 2.2 a	33.4 ± 0.6 a	12.2 ± 0.6 a	8.0 ± 0.1 a
	125	55.6 ± 1.7 b	27.3 ± 0.4 b	9.6 ± 0.2 b	6.7 ± 0.1 b
	62.5	23.8 ± 2.5 c	11.4 ± 0.4 c	7.3 ± 0.4 c	5.3 ± 0.4 c
[2] Control	-	-	3.9 ± 0.2 d	4.3 ± 0.3 d	3.6 ± 0.2 d
pyraclostrobin	250	69.4 ± 1.7 a	25.8 ± 0.93 a	10.4 ± 0.3 a	7.6 ± 0.2 a
	125	43.1 ± 5.4 b	21.6 ± 0.95 b	7.8 ± 0.2 a	6.5 ± 0.2 b
	62.5	18.8 ± 2.2 c	7.9 ± 0.12 c	6.4 ± 0.6 b	4.2 ± 0.1 c
[2] Control	-	-	3.9 ± 0.2d	4.3 ± 0.3 c	3.6 ± 0.1d

[1] Values in the column represent the mean ± standard error (SE) of three repeated experiments. Values followed by different letters are significantly different according to the least significant difference test at $p = 0.05$. [2] Control: Not treated with fungicide.

4. Discussion

Soybean is a pivotal food crop and oil crop in China, of which Heilongjiang Province is the main soybean-producing area [67]. Soybean root rot, a soil-borne ailment, affects the entire soybean production lifecycle and severely curtails soybean yields globally [68]. However, distinguishing *F. asiaticum* from other *Fusarium* species using traditional morphological inspections or molecular analysis relying on rDNA-ITS sequencing proves challenging [69]. To ensure the accuracy of identification, the translation elongation factor 1-α (*Tef1*) and *β*-tubulin (*Tub2*) genes of representative isolates can be amplified and sequenced, as was performed in the current study. *Fusarium asiaticum* has been reported to infect soybeans and cause root rot in southwest China [15], but it is the first identified pathogen of soybean root rot in northeastern China. There are significant differences in climate and soil between the two places. Thus, characterizing *F. asiaticum* is crucial for understanding the etiology of the disease, including its occurrence and prevalence, as well as developing more scientific and appropriate prevention strategies.

There are many inoculation methods for determining the pathogenicity of *Fusarium* spp., including the root injury inoculation method, root-dipping inoculation method, and root-burying method using sorghum grains with fungal hyphae [70–72]. The inoculation amount, investigation time, and investigation standards also differ among these methods. Because the root embedding method is simple, fast, and more closely reflects natural disease infection, this inoculation method was selected in the current study.

A previous study showed that the optimum temperature and pH for *F. asiaticum* were 20–25 °C and 7.0–9.0, respectively. Soybean root rot occurred during the entire soybean growth period. The optimum growth conditions for *F. asiaticum* causing soybean root rot were temperatures of 20–25 °C and pH between 7.0 and 9.0, which was consistent with temperature and soil alkalinity conditions in Northeast China. Our results differed from other studies to some extent, which may be due to the different environmental conditions in which the disease occurs. Therefore, this may be one of the reasons why these isolates may seriously harm soybean production in Northeast China.

In the host range determination, *F. asiaticum* isolates infected black soybean, French bean, white hyacinth bean, mung bean, and adzuki bean roots, but did not infect rice, peanut, corn, and oat roots. It has been reported in China that *F. asiaticum* infection caused panicle rot in foxtail millet [*Setaria italica* (L.) P. Beauv.], stem rot in Ligusticum chuanxiong, seedling blight in maize, fruit rot in melon (*Cucumis melo* L.), and Fusarium head blight (FHB) in wheat (*Triticum aestivum* L.) [73–76]. *Fusarium asiaticum* has a relatively wide geographical distribution and host range, which can lead to significant yield losses. The non-host crops can disrupt the disease cycle by reducing the pathogen's population in the soil. By alternating soybean cultivation with non-host crops, the accumulation of *F. asiaticum* in the soil can be minimized, thus reducing the risk of root rot in subsequent soybean crops. However, the selection of non-host crops depends on various factors, such as soil type, climate, and local agricultural practices. Further research is still being conducted to optimize the combination of non-host crops in different regions to achieve the best disease-control and yield-improvement effects.

Currently, the most efficient way to mitigate soybean root rot caused by *Fusarium* spp. is soybean seed coating with appropriate fungicides [77]. In this study, four chemical fungicides—pyraclostrobin, prochloraz, fludioxonil, and pyraclostrobine—were selected. Results indicated that fludioxonil exerted the strongest inhibitory effect on the growth of *F. asiaticum*, followed by pyraclostrobin, based on their sensitivities to the selected fungi. In the greenhouse pot experiment, 250 $\mu g \cdot mL^{-1}$ fludioxonil reduced disease incidence by 73.8% and improved soybean-seedling quality. However, Qiu et al. (2018) found that four field strains of *F. asiaticum* were highly resistant to fludioxonil, the EC_{50} values ranging from 80 to > 400 $\mu g \cdot mL^{-1}$ [78]. In the present study, *F. asiaticum* isolates from diseased soybean roots showed high sensitivity to fludioxonil. *F. asiaticum* is a newly emerged pathogen causing soybean root rot in northeast China; the isolates have not developed resistance to fludioxonil. Thus, fludioxonil holds potential for controlling soybean root rot caused by *F. asiaticum* in northeast China. Nevertheless, further research is advisable to precisely determine the appropriate application method and timing.

5. Conclusions

To the best of our knowledge, this study is the first to delve into the effects of *Fusarium asiaticum* on soybean root rot in northeast China. Our results show that *F. asiaticum* has a broad host range and can cause root rot, thus posing a potential risk to regional crop production. Employing intercropping with non-host plants, combined with the application of fludioxonil, can effectively control the soybean root rot caused by *F. asiaticum*. Therefore, when devising advanced strategies for soybean disease management, it is crucial to conduct in-depth investigations into the occurrence of this disease. This not only aids in better understanding the disease mechanism but also facilitates the development of more targeted and effective control measures. By comprehensively examining disease occurrence and considering factors such as soil conditions, climate, and crop growth cycles, we can optimize the application of control methods like intercropping and fungicide use. This comprehensive approach will contribute to the sustainable development of soybean production in northeast China, ensuring both high yields and good quality.

Author Contributions: Conceptualization, Y.L. and J.L.; methodology, Y.L. and J.D.; software, J.L., W.C., Q.Z., Z.R., L.L., and L.S.; validation, J.L., W.C., Q.Z., Z.R., L.L., and L.S.; formal analysis, Y.L. and J.D.; investigation, J.L., W.C., Q.Z., Z.R., L.L., and L.S.; resources, Y.L. and J.D.; data curation, J.L., W.C., Q.Z., Z.R., L.L., and L.S.; writing—original draft preparation, J.L., W.C., and Q.Z.; writing—review and editing, Y.L. and J.L.; funding acquisition, Y.L. and J.D. All authors have read and agreed to the published version of the manuscript.

Funding: This study was supported by the Key Project of the Heilongjiang Provincial Natural Science Foundation (ZD2023C001), the National Natural Science Foundation of China (32372048), the China Agriculture Research System of MOF and MARA (CARS04), the Heilongjiang Province Seed Industry Innovation and Development Project, the Heilongjiang Collaborative Innovation and Extension System of Modern Agricultural Industry Technology of Forage and Feed, and the Talent Introduction Project of Northeast Agricultural University (54960312).

Data Availability Statement: The datasets generated and/or analyzed in the current study are available from the corresponding author upon reasonable request.

Conflicts of Interest: The authors declare no conflicts of interest.

Appendix A

Table A1. Sequences used for concatenated alignment.

Strains	ITS Regions	*TEF* Gene	*GPD* Gene
Fusarium asiaticum HL9	OQ061210.1	OQ378361.1	OQ378358.1
F. asiaticum HL15	OQ061466.1	OQ378360.1	OQ378363.1
F. asiaticum HL25	OQ061472.1	OQ378362.1	OQ378359.1
F. asiaticum HL35	OM967192.1	ON011079.1	ON011080.1
F. asiaticum	OM100564.1	LC500693.1	KM062027.1
F. oxysporum	MH221085.1	KY123890.1	LC592361.1
F. subglutinans	KY318486.1	KF467375.1	OK000516.1
F. verticillioides	KX385055.1	KF467376.1	OK000520.1
F. mundagurra	MZ379241.1	MZ399212.1	MZ399215.1
F. verrucosum	KM231812.1	KM231940.1	KM232077.1
F. proliferatum	GU074010.1	KF467371.1	GU338455.1
F. sambucinum	DQ132833.1	KM231941.1	KF896804.1
F. illudens	KM231806.1	KM231934.1	KM232068.1
F. graminearum	JX162395.1	MW620072.1	OM048104.1
Alternaria alternata	MK351431.1	MT178330.1	MN607983.1

References

1. Patil, G.; Vuong, T.D.; Kale, S.; Valliyodan, B.; Deshmukh, R.; Zhu, C.S.; Wu, X.L.; Bai, Y.H.; Yungbluth, D.; Lu, F.; et al. Dissecting genomic hotspots underlying seed protein, oil, and sucrose content in an interspecific mapping population of soybean using high-density linkage mapping. *Plant Biotechnol. J.* **2018**, *16*, 1939–1953. [CrossRef]
2. Tan, Z.Y.; Kang, Z.; Huang, H.N.; Chen, H.B.; Gong, G.S.; Yong, T.W.; Yang, W.; Chang, X.L. Development of 8% uniconzole·propiconazole·abamectin suspension seed coating agent and its control effect against Fusarium root rot of soybean. *J. Nucl. Agric.* **2020**, *34*, 954–962.

3. Li, Y.G.; Zhao, T.X.; Khuong Gia, H.H.; Xu, L.K.; Liu, J.X.; Li, S.X.; Huang, H.W.; Ji, P.S. Pathogenicity and genetic diversity of *Fusarium oxysporum* causing soybean root rot in northeast China. *J. Agric. Sci.* **2018**, *10*, 13–24. [CrossRef]
4. Li, N.N.; Zhou, Q.X.; Chang, K.F.; Yu, H.T.; Hwang, S.F.; Conner, R.L.; Strelkov, S.E.; McLaren, D.L.; Turnbull, G.D. Occurrence, pathogenicity and species identification of *Pythium* causing root rot of soybean in Alberta and Manitoba, Canada. *Crop Prot.* **2019**, *118*, 36–43. [CrossRef]
5. Roth, M.G.; Webster, R.W.; Mueller, D.S.; Chilvers, M.I.; Faske, T.R.; Mathew, F.M.; Bradley, C.A.; Damicone, J.P.; Kabbage, M.; Smith, D.L. Integrated management of important soybean pathogens of the United States in changing climate. *J. Integr. Pest. Manag.* **2020**, *11*, 17. [CrossRef]
6. Han, S.Y.; Chen, J.X.; Zhao, Y.J.; Cai, H.S.; Guo, C.H. *Bacillus subtilis* HSY21 can reduce soybean root rot and inhibit the expression of genes related to the pathogenicity of *Fusarium oxysporum*. *Pestic. Biochem. Phys.* **2021**, *178*, 104916. [CrossRef] [PubMed]
7. Xue, A.G.; Cober, E.; Voldeng, H.D.; Cober, E.; Voldeng, H.D.; Babcock, C.; Clear, R.M. Different aggressiveness in isolates of *Fusarium graminearum* and *Fusarium pseudograminearum* causing root rot of soybean. *Can. J. Plant Pathol.* **2006**, *28*, 369.
8. Chang, K.F.; Hwang, S.F.; Conner, R.L.; Ahmed, H.U.; Zhou, Q.; Turnbull, G.D.; Strelkov, S.E.; McLaren, D.L.; Gossen, B.D. First report of *Fusarium proliferatum* causing root rot in soybean (*Glycine max* L.) in Canada. *Crop Prot.* **2015**, *67*, 52–58. [CrossRef]
9. Abdelmagid, A.; Hafez, M.; Soliman, A.; Adam, L.R.; Daayfe, F. First report of *Fusarium sporotrichioides* causing root rot of soybean in Canada and detection of the pathogen in host tissues by PCR. *Can. J. Plant Pathol.* **2021**, *43*, 527–536. [CrossRef]
10. Detranaltes, C.; Jones, C.R.; Cai, G. First report of *Fusarium fujikuroi* causing root rot and seedling elongation of soybean in Indiana. *Plant Dis.* **2021**, *105*, 3762. [CrossRef]
11. Naeem, M.; Munir, M.; Li, H.J.; Ali Raza, M.; Song, C.; Wu, X.L.; Irshad, G.; Bin Khalid, M.H.; Yang, W.Y.; Chang, X.L. Transcriptiona1 responses of *Fusarium graminearum* interacted with soybean to cause root rot. *J. Fungi.* **2021**, *7*, 422. [CrossRef] [PubMed]
12. Jang, Y.; Yi, H.; Maharjan, R.; Jeong, M.; Yoon, Y. First report of root rot caused by *Fusarium armeniacum* on soybean in Korea. *Plant Dis.* **2022**, *106*, 1306. [CrossRef] [PubMed]
13. Detranaltes, C.; Saldanha, M.; Scofield Steven, R.; Cai, G.H. First report of *Fusarium commune* causing root rot of soybean seedlings in Indiana. *Plant Dis.* **2022**, *106*, 3216. [CrossRef] [PubMed]
14. Yan, H.; Nelson Jr, B. Effects of soil type, temperature and moisture on development of Fusarium root rot of soybean by *Fusarium solani* (FSSC 11) and *Fusarium tricinctum*. *Plant Dis.* **2022**, *106*, 2974–2983. [CrossRef] [PubMed]
15. Chang, X.L.; Yan, L.; Naeem, M.; Khaskheli, M.I.; Zhang, H.; Gong, G.S.; Zhang, M.; Song, C.; Yang, W.Y.; Liu, T.G.; et al. Maize/Soybean relay strip intercropping reduces the occurrence of Fusarium root rot and changes the diversity of the pathogenic *Fusarium* species. *Pathogens* **2020**, *9*, 211. [CrossRef] [PubMed]
16. Lerch, E.; Arritt, N.; Dorrance, A.E.; Robertson, A.E. Quantitative Trait Loci (QTL) conferring resistance in soybean to root rot caused by *Pythium oopapillum*. In Proceedings of the 2017 APS Annual Meeting, San Antonio, TX, USA, 5–9 August 2017.
17. Grijalba, P.E.; Ridao, A.D.; Steciow, M. Damping off on soybean (*Glycine max*) caused by *Pythium aphanidermatum* in Buenos Aires Province (Argentina). *Rev. Fac. Cienc. Agrar.* **2020**, *52*, 282–288.
18. Kumar, S.; Rajput, L.S.; Ramteke, R.; Nataraj, V.; Ratnaparkhe, M.B.; Maheshwari, H.S.; Shivakumar, M. First report of root rot and damping-off disease in soybean (*Glycine max*) caused by *Pythium deliense* in India. *Plant Dis.* **2021**, *105*, 2022. [CrossRef]
19. Liu, B.; Shen, W.S.; Wei, H.; Smith, H.; Louws, F.J.; Steadman, J.R.; Correl, J.C. Rhizoctonia communities in soybean fields and their relation with other microbes and nematode communities. *Eur. J. Plant Pathol.* **2016**, *144*, 671–686. [CrossRef]
20. Shehata, M.A.; Pflege, F.L.; Davis, D.W. Response of susceptible and moderately resistant pea genotypes to interaction between rhizoctonia and three other stem and root rot pathogens. *Plant Dis.* **1983**, *67*, 1146–1148. [CrossRef]
21. Anderson, T.R. Fungi isolated from stems and roots of soybean in Ontario. *Can. Plant Dis. Sur.* **1987**, *67*, 3–5.
22. Li, S.; Hartman, G.L. First report of *Stachybotrys chartarum* causing soybean root rot. *Plant Dis.* **2000**, *84*, 100. [CrossRef] [PubMed]
23. Susilowati, A.; Wahyudi, A.T.; Lestari, Y.; Suwanto, A.; Wiyono, S. Potential *Pseudomonas* isolated from soybean rhizosphere as biocontrol against soilborne phytopathogenic fungi. *Hayati J. Biosci.* **2011**, *18*, 51–56. [CrossRef]
24. Detranaltes, C.; Cai, G. First report of *Mycoleptodiscus terrestris* causing root rot of soybean in Indiana. *Plant Dis.* **2020**, *105*, 1194. [CrossRef] [PubMed]
25. Mattupalli, C.; Cuenca, F.P.; Shiller, J.B.; Watkins, T.; Hansen, K.; Garzon, C.D.; Marek, S.M.; Young, C.A. Genetic diversity of *Phymatotrichopsis omnivora* based on mating type and microsatellite markers reveal heterothallic mating system. *Plant Dis.* **2022**, *106*, 2105–2116. [CrossRef]
26. Xin, H.P.; Ma, H.Q.; Liu, J.R.; Zhang, Y.P.; Liu, Y.C.; Zhang, X.D. A preliminary study on epidemiology and control of disease. *Soybean Sci.* **1987**, 189–196.
27. Li, B.Y.; Ma, S.M. Pathogen species and antigen screening of soybean root rot. *J. Plant Prot.* **2000**, *27*, 91–92.
28. Xing, A.; Wen, J.Z.; Lv, G.Z.; Sun, X.D. Isolation and identification of *Fusarium* spp. on soybean root rot plants in Heilongjiang Province. *J. Northeast Agri. Univ.* **2009**, *40*, 5–9.

29. Wang, X.Y.; Wen, J.Z. Analysis on species and pathogenicity of Fusarium sojae root rot in three northeastern provinces. *Chin. J. Oil Crop* **2011**, *33*, 391–395.
30. Bai, L.Y.; Zhang, Q.D.; Li, B.; Guo, Q.Y. Identification and pathogenicity determination of the pathogenic *Fusarium* of soybean root rot in the altay region of Xinjiang. *Xinjiang Agric. Sci.* **2009**, *46*, 543–548.
31. Geng, X.B.; Wang, C.L.; Huang, M.H.; Li, Y.G. Identification of soybean stem rot pathogen causing soybean seedling root rot. *Plant Prot.* **2015**, *41*, 127–129.
32. Yang, X.H.; Gu, X.; Zhao, H.H.; Yao, L.L.; Liu, W.; Shen, H.B.; Zhang, Y.; Liu, L.J.; Ding, J.J. Investigation report on soybean root rot in Sanjiang Plain Area. *Chin. Agron. Bull.* **2015**, *31*, 113–116.
33. Sugimoto, T.; Watanabe, K.; Yoshida, S.; Aino, M.; Matsuyama, M.; Maekawa, K.; Irie, K. The effects of inorganic elements on the reduction of Phytophthora stem rot disease of soybean, the growth rate and zoospore release of *Phytophthora sojae*. *J. Phytopathol.* **2007**, *155*, 97–107. [CrossRef]
34. Lin, F.; Li, W.; Mccoy, A.G.; Gao, X.; Collins, P.J.; Zhang, N.; Wen, Z.X.; Cao, S.Z.; Wani, S.H.; Gu, C.H.; et al. Molecular mapping of quantitative disease resistance loci for soybean partial resistance to *Phytophthora sansomeana*. *Theor. Appl. Genet.* **2021**, *134*, 1977–1987. [CrossRef] [PubMed]
35. Ma, S.M. Pathogenic pathogen categories distribution and germplasm resistance identification of soybean root rot in Heilongjiang Province. *Bull. Chin. Agric. Sci.* **2012**, *28*, 230–235.
36. Wang, K.; Liu, Y.; Hao, P.H.; Xia, Y.H.; Sun, B.J.; Li, H.L.; Li, Y.U. Occurrence of *Pratylenchus coffeae* causing root rot of soybean in Shandong Province of China. *Plant Dis.* **2021**, *105*, 1227. [CrossRef]
37. Shen, C.Y.; Su, Y.C. Discovery and preliminary study of *Phytophthora* soybean in China. *Acta Phytopathol. Sin.* **1991**, *4*, 60.
38. Zhang, L.; Geng, X.B.; Wang, C.L.; Li, Y.G. Identification and virulence of *Fusarium* spp. causing soybean root rot in Heilongjiang Province. *Plant Prot.* **2014**, *40*, 165–168.
39. Du, Y.X.; Shi, N.N.; Ruan, H.C.; Lian, J.F.; Gan, L.; Chen, F.R. Study on pathogenic fungi causing soybean root rot in Yinchuan and field disease control efficiency of seed coating. *Chin. Agric. Sci. Bull.* **2021**, *37*, 103–109.
40. Broders, K.D.; Lipps, P.E.; Paul, P.A.; Dorrance, A.E. Evaluation of *Fusarium graminearum* associated with corn and soybean seed and seedling disease in Ohio. *Plant Dis.* **2007**, *91*, 1155–1160. [CrossRef]
41. Dorrance, A.; McClure, S.; Martin, S. Effect of partial resistance on phytophthora stem rot incidence and yield of soybean in Ohio. *Plant Dis.* **2003**, *87*, 308–312. [CrossRef]
42. Xi, X.D.; Fan, J.L.; Yang, X.Y.; Liang, Y.; Zhao, X.L.; Wu, Y.H. Evaluation of the anti-oomycete bioactivity of rhizosphere soil-borne isolates and the biocontrol of soybean root rot caused by *Phytophthora sojae*. *Biol. Control* **2022**, *166*, 104818. [CrossRef]
43. Zhang, C.J.; Liao, S.Q.; Song, H.; Zhao, X.; Han, Y.P.; Liu, Q.; Li, W.B.; Wu, X.X. Identification for resistance to root rot caused by *Fusarium Oxysporum* in soybean germplasm and physiological analysis. *Soybean Sci.* **2017**, *36*, 441–446.
44. Wang, C.L. The Analysis and Genetic Diversity of the 48 Soybean Varities Against Three Species of Fusaium Causing Soybean Root Rot. Master's Thesis, Northeast Agricultural University, Harbin, China, 2016.
45. Díaz Arias, M.M.; Leandro, L.F.; Munkvold, G.P. Aggressiveness of *Fusarium* Species and impact of root infection on growth and yield of soybeans. *Phytopathology* **2013**, *103*, 822–832. [CrossRef] [PubMed]
46. Hegde, N.P. Evaluating Chemical Seed Treatments for Fusarium Root Rot Control in Dry Beans and Field Peas. Master's Thesis, North Dakota State University, Fargo, ND, USA, 2014.
47. Degani, O.; Kalman, B. Assessment of commercial fungicides against onion (*Allium cepa*) basal rot disease caused by *Fusarium oxysporum* f. sp. cepae and *Fusarium acutatum*. *J. Fungi* **2021**, *7*, 235. [CrossRef] [PubMed]
48. Zhang, Y.Z.; Li, Z.; Man, J.; Xu, D.; Wen, L.; Yang, C.; Xu, Q.; Jiang, Q.T.; Chen, G.Y.; Deng, M.; et al. Genetic diversity of field *Fusarium asiaticum* and *Fusarium graminearum* isolates increases the risk of fungicide resistance. *Phytopathol. Res.* **2023**, *5*, 51. [CrossRef]
49. Li, W.Q.; Huang, W.; Zhou, J.; Wang, J.; Liu, J.; Li, Y. Evaluation and control of *Alternaria alternata* causing leaf spot in soybean in Northeast China. *J. Appl. Microbiol.* **2023**, *134*, lxad004. [CrossRef]
50. Shi, J.L.; Li, Y.Q.; Hu, K.M.; Ren, J.G.; Liu, H.M. Isolation and identification of pathogens from rotted root of *Pinellia ternata* in Guizhou Province. *J. Microbiol. Chin.* **2015**, *42*, 289–299.
51. Chang, X.L.; Dai, H.; Wang, D.P.; Zhou, H.H.; He, W.Q.; Fu, Y.; Ibrahim, F.; Zhou, Y.; Gong, G.S.; Shang, J.; et al. Identification of *Fusarium* species associated with soybean root rot in Sichuan Province, China. *Eur. J. Plant Pathol.* **2018**, *151*, 563–577. [CrossRef]
52. Wang, C.L.; Geng, X.B.; Huang, M.H.; Sun, L.P.; Li, Y.G. A Method to determine pathogenicity of *Fusarium oxysporum* Causing root rot of soybean at seedling stage. *Jilin Agr. Sci.* **2015**, *40*, 69–72.
53. Wei, J.C. *Handbook of Fungus Identification*; Shanghai Scientific: Shanghai, China, 1979.
54. White, T.J.; Bruns, T.D.; Lee, S.B.; Taylor, J.W.; Innis, M.A.; Gelfand, D.H.; Sninsky, J. Amplification and direct sequencing of fungal ribosomal RNA genes for phylogenetics. Pcr Protocols: A Guide to Methods and Application. *PCR Protoc. Guide Methods Appl.* **1990**, *18*, 315–322.

55. Carbone, I.; Kohn, L.M. A method for designing primer sets for speciation studies in filamentous ascomycetes. *Mycologia* **1999**, *91*, 553–556. [CrossRef]
56. Glass, N.L.; Donaldson, G.C. Development of primer sets designed for use with the PCR to amplify conserved genes from filamentous ascomycetes. *Appl. Environ. Microbiol.* **1995**, *61*, 1323–1330. [CrossRef]
57. Zhang, D.; Gao, F.L.; Jakovlic, I.; Zou, H.; Zhang, J.; Li, W.X.; Wang, G.T. PhyloSuite: An integrated and scalable desktop platform for streamlined molecular sequence data management and evolutionary phylogenetics studies. *Mol. Ecol. Resour.* **2020**, *20*, 348–355. [CrossRef] [PubMed]
58. Ronquist, F.; Teslenko, M.; Mark, P.; van der Ayres, D.L.; Darling, A.; Höhna, S.; Larget, B.; Liu, L.; Suchard, M.A.; Huelsenbeck, J.P. MrBayes 3.2: Efficient bayesian phylogenetic inference and model choice across a large model space. *Syst. Biol.* **2012**, *61*, 539–542. [CrossRef]
59. Vasi, T.; Vojinovi, U.; Ujovi, S.; Krnjaja, V.; Stevi, M. In vitro toxicity of fungicides with different modes of action to alfalfa anthracnose fungus, *Colletotrichum destructivum*. *J. Environ. Sci. Health A* **2019**, *54*, 1–8.
60. Zhang, C.Q.; Liu, Y.H.; Wu, H.M.; Xu, B.C.; Sun, P.L.; Xu, Z.H. Baseline sensitivity of *Pestalotiopsis microspora*, which causes black spot disease on chinese hickory (*Carya cathayensis*), to pyraclostrobin. *Crop Prot.* **2012**, *42*, 256–259. [CrossRef]
61. Jiang, N.; Hu, F.Y.; Ye, Y.F. Pathogen identification of a new disease in Siraitia grosvenorii and screening of effective fungicides. *Plant Prot.* **2015**, *41*, 173–177.
62. Lehner, M.S.; Paula Júnior, T.J.; Silva, R.A.; Vieira, R.F.; Carneiro, J.E.S.; Schnabel, G.; Mizubuti, E.S.G. Fungicide sensitivity of *Sclerotinia sclerotiorum*: A thorough assessment using discriminatory dose, EC_{50}, high-resolution melting analysis, and description of new point mutation associated with thiophanate-methyl resistance. *Plant Dis.* **2015**, *99*, 1537–1543. [CrossRef]
63. Ji, X.; Li, J.; Meng, Z.; Zhang, S.; Dong, B.; Qiao, K. Synergistic effect of combined application of a new fungicide fluopimomide with a biocontrol agent *Bacillus methylotrophicus* TA-1 for management of gray mold in tomato. *Plant Dis.* **2019**, *103*, 1991–1997. [CrossRef] [PubMed]
64. O'Donnell, K.; Todd, J.W.; David, M.G.; Kistler, H.C.; Aoki, T. Genealogical concordance between the mating type locus and seven other nuclear genes supports formal recognition of nine phylogenetically distinct species within the *Fusarium graminearum* clade. *Fungal Genet. Biol.* **2004**, *41*, 600–623. [CrossRef] [PubMed]
65. Leslie, J.F.; Summerell, B.A.; Bullock, S. *The Fusarium Laboratory Manual*; Blackwell Publishing: Ames, IA, USA, 2006; p. 176.
66. Xing, H.Q.; Wang, C.M.; Jin, S.L.; Zhou, T.W.; Guo, C. Isolation and identification of *Fusarium asiaticum* causing *Lathyrus sativus* root rot. *Acta Agrestia Sin.* **2021**, *29*, 1350–1356.
67. Xiao, J.L.; Wang, G.J.; Ming, Z.; Jing, Y.; Wen, L.; Bi, Y.; Wang, L.; Lai, Y.C.; Shu, X.T.; Wang, Z. Effect of cultivation pattern on the light radiation of group canopy and yield of spring soybean (*Glycine Max* L. Merrill). *American J. Plant Sci.* **2013**, *4*, 1204–1211. [CrossRef]
68. Yang, X.B.; Ruff, R.L.; Meng, X.Q. Race of *Phytophthora sojae* in Iowa sobean fields. *Plant Dis.* **1996**, *80*, 1418–1420. [CrossRef]
69. Wu, W.X.; Liu, Y.; Huang, X.Q.; Zhang, L.; Zhou, X.Q.; Liu, H.Y. Establishment and application of rapid molecular detection for *Fusarium oxysporum*. *Acta Pratacul. Sin.* **2016**, *25*, 109–115.
70. Yin, Y.; Liu, X.; Li, B.; Ma, Z. Characterization of sterol demethylation inhibitor-resistant isolates of *Fusarium asiaticum* and *F. graminearum* collected from wheat in China. *Phytopathology* **2009**, *99*, 487–497. [CrossRef] [PubMed]
71. Zhang, H.; Zhang, Z.; Lee, T.; Chen, W.Q.; Xu, J.; Xu, J.S.; Yang, L.; Yu, D.; Waalwijk, C.; Feng, J. Population genetic analyses of *Fusarium asiaticum* populations from barley suggest a recent shift favoring 3ADON producers in southern China. *Phytopathology.* **2010**, *100*, 328–336. [CrossRef]
72. Chang, X.L.; Naeem, M.; Li, H.J.; Yan, L.; Liu, T.G.; Liu, B.; Zhang, H.; Khaskheli, M.I.; Gong, G.S.; Zhang, M.; et al. First report of *Fusarium asiaticum* as a causal agent for seed decay of soybean (*Glycine max*) in Sichuan, China. *Plant Dis.* **2020**, *104*, 1542. [CrossRef]
73. Li, B.J.; Chen, Q.H.; Lan, C.Z.; Wang, N.N.; Wang, Y.C.; Weng, Q.Y. Indenfication and pathogenicity test of the pathogens causing soybean root in Fujian. *Fujian J. Agricl. Sci.* **2011**, *26*, 798–803.
74. Yang, S.; Wang, J.S.; Ma, Z.C.; Wang, Y.C.; Wang, K.R. Isolation and identification of *Fusarium proliferatum* causing soybean root rot and its biological characterization. *J. Plant Prot.* **2012**, *39*, 187–188.
75. Li, Y.G.; Zhang, X.; Zhang, R.; Liu, J.X.; Ali, E.; Ji, P.; Pan, H.Y. Occurrence of seedling blight caused by *Fusarium tricinctum* on rice in China. *Plant Dis.* **2019**, *103*, 1789. [CrossRef]
76. Kong, F.X.; Zhang, H.J.; Liu, Z.; Chen, G.Q.; Xu, J. First report of panicle rot caused by *Fusarium asiaticum* on *Foxtail millet* in China. *Plant Dis.* **2022**, *106*, 1062. [CrossRef] [PubMed]

77. Nyandoro, R.; Chang, K.F.; Hwang, S.F.; Ahmed, H.U.; Turnbull, G.D.; Strelkov, S.E. Management of root rot of soybean in Alberta with fungicide seed treatments and genetic resistance. *Can. J. Plant Sci.* **2019**, *99*, 499–509. [CrossRef]
78. Qiu, J.B.; Yu, M.Z.; Yin, Q.; Xu, J.H.; Shi, J.R. Molecular characterization, fitness, and mycotoxin production of *Fusarium asiaticum* strains resistant to fludioxonil. *Plant Dis.* **2018**, *102*, 1759–1765. [CrossRef]

Disclaimer/Publisher's Note: The statements, opinions and data contained in all publications are solely those of the individual author(s) and contributor(s) and not of MDPI and/or the editor(s). MDPI and/or the editor(s) disclaim responsibility for any injury to people or property resulting from any ideas, methods, instructions or products referred to in the content.

agronomy

MDPI

Article

High-Throughput Sequence Analysis of Microbial Communities of Soybean in Northeast China

Yuanyuan Wang [1], Qingyao Bai [1], Fanqi Meng [1], Wei Dong [2], Haiyan Fan [2], Xiaofeng Zhu [2], Yuxi Duan [2] and Lijie Chen [2,*]

[1] College of Bioscience and Biotechnology, Shenyang Agricultural University, Shenyang 110866, China; wyuanyuan1225@syau.edu.cn (Y.W.); 15141554915@163.com (Q.B.); 17302487927@163.com (F.M.)
[2] Nematology Institute of Northern China, Shenyang Agricultural University, Shenyang 110866, China; dongwei9501@163.com (W.D.); fanhaiyan2017@syau.edu.cn (H.F.); syxf2000@syau.edu.cn (X.Z.); duanyx@syau.edu.cn (Y.D.)
* Correspondence: chenlj-0210@syau.edu.cn

Abstract: Soybean, an essential oil crop in China, has witnessed accelerated seed transfer domestically and abroad in recent years. Seed carriage has emerged as a major route for the dissemination of soybean diseases. In this study, 14 soybean cultivars from three northeastern provinces were collected and examined for seed-borne microorganisms using traditional detection technology and high-throughput sequencing technology. Through traditional detection techniques, a total of six genera of bacteria and seventeen genera of fungi were isolated from the test varieties. The quantity and types of microorganisms on the seed surface were greater than those on the seed coat and within the seed, while the seed coat and internal seed contained fewer microorganisms. The dominant fungal genera were *Cladosporium*, *Fusarium*, *Aspergillus*, and *Alternaria*, accounting for 21.23%, 17.45%, 15.57%, and 11.56% of the genera, respectively. The dominant bacterial genera were *Pseudomonas*, *Sphingomonas*, and *Pantoea*, accounting for 37.46%, 17.29%, and 15.27% of the genera, respectively. The dominant genera obtained through traditional seed-carrying assay techniques were also dominant in high-throughput sequencing. However, some dominant genera obtained through high-throughput sequencing were not isolated by traditional methods. High-throughput sequencing analysis revealed that soybean seeds from Jilin Province had the highest abundance of seed-borne fungi, followed by seeds from Liaoning Province and Heilongjiang Province. Jilin Province also had the highest abundance of seed-borne bacteria, followed by Heilongjiang Province and Liaoning Province. The isolation and identification of microorganisms on soybean seeds provide a scientific basis for seed quarantine treatment and disease control, which is of great significance for soybean production in China.

Keywords: soybean; microbial diversity; conventional seed-borne microorganism detection technology; high-throughput sequencing technology

1. Introduction

Soybean (*Glycine max* (L.) Merr), a significant grain and oil crop globally, supplies humans with high-quality plant protein and oil, playing a crucial role in ensuring food production safety [1]. As an agricultural production means, high-quality seeds directly impact production [2]. Seeds are associated with diverse microorganisms during their development, harvesting, and storage stages [3]. Some of these microorganisms can cause seed discoloration and decay, and some can adhere to the seed surface or invade seed tissues, facilitating the spread of diseases across growing seasons and locations, having a significant negative impact on germination, seedling growth, and adult plant health [4,5].

The traditional identification of fungi in seed-carrying detection mainly classifies fungi based on their morphological, growth, physiological, and biochemical characteristics. However, numerous fungal species may not be easily detected, and their development, physiological, and biochemical characteristics are susceptible to environmental changes [6]. Due to the limitations of the culture medium used for isolation, it is difficult to isolate all types of bacteria, and identification is also challenging [7]. Consequently, it is challenging to fully identify seed-borne bacteria using traditional methods [8]. With the advancement of molecular biology technology, nucleic acid sequence analysis has become prevalent in microbial classification and identification. Currently, commonly used techniques include 18S rDNA, Internal Transcribed Spacer (ITS), and 16S rDNA [9,10]. 18S rDNA and ITS are suitable for fungal identification, and 16S rDNA can reveal the characteristic nucleic acid sequences of biological species and is considered the most suitable indicator for bacterial phylogeny and classification identification [11]. In recent years, powerful databases and user-friendly software for high-throughput sequencing (NGS) platforms (including SOLiD, Illumina, and 454 sequencing) have contributed to a deeper and more comprehensive analysis of complex microbial communities [12]. To date, DNA-NGS technology has been successfully applied to the analysis of microbial communities in various samples, such as soil, air, water, intestines, leaves, roots, and fruits, and is widely used in agricultural research on plant rhizosphere microorganisms [13]. The plants involved include Arabidopsis thaliana, rice, potato, tobacco, soybean, etc. [14]. However, few studies have been conducted to understand the changes in microbial communities in seeds.

This experiment is based on the use of high-throughput sequencing technology and traditional seed-carrying detection technology to study the diversity of microbial communities on soybean seeds and aims to clarify the seed-carrying situation in different regions and provide references for soybean seed production, safe transportation, and disease control.

2. Materials and Methods

2.1. Soybean Cultivars

A total of 14 test soybean seed varieties collected from different locations were provided by the Comprehensive Soybean Test Station in Northeast China in Nov 2023. Detailed information about the soybean seeds is presented in Table 1. All samples were sealed in Ziplock bags and stored under cold storage at 4 °C.

Table 1. The locations and codes of different soybean varieties.

Location	Soybean Cultivars	Abbreviation
Liaoning	Tiefeng 31	L.1
	Liaodou 21	L.2
	Liaodou 15	L.3
Jilin	Huamidou 30	J.1
	Jinong 38	J.2
	Dedou 10	J.3
Heilongjiang	Nongqing Bean 20	H.1
	Heinong 64	H.2
	Kangxian No. 9	H.3
	Beidou 17	H.4
	Dongsheng 9	H.5
	Heihe 43	H.6
	Qinong 10	H.7
	Longken 336	H.8

2.2. Traditional Detection of Seed Carrier Microbes

In the present study, seed carrier microbes included fungi and bacteria on the seed surface, seed coat, and seed interior.

Isolation of seed surface microbes: Twenty grains of different soybean varieties were randomly selected. The grains were soaked in 20 mL of sterile distilled water for 30 min and centrifuged at 12,000 rpm for 10 min. A 100 μL volume of liquid from the bottom of the centrifuge tube was transferred and spread uniformly on a Petri dish containing potato glucose AGAR (PDA) and beef extract peptone AGAR (NA); each treatment was repeated 4 times. Sterile distilled water was used as a control under the same conditions. The Petri dishes were incubated at 28 °C.

Isolation of seed coat and interior microbes: The seed coat and seed embryo were separated and soaked in 0.5% NaClO for 10 min, washed 3 times with sterile distilled water, and then air-dried [15]. The dried seed coat and seed embryo were placed on PDA and NA media, with 5 grains or 5 tissue blocks per dish, and each treatment was repeated 4 times [16]. The Petri dishes were incubated at 28 °C.

2.3. Identification of Fungi

After culturing for 3–5 d, the colonies were isolated according to morphology, color, and other apparent characteristics and then cultured at 28 °C until most of the colonies produced spores. Single spores were isolated from the colonies and cultured for final identification. Isolates were identified according to their phenotypic characteristics, including the culture-based characteristics of fungi. Fungal DNA isolation was carried out using the TIANamp Genomic DNA Kit (TIANGEN Biotech Beijing Co. Ltd., Beijing, China). Fungal rDNA ITS fragments were sequenced and compared to assist with identification [17]. The fungal ITS universal primer pairs ITS1 (5′-TCCGTAGGTGAACCTGCGG-3′) and ITS4 (5′-TCCTCCGCTTATTGATATGC-3′) were utilized [18].

2.4. Identification of Bacteria

After culturing for 2–3 days, morphologically unique bacterial colonies were selected from each plate, streaked on fresh plates for purification, and then cultured in NA medium for identification. Physiological and biochemical tests were performed to identify bacteria, according to Gerhardt and Murray [19]. Isolates were further identified via phylogenetic analysis of 16S rRNA gene sequences. To extract genomic DNA, standard techniques were used [10,20,21]. The 16S rRNA sequences were amplified with the primer pair 27F (5′-AGAGTTTGATCATGGCTCAG-3′) and 1492R (5′-ACGGTTACCTTGTTACGACTT-3′) [22,23]. The PCR products were inserted into the PGEM-T vector (Promega, Madison, WI, USA). The plasmid was extracted and sequenced using the Genewiz Biotechnology Co., Ltd. (Suzhou, China). The 16S rRNA gene sequence was identified using the GenBank database.

2.5. Microbial DNA Extraction and Polymerase Chain Reaction

Seeds from 14 different soybean varieties were soaked in a 5% NaClO solution for 5 min and then rinsed with autoclaved distilled water after the bleach was drained. Once seed surface sterility was confirmed, 150 seeds were ground gently in an autoclaved mortar using 0.5 mL of 50 mM Na_2HPO_4 buffer per gram of dry seed weight [24]. DNeasy PowerSoil DNA Kit (Qiagen, Germany) was utilized to extract total DNA from the ground seed suspension (100 μL). PCR was performed using the specific primers 341F (5′-CCTAYGGGRBGCASCA) and 806R (5′-GGACTACNNGGGTATCTAAT) and ITS5-1737 (5′-GGAAGTAAAAGTCGTAACAAGG-3′) and /ITS2-2043 (5′-GCTGCGTTCTTCAT CGATGC-3′) to amplify the 16S rRNA and ITS genes [25]. Amplification was carried out using the following program: an initial denaturation at 98 °C for 1 min, followed by 30 cycles of amplification at 98 °C for 10 s, 50 °C for 30 s, and 72 °C for 30 s, and a final extension at 72 °C for 5 min. Amplicons were gel-purified using the gel recovery kit (Gene JET, Thermo Scientific, Waltham, MA, USA) and sequenced using Ion S5TMXL at the Beijing Novo Gene Technology Co., Ltd., Beijing, China.

2.6. Bioinformatics Analysis

Bacterial species annotation was conducted using the Mothur method and the SSUrRNA database [26] of SILVA132 (with a set threshold of 0.8–1) [27]. Species annotation analysis of fungi was performed using the blast method in Qiime software [28] and the Unit (Version 7.2) database [29] for fungi, which were analyzed at each taxonomic level. The MUSCLE software (Version 3.8.31) was used for fast multi-sequence alignment to obtain the phylogenetic relationships of all OTU sequences [30]. Finally, data for each sample were homogenized.

The OTU abundance information was normalized using the sequence number. The alpha diversity was subsequently analyzed using the normalized data.

Alpha diversity was employed to analyze species diversity in a sample using five indices: observed species, Chao1, Shannon, Simpson, and ACE. All of these indices were calculated with QIIME (Version 1.7.0).

A Petal map was created according to the OTU analysis results to visually display the species distinctions of the microbes in the sample. The top 10 species with the highest abundance at each classification level were selected based on the species annotation results, and a cumulative columnar plot of relative species abundance was generated. Rarefaction and rank abundance curves were generated, and NMDS Analysis was conducted using R software (Version 3.2.2).

3. Results

3.1. Analysis of the Diversity of Microorganisms Carried by Soybean Seeds Using Traditional Detection Technique

3.1.1. Fungal Diversity on Soybean Seeds

A total of 17 genera of fungi were isolated from the tested soybean seeds. More fungal species were carried on the surface of the seeds than in the interior of the seeds and on the seed coat. The L.3 sample carried the highest number of fungi, and the H.2 sample carried the least number of fungi (Figure 1). The number of fungi carried by seeds in Heilongjiang Province was relatively lower.

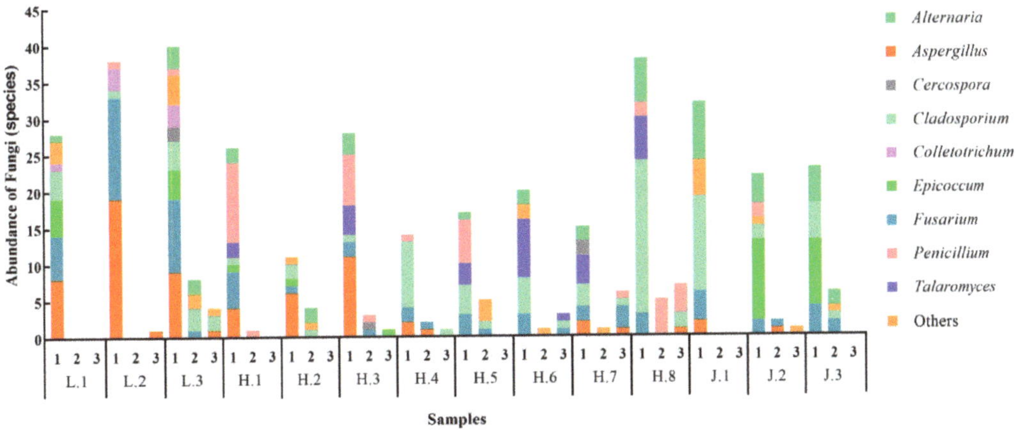

Figure 1. Distribution of seed-borne fungi on 14 soybean varieties in northeast China. 1 represents the fungi detected on the seed surface, 2 represents the fundi detected on the seed coat, and 3 represents the fungi detected on the interior of the seeds.

3.1.2. Bacterial Diversity on Soybean Seeds

A total of six genera of bacteria were identified in soybean seeds; the dominant genera were *Pseudomonas*, *Sphingomonas*, *Pantoea*, and *Bacillus*, with frequencies of 37.46%, 17.29%, 15.28%, and 7.49%, respectively. The distribution of bacteria on soybean seeds was basically the same as that of fungi. The number of bacteria on the seed surface was higher than on the seed coat and in the seed embryo. Soybean seeds from Jilin had a greater diversity of bacteria (Figure 2).

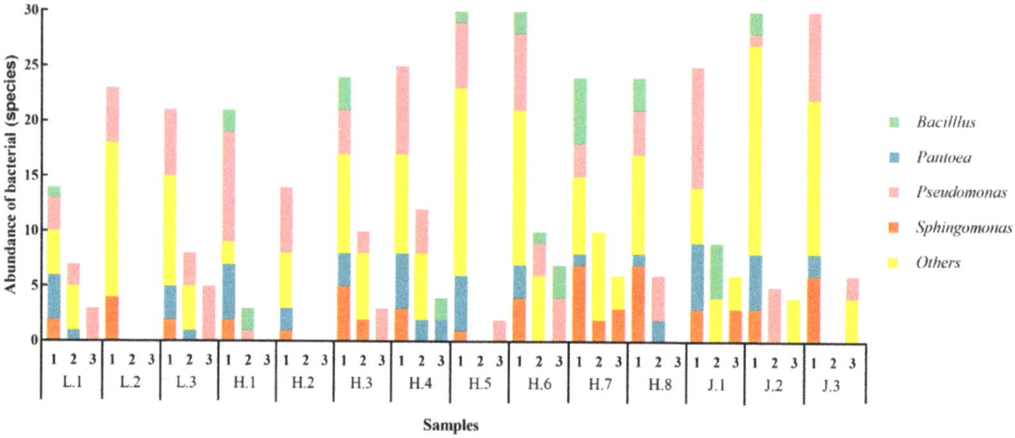

Figure 2. Distribution of seed-borne bacteria on 14 soybean varieties in northeast China. 1 represents bacteria detected on the seed surface, 2 represents bacteria detected on seed coats, and 3 represents bacteria detected in the interior of seeds.

3.2. Microbial Diversity of Soybean Seeds Based on High-Throughput Sequencing Technology

3.2.1. Analysis of Fungi Using ITS Sequencing

The Alpha indices and ITS sequencing information for the 14 soybean seed varieties are shown in Table 2. A total of 1,179,748 original data were obtained from the ITS region. The average number of original data from soybean seeds sampled from Liaoning, Jilin, and Heilongjiang Provinces were 83,088, 84,972, and 84,446, respectively. Analyzing the relationship between Operational Taxonomic Units (OTUs) obtained from different samples and represented using the OTU Flower figure (Figure 3A) showed that the number of OTUs shared by the 14 soybean seed samples was 140.

Table 2. Alpha indices and sequence information for fungi on soybean seeds.

Sample Name	Raw Reads	Observed Species	Shannon	Simpson	Chao1	ACE
L.1	81,918	421	4.464	0.852	458.558	459.703
L.2	84,885	606	5.793	0.934	800.130	669.691
L.3	82,460	332	3.047	0.732	367.646	370.855
H.1	85,474	484	4.902	0.875	505.121	512.919
H.2	84,255	493	5.339	0.888	522.647	537.702
H.3	83,843	476	4.715	0.860	511.019	514.682
H.4	85,262	492	4.863	0.882	538.250	543.746
H.5	81,924	485	4.755	0.832	515.611	519.778
H.6	84,103	390	4.125	0.838	447.468	444.756
H.7	82,806	450	4.593	0.887	497.788	493.661

Table 2. *Cont.*

Sample Name	Raw Reads	Observed Species	Shannon	Simpson	Chao1	ACE
H.8	87,901	426	4.917	0.913	462.167	468.367
J.1	84,377	525	4.564	0.839	554.500	555.380
J.2	85,514	520	5.390	0.905	535.769	537.281
J.3	85,026	504	4.459	0.831	534.000	539.290

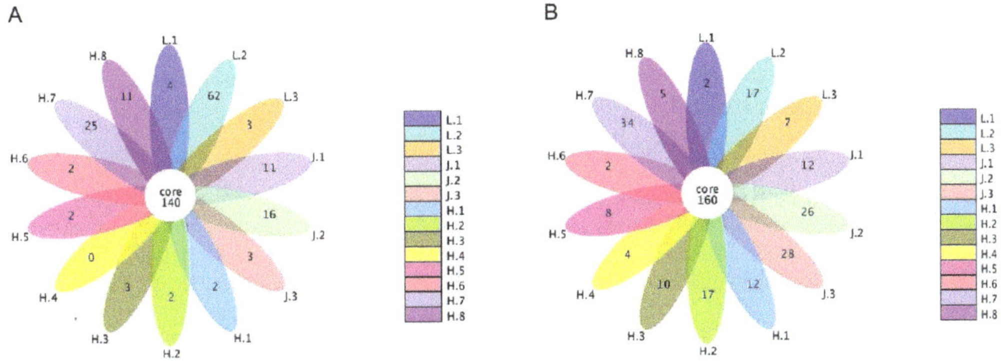

Figure 3. OTU flower figures of soybean seed microbial carriers. (**A**) Fungal flower figure; (**B**) bacterial flower figure.

3.2.2. Analysis of Bacteria Based on 16S rDNA Sequences

Alpha indices based on 16S rDNA sequencing of the 14 soybean seed samples are detailed in Table 3. A total of 1,166,335 original data were obtained from the 16S rDNA region. The average number of original data for soybean seed samples from Liaoning, Jilin, and Heilongjiang Provinces were 83,732, 87,194, and 81,645, respectively. The results from the OTU flower figure (Figure 3B) showed that 160 OTUs were shared by the 14 soybean seed samples. Sample H.7 had the highest number of unique OTUs (34), while samples L.1 and H.6 had the lowest number of unique OTUs (2 each).

Table 3. Alpha indices and sequence information for bacteria on soybean seeds.

Sample Name	Raw Reads	Observed Species	Shannon	Simpson	Chao1	ACE
L.1	76,671	549	5.989	0.960	710.803	672.319
L.2	86,246	660	6.310	0.953	811.076	795.754
L.3	88,279	439	4.767	0.906	566.147	565.411
H.1	87,645	751	5.934	0.941	918.632	918.984
H.2	86,243	728	6.050	0.954	790.569	827.661
H.3	81,837	672	5.336	0.911	861.000	842.736
H.4	88,130	665	5.738	0.939	829.478	853.429
H.5	87,417	629	5.969	0.960	853.438	814.259
H.6	81,822	567	5.436	0.932	721.670	758.503
H.7	81,953	699	6.458	0.965	814.636	832.801
H.8	58,510	442	4.408	0.831	457.000	477.162
J.1	88,793	742	6.245	0.955	919.300	919.734
J.2	86,339	842	6.259	0.939	929.345	943.371
J.3	86,450	788	6.803	0.965	833.200	854.305

3.3. *Alpha Diversity of Microorganisms Carried by Soybean Seeds*

3.3.1. Alpha Diversity of Fungi

To analyze the richness of the fungal community structure of the 14 soybean seeds from northeast China, the microbial diversity index was analyzed at a 97% similarity threshold using the chao1 index, observed species index, Shannon index, ACE index, and Simpson index (Table 2). Among the 14 soybean seed samples, sample L.2 had the largest Chao1 and Shannon index values and contained the most observed species, indicating a more abundant fungal community. Conversely, sample L.3 had the smallest Chao1 and Shannon index values and contained the fewest observed species, suggesting a relatively harmonious fungal community.

Rarefaction and rank abundance curves were also employed for alpha diversity analysis. The rank abundance curve (Figure 4) indicated a more even distribution and higher abundance of fungi associated with soybean seeds from Jilin Province, followed by Liaoning Province and Heilongjiang Province. In the rarefaction curve in Figure 4, the number of observed OTUs in soybean samples from Jilin Province was more than 500, while Liaoning Province had the lowest OTU count, at 400.

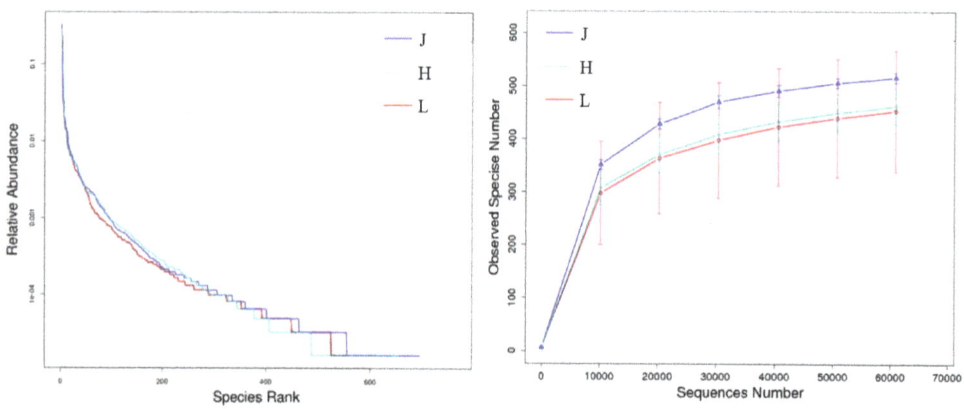

Figure 4. Rank abundance and rarefaction curves of fungi on soybean seeds. (**Left**) Rank abundance curve; (**Right**) rarefaction curve.

3.3.2. Alpha Diversity of Bacteria

Among the 14 soybean seed samples, samples J.2 and J.3 had relatively large Chao1 and Shannon index values (Table 3) and contained more observed species, suggesting a more abundant bacterial community. From the rank abundance curve, it can be seen that the distribution and abundance of bacteria in the three provinces were essentially the same as those of fungi (Figure 5). The number of observed OTUs in soybean samples from Jilin Province was about 800, while the number of OTUs in samples from Liaoning Province was about 500. Overall, soybean seeds from Jilin Province were more abundant in bacteria, followed by seeds from Heilongjiang Province and Liaoning Province.

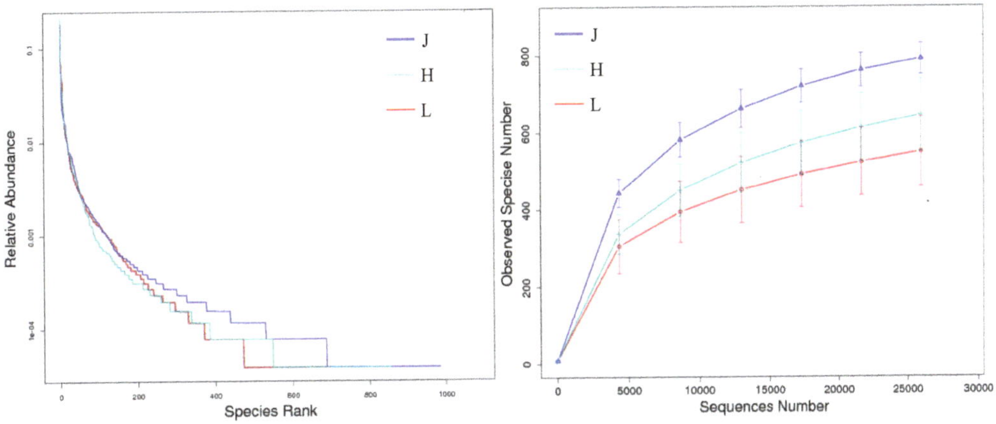

Figure 5. Rank abundance and rarefaction curves of bacteria on soybean seeds. (**Left**) Rank abundance curve; (**Right**) rarefaction curve.

3.4. *Characteristics of Microbial Communities Carried by Soybean Seeds*

3.4.1. Characteristics of Fungal Communities

A total of 185 species of fungi were detected and identified in soybean seed samples from northeast China. The distribution and proportions of the 10 dominant genera with maximum abundance at each taxonomic level are shown in Figure 6. *Cladosporium* and *Boeremia* had the greatest relative abundances among the top ten genera in this study.

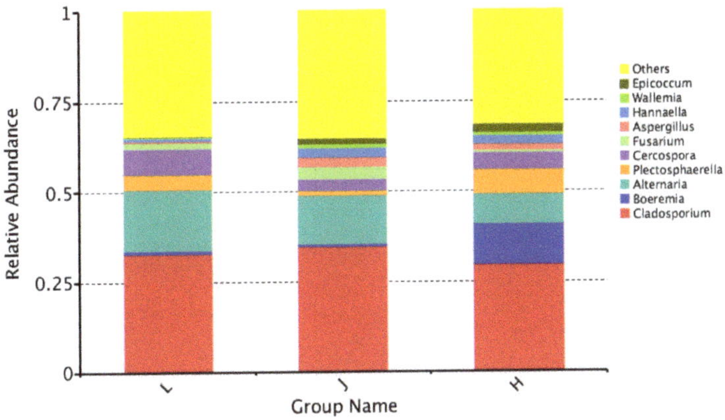

Figure 6. Relative abundance of the top ten genera of fungi on soybean seeds.

3.4.2. Characteristics of Bacterial Communities

A total of 273 species of bacteria were detected and identified in soybean seed samples from northeast China. The distribution and proportions of the top 10 dominant genera with maximum abundance are shown in Figure 7. *Methylobacterium* and *Sphingomonas* had the greatest relative abundances among the top ten genera in this study.

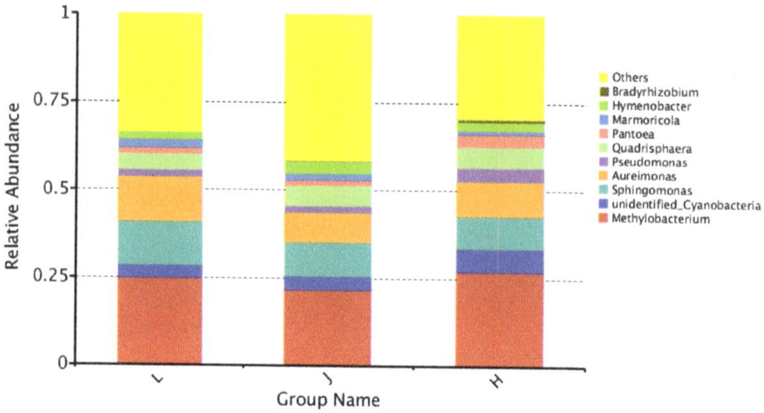

Figure 7. Relative abundance of the top ten genera of bacteria on soybean seeds.

3.5. *Changes in Related Microbial Communities Carried by Soybean Seeds in Northeast China*

The results of Non-Metric Multi-Dimensional Scaling (NMDS) showed that there were inter-group and intra-group differences between microbial species associated with soybean seeds from Liaoning, Jilin, and Heilongjiang Provinces (Figure 8).

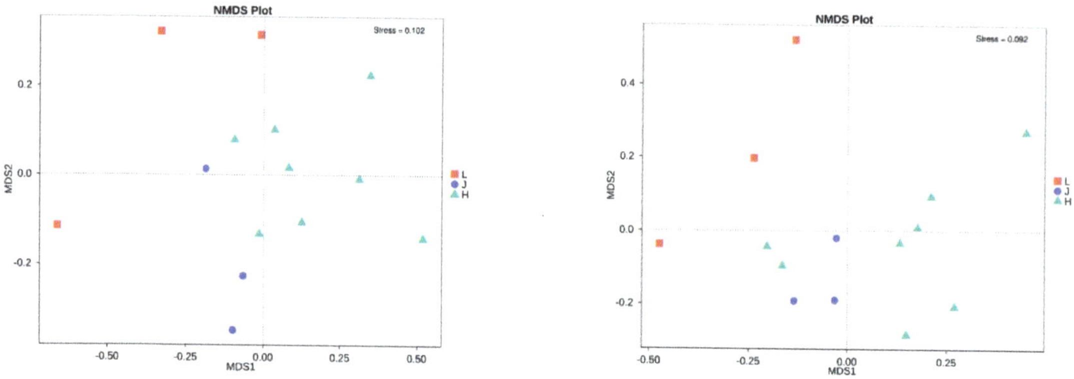

Figure 8. NMDS analysis of soybean seed carrier microbes (**left**: fungi; **right**: bacteria).

3.5.1. Changes in Fungal Communities

Analysis of similarities (ANOSIM) was used to further detect changes in fungal composition. The statistical analysis score for Liaoning and Jilin samples was r = 0.07407, $p = 0.4$ ($p < 0.05$, significant level). At the genus level (Figure 9), there were significant differences in *Penicillium* and *Hannaella* between the Jilin and Liaoning groups, in *Plectosphaerella* between Heilongjiang and Jilin groups, and in *Epicoccum* and *Symmetrospora* between Heilongjiang and Liaoning groups.

3.5.2. Changes in Related Bacterial Communities

The statistical analysis score for the Liaoning and Jilin samples was r = 0, $p = 0.5$ ($p < 0.05$, significant level). However, at the genus level (Figure 10), there were significant differences in *Roseomonas*, *Nakamurella*, and *Segetibacter* between the Heilongjiang and Jilin groups and in *Spirosoma* and *Kineococcus* between the Heilongjiang and Liaoning groups.

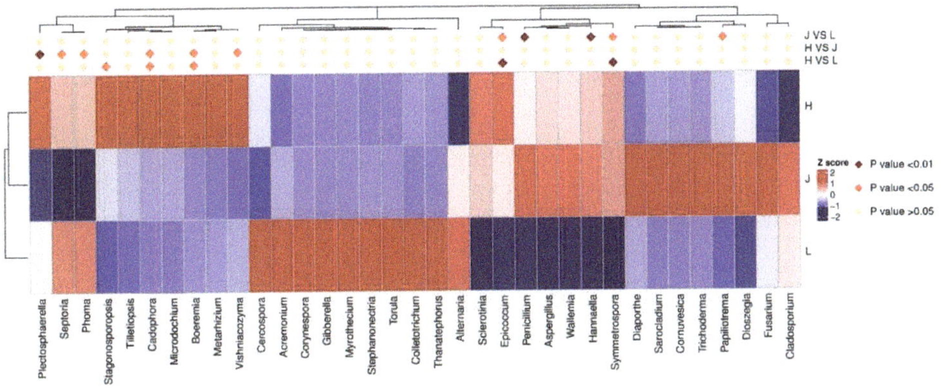

Figure 9. Heatmap of species annotations showing differences between groups of fungi on soybean seeds.

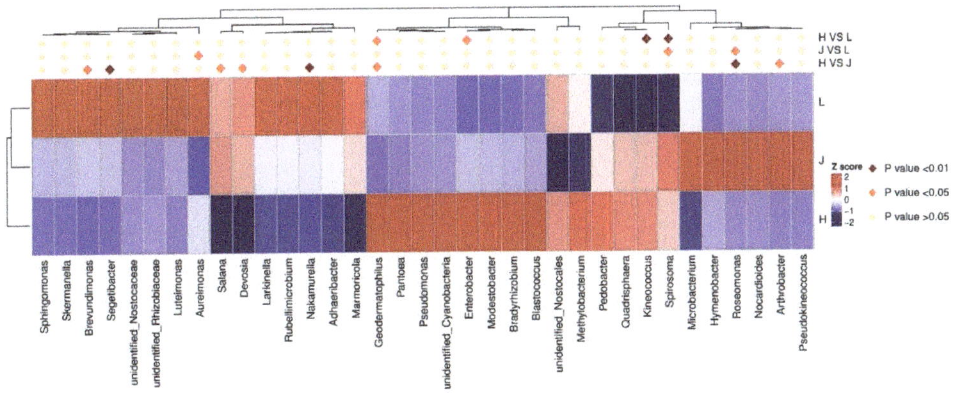

Figure 10. Heatmap of species annotations showing differences between groups of bacteria on soybean seeds.

4. Discussion

Plants are colonized by numerous microorganisms, forming intricate plant microbial communities [31]. Plant-associated microorganisms, mainly comprising bacteria and fungi, have the potential to influence the primary functions of the host [32]. A wide variety of microorganisms directly or indirectly affect seeds during their development, harvesting, and storage [33]. Some microbes can induce seed coloration and decomposition, while others can adhere to the seed surface or penetrate internal tissues, facilitating the transmission of diseases across consecutive growing seasons and different locations [4, 34]. Therefore, it is imperative to investigate the diversity of the microbial composition of soybean seeds to clarify the soybean seed-carrying situation in different regions and provide references for soybean seed production, safe transportation, and disease control. In the present study, traditional seed bacteria detection technology and high-throughput sequencing technology were compared to analyze bacteria on soybean seeds in northeast China.

Comparison of two isolation methods: Traditional fungal isolation and identification revealed that dominant genera such as *Cladosporium* (21.23%), *Fusarium* (17.45%), *Aspergillus* (15.57%), and *Alternaria* (11.56%) were also dominant in high-throughput sequencing identification, accounting for 31.40%, 1.25%, 1.61%, and 11.15% of the fungi, respectively. Other dominant genera, *Cercospora* (1.42%) and *Epicoccum* (1.89%), accounted for 4.77% and 1.62% of the fungi, respectively, in high-throughput sequencing. *Boeremia* (4.20%) and *Hannaella* (1.23%) were not found in traditional fungal isolates. The dominant genera identified through traditional bacterial isolation, namely *Pseudomonas* (37.46%), *Sphingomonas* (17.29%), and *Pantoea* (15.27%), accounted for 3.06%, 9.97%, and 2.81% of the bacteria, respectively, in high-throughput sequencing identification. Other dominant genera identified by high-throughput sequencing were not found in traditional bacterial isolation. Different separation methods lead to different separation ratios.

The analysis of microbial communities by high-throughput sequencing technology showed that there were differences in the composition of microbial communities between groups and within groups. In pairwise comparisons of three regions, there was no significant difference in community structure between groups; however, at the genus level, the differences between some genera between the groups were significant or even extremely significant. The results indicate that there were significant differences in soybean seed bacteria carriage in northeast China.

The detection of seed diseases is a crucial part of crop disease control and prevention throughout the introduction, reproduction, and acquisition stages [35]. Seed health testing is used to assess and determine the specific varieties and quantities of pathogens present in the seeds [36]. The quality of seeds has a significant impact on the full realization of their yield and crop value potential [37]. The traditional seed carrier detection technology is suitable for detecting the carrier site and each part of the seed and is suitable for most seed carrier detections. This method can be used for the study of seed health, species-borne diseases, and target pathogens, and it can identify and preserve all isolated microorganisms [38]. However, one drawback is that the cultivated colonies may exhibit interconnectivity or even overlap, thus posing challenges in terms of their separation. Identifying certain isolated bacteria might be challenging due to the susceptibility of a small number of slow-growing bacteria to competition and the influence of dominant bacteria. Moreover, these bacteria may exhibit minimal or invisible colony growth [39]. During this study, due to the limitations of the medium, it was not possible to detect all microorganisms in the tested seed samples. High-throughput sequencing technology offers the advantage of directly obtaining genomic DNA from the sample, which contains both cultured and non-cultured microbial genes. The absence of separation significantly reduces expenses. Using a large amount of data, it is possible to obtain the entire sample's microbiological information in order to streamline the taxonomic examination of microorganisms. Without a strain, we cannot perform further tests; hence, in the context of seed bacterial identification, it is possible to employ both of these detection approaches concurrently, thereby ensuring that the test outcomes are complementary.

5. Conclusions

Whether the dominant genera of *Fusarium*, *Alternaria*, *Urospora*, and *Pseudomonas* can cause soybean diseases still needs to be verified using Koch's rule. In this experiment, traditional seed bacteria detection technology and high-throughput sequencing technology were combined to detect the bacteria-carrying status and health status of soybean seeds in northeast China to provide a theoretical basis for safe transportation and disease control in soybean seed production areas and to achieve accurate inspection purposes.

Author Contributions: Conceptualization, Y.W., L.C. and Y.D.; methodology, F.M. and W.D.; software, Q.B. and X.Z.; validation, F.M. and Y.W.; formal analysis, F.M.; investigation, W.D. and Q.B.; resources, Y.W., Q.B. and H.F.; data curation, W.D. and Y.D.; writing—original draft preparation, Y.W. and H.F.; writing—review and editing, L.C.; supervision, Y.D.; project administration, L.C.; funding acquisition, Y.W. and X.Z. All authors have read and agreed to the published version of the manuscript.

Funding: This work was supported by the National Key R&D Program of China (2023YFD1400400) and the Central Guidance for Local Scientific and Technological Development Funds (2024ZY0018).

Data Availability Statement: The original contributions presented in the study are included in the article; further inquiries can be directed to the corresponding author.

Conflicts of Interest: The authors declare no conflicts of interest.

References

1. Colletti, A.; Attrovio, A.; Boffa, L.; Mantegna, S.; Cravotto, G. Valorisation of by-products from soybean (*Glycine max* (L.) Merr.) processing. *Molecules* **2020**, *25*, 2129. [CrossRef]
2. Kockelmann, A.; Tilcher, R.; Fischer, U. Seed production and processing. *SugarTech* **2010**, *12*, 267–275. [CrossRef]
3. Martín, I.; Gálvez, L.; Guasch, L.; Palmero, D. Fungal Pathogens and Seed Storage in the Dry State. *Plants* **2022**, *11*, 3167. [CrossRef] [PubMed]
4. Barret, M.; Guimbaud, J.F.; Darrasse, A.; Jacques, M.A. Plant microbiota affects seed transmission of phytopathogenic microorganisms. *Mol. Plant Pathol.* **2016**, *17*, 791–795. [CrossRef] [PubMed]
5. Javaid, M.M.; Mahmood, A.; Alshaya, D.S.; AlKahtani, M.D.F.; Waheed, H.; Wasaya, A.; Khan, S.A.; Naqve, M.; Haider, I.; Shahid, M.A.; et al. Influence of environmental factors on seed germination and seedling characteristics of perennial ryegrass (*Lolium perenne* L.). *Sci. Rep.* **2022**, *12*, 9522. [CrossRef] [PubMed]
6. Long, R.L.; Gorecki, M.J.; Renton, M.; Scott, J.K.; Colville, L.; Goggin, D.E.; Commander, L.E.; Westcott, D.A.; Cherry, H.; Finch-Savage, W.E. The ecophysiology of seed persistence: A mechanistic view of the journey to germination or demise. *Biol. Rev.* **2015**, *90*, 31–59. [CrossRef] [PubMed]
7. Gitaitis, R.; Walcott, R. The epidemiology and management of seedborne bacterial diseases. *Annu. Rev. Phytopathol.* **2007**, *45*, 371–397. [CrossRef] [PubMed]
8. Berg, T.; Tesoriero, L.; Hailstones, D. PCR-based detection of *Xanthomonas campestris* pathovars in *Brassica* seed. *Plant Pathol.* **2005**, *54*, 416–427. [CrossRef]
9. Franco-Duarte, R.; Černáková, L.; Kadam, S.; Kaushik, K.S.; Salehi, B.; Bevilacqua, A.; Corbo, M.R.; Antolak, H.; Dybka-Stępień, K.; Leszczewicz, M.; et al. Advances in chemical and biological methods to identify microorganisms—From past to present. *Microorganisms* **2019**, *7*, 130. [CrossRef] [PubMed]
10. Simonin, M.; Briand, M.; Chesneau, G.; Rochefort, A.; Marais, C.; Sarniguet, A.; Barret, M. Seed microbiota revealed by a large-scale meta-analysis including 50 plant species. *New Phytol.* **2022**, *234*, 1448–1463. [CrossRef] [PubMed]
11. Jiang, F.; Ruan, Y.; Chen, X.-H.; Yu, H.-L.; Cheng, T.; Duan, X.-Y.; Liu, Y.-G.; Zhang, H.-Y.; Zhang, Q.-Y. Metabolites of pathogenic microorganisms database (MPMdb) and its seed metabolite applications. *Microbiol. Spectr.* **2024**, *12*, e02342-23. [CrossRef] [PubMed]
12. Xie, Y. Developmental Paths Differentiation and Genome Analysis of the Basidiomycota, *Coprinopsis cinerea*. ProQuest Dissertations & Theses, Hong Kong University of Science and Technology, Hong Kong, 2020.
13. Olimi, E.; Kusstatscher, P.; Wicaksono, W.A.; Abdelfattah, A.; Cernava, T.; Berg, G. Insights into the microbiome assembly during different growth stages and storage of strawberry plants. *Environ. Microbiome* **2022**, *17*, 21. [CrossRef] [PubMed]
14. Razzaq, A.; Saleem, F.; Kanwal, M.; Mustafa, G.; Yousaf, S.; Arshad, H.M.I.; Hameed, M.K.; Khan, M.S.; Joyia, F.A. Modern Trends in Plant Genome Editing: An Inclusive Review of the CRISPR/Cas9 Toolbox. *Int. J. Mol. Sci.* **2019**, *20*, 4045. [CrossRef]
15. Hosseini, P.; Matthews, B.F. Regulatory interplay between soybean root and soybean cyst nematode during a resistant and susceptible reaction. *BMC Plant Biol.* **2014**, *14*, 300. [CrossRef] [PubMed]
16. Roy, K.W.; Baird, R.E.; Abney, T.S. A review of soybean (*Glycine max*) seed, pod, and flower mycofloras in North America, with methods and a key for identification of selected fungi. *Mycopathologia* **2001**, *150*, 15–27. [CrossRef]
17. Azerang, P.; Khalaj, V.; Kobarfard, F.; Owlia, P.; Sardari, S.; Shahidi, S. Molecular characterization of a fungus producing membrane active metabolite and analysis of the produced secondary metabolite. *Iran. Biomed. J.* **2019**, *23*, 121–128. [CrossRef]
18. Manter, D.K.; Vivanco, J.M. Use of the ITS primers, ITS1F and ITS4, to characterize fungal abundance and diversity in mixed-template samples by qPCR and length heterogeneity analysis. *J. Microbiol. Methods* **2007**, *71*, 7–14. [CrossRef]
19. Gerhardt, P.; Murray, R.G.E.; Costilow, R.; Nester, E.; Wood, W.; Krieg, N.; Phillips, G. *Manual of Methods for General Bacteriology*; American Society for Microbiology: Washington, DC, USA, 1981.
20. Sambrook, J.; Fritsch, E.F.; Maniatis, T. *Molecular Cloning: A Laboratory Manual*; Cold Spring Harbor Laboratory Press: Cold Spring Harbor, NY, USA, 1989.
21. Zhao, J.; Liu, D.; Wang, Y.; Zhu, X.; Xuan, Y.; Liu, X.; Fan, H.; Chen, L.; Duan, Y. Biocontrol potential of *Microbacterium maritypicum* Sneb159 against *Heterodera glycines*. *Pest Manag. Sci.* **2019**, *75*, 3381–3391. [CrossRef] [PubMed]
22. Edwards, U.; Rogall, T.; Blöcker, H.; Emde, M.; Böttger, E.C. Isolation and direct complete nucleotide determination of entire genes. Characterization of a gene coding for 16S ribosomal RNA. *Nucleic Acids Res.* **1989**, *17*, 7843–7853. [CrossRef]

23. Lane, D. 16S/23S rRNA sequencing. In *Nucleic Acid Techniques in Bacterial Systematics*; Modern Microbiological Methods; John Wiley & Sons: New York, NY, USA, 1991; pp. 115–175.
24. Johnston Monje, D.; Raizada, M.N. Conservation and diversity of seed associated endophytes in *Zea* across boundaries of evolution, ethnography and ecology. *PLoS ONE* **2011**, *6*, e20396. [CrossRef] [PubMed]
25. Lima, H.S.; Mancine, N.; Peruchi, G.B.; Francisco, C.S.; Wang, N.; de Souza, R.S.C.; Armanhi, J.S.L.; Della Coletta-Filho, H. Microbial community of cultivated and uncultivated citrus rhizosphere microbiota in Brazil. *Sci. Data* **2024**, *11*, 1294. [CrossRef] [PubMed]
26. Klindworth, A.; Pruesse, E.; Schweer, T.; Peplies, J.; Quast, C.; Horn, M.; Glöckner, F.O. Evaluation of general 16S ribosomal RNA gene PCR primers for classical and next-generation sequencing-based diversity studies. *Nucleic Acids Res.* **2013**, *41*, e1. [CrossRef]
27. Wang, Q.; Garrity, G.M.; Tiedje, J.M.; Cole, J.R. Naive Bayesian classifier for rapid assignment of rRNA sequences into the new bacterial taxonomy. *Appl. Environ. Microbiol.* **2007**, *73*, 5261–5267. [CrossRef] [PubMed]
28. Altschul, S.F.; Gish, W.; Miller, W.; Myers, E.W.; Lipman, D.J. Basic local alignment search tool. *J. Mol. Biol.* **1990**, *215*, 403–410. [CrossRef]
29. Kõljalg, U.; Nilsson, R.H.; Abarenkov, K.; Tedersoo, L.; Taylor, A.F.S.; Bahram, M.; Bates, S.T.; Bruns, T.D.; Bengtsson-Palme, J.; Callaghan, T.M.; et al. Towards a unified paradigm for sequence-based identification of fungi. *Mol. Ecol.* **2013**, *22*, 5271–5277. [CrossRef]
30. Edgar, R.C. UPARSE: Highly accurate OTU sequences from microbial amplicon reads. *Nat. Methods* **2013**, *10*, 996–998. [CrossRef]
31. Bever, J.D.; Platt, T.G.; Morton, E.R. Microbial population and community dynamics on plant roots and their feedbacks on plant communities. *Annu. Rev. Microbiol.* **2012**, *66*, 265–283. [CrossRef] [PubMed]
32. Carpentieri-Pipolo, V.; de Almeida Lopes, K.B.; Degrassi, G. Phenotypic and genotypic characterization of endophytic bacteria associated with transgenic and non-transgenic soybean plants. *Arch. Microbiol.* **2019**, *201*, 1029–1045. [CrossRef]
33. Abdulmumeen, H.A.; Risikat, A.N.; Sururah, A.R. Food: Its preservatives, additives and applications. *Int. J. Chem. Biochem. Sci.* **2012**, *1*, 36–47.
34. Barret, M.; Briand, M.; Bonneau, S.; Préveaux, A.; Valière, S.; Bouchez, O.; Hunault, G.; Simoneau, P.; Jacques, M.-A. Emergence shapes the structure of the seed microbiota. *Appl. Environ. Microbiol.* **2015**, *81*, 1257–1266. [CrossRef]
35. Munkvold, G.P. Seed pathology progress in academia and industry. *Annu. Rev. Phytopathol.* **2009**, *47*, 285–311. [CrossRef] [PubMed]
36. Mancini, V.; Murolo, S.; Romanazzi, G. Diagnostic methods for detecting fungal pathogens on vegetable seeds. *Plant Pathol.* **2016**, *65*, 691–703. [CrossRef]
37. Sevik, M.A.; Arli-Sokmen, M. Estimation of the effect of Tomato spotted wilt virus (TSWV) infection on some yield components of tomato. *Phytoparasitica* **2012**, *40*, 87–93. [CrossRef]
38. ElMasry, G.; ElGamal, R.; Mandour, N.; Gou, P.; Al-Rejaie, S.; Belin, E.; Rousseau, D. Emerging thermal imaging techniques for seed quality evaluation: Principles and applications. *Food Res. Int.* **2020**, *131*, 109025. [CrossRef] [PubMed]
39. Li, X.; Gu, A.; Zhang, X.; Xu, Y.; Luo, L.; Li, J. Comparative analysis of three kinds of testing methods for rice seed borne fungi. *Food Res. Int.* **2011**, *11*, 43–51.

Disclaimer/Publisher's Note: The statements, opinions and data contained in all publications are solely those of the individual author(s) and contributor(s) and not of MDPI and/or the editor(s). MDPI and/or the editor(s) disclaim responsibility for any injury to people or property resulting from any ideas, methods, instructions or products referred to in the content.

agronomy

MDPI

Article

Monitoring of Soybean Bacterial Blight Disease Using Drone-Mounted Multispectral Imaging: A Case Study in Northeast China

Weishi Meng [1], Xiaoshuang Li [1], Jing Zhang [2], Tianhao Pei [1] and Jiahuan Zhang [1,3,*]

1 College of Plant Protection, Jilin Agricultural University, Changchun 130118, China
2 Changbaishan Customs, People's Republic of China, Antu 133613, China
3 Key Laboratory of Soybean Disease and Pest Control, Ministry of Agriculture and Rural Affairs, Changchun 130118, China
* Correspondence: zhjh63@126.com

Abstract: Soybean bacterial blight disease is a threat to soybean production. Multispectral technology has shown good potential in detecting this disease and can overcome the limitations of traditional methods. The aim of this study was to perform field monitoring of the dynamics of this disease in Northeast China in 2022. The correlation between the soybean chlorophyll content index (CCI) and disease grade was obtained using artificial inoculation of the pathogen. The correlation between the soybean CCI, disease grade, green normalized difference vegetation index (GNDVI), and soybean yield was analyzed using a drone-mounted spectrometer platform for image acquisition and preprocessing. The soybean CCI was negatively correlated with the disease grade. The GNDVI declined with disease progression, which allowed for an indirect determination of the disease grade. The soybean yield loss was significant at disease grade 4 for soybean bacterial blight disease. The random forest regression model was more accurate than the regression model in estimating the yield based on the GNDVI. Therefore, the GNDVI could be used to survey the disease class and estimate the yield using the random forest model. This study provides support for field trials of drone-mounted multispectral equipment. This surveillance approach holds the potential to bring about precision plant protection in the future.

Keywords: soybean bacterial blight disease; multispectroscopy; chlorophyll content index; GNDVI; disease monitoring

1. Introduction

Soybeans are a crucial crop globally, playing a significant role in food production and the agricultural economy. However, they are constantly threatened by various diseases [1], including soybean bacterial blight disease, which is caused by infestation of *Pseudomonas savastanoi* pv. *glycinea* [2–4]. This disease occurs in all soybean production areas in both northern and southern China, with more severe cases in the north. The disease used to occur mainly in the Huang-Hua-Hai region in China and was one of the major diseases of the soybean [5]. In recent years, due to climate warming, it has become more and more common in the main soybean-producing areas in Northeast China, and the degree of damage has gradually increased [6]. When severe, it can reduce soybean yields [7]. In Northeast China, the onset of the disease generally occurs during the soybean flowering stage up to the full grain stage (in July–August), especially under cool and humid climatic conditions [8,9]. The disease mainly affects leaves, but also petioles, stems, and bean pods

https://doi.org/10.3390/agronomy15040921

(Figure 1). Leaf spots in the early stage are green, polygonal, and water-soaked, which then turn yellow to light brown and then reddish brown to blackish brown. The edge of the spot has a clear yellow-green halo and white bacterial pus overflows from the back of the spots. The spots often converge to form a large spot, causing some or all the leaves to turn yellow and die, resulting in early defoliation [10,11]. As soybean canopy leaves are the key part of the plant for photosynthesis, their growth, and death due to disease directly affect the final soybean quality and yield [12].

Figure 1. Soybean bacterial blight disease in a field. (**A**) Early stage. (**B**) Mid stage. (**C**) Late stage. Red circles indicate disease spots (Photographed by Jiahuan Zhang).

Conventional detection methods usually involve manual visual inspection; however, using these methods, the disease can only be observed after a period of crop infestation and therefore, they cannot be used for prevention [13]. To overcome these limitations and achieve early disease and pest detection, new technologies are needed. Unmanned aerial vehicles (UAVs) are aircraft that can be operated from miles away without the presence of a pilot [14]. UAVs can be equipped with various types of auxiliary sensors to obtain and quantify environmental parameters in the vicinity of the drone, which makes them useful in many fields [15]. In recent years, UAVs have been widely used in the field of low-altitude and low-speed remote sensing of farmland [16]. UAVs are used for remote sensing because of their high flexibility, low operating costs and ability to capture high-resolution images [17]. Moreover, multispectroscopy has been widely used in land-use planning, agricultural production, environmental protection, and other research fields because of its advantages such as a fast detection speed and the ability of on-site detection [18–21]. The potential of using UAVs and spectroscopic techniques to monitor soybean diseases and pests has been demonstrated in a few studies [22]. For example, near-infrared reflectance, detected from UAV-based multispectral imagery, decreased with increasing soybean aphid (*Aphis glycines*) populations in open-field trials when soybean aphid populations were above the economic threshold [23]. UAVs and cell phones were also used to capture images of caterpillars and *Diabrotica speciosa* (cucurbit beetle) [24]. In another study, UAV-acquired images were used to visualize the spatial and time series variation in an area damaged by red crown rot (a soil-borne disease of soybean) [25]. Tetila et al. (2017) proposed a computer vision system to track soybean foliar diseases in the field using images captured by an unmanned aerial vehicle (model DJI Phantom 3), which can aid experts and farmers in monitoring diseases in soybean fields [26]. Nagasubramanian et al. (2019) deployed a novel 3D deep convolutional neural network to identify charcoal rot disease in soybean stems [27]. Liu et al. (2023) discussed the use of hyperspectral analysis for classifying soybean diseases; they found that the PCA-SI combination method had a significantly better classification accuracy and could effectively distinguish between healthy and diseased soybean leaves [28]. Therefore, multispectroscopy image surveys of canopy leaves could be used to monitor soybean growth and disease for targeted field management.

The features of soybean diseases in fields are highly complex and multiple similar symptoms are usually mixed together [29,30]. The resolution of monitoring tools is also not sufficient for detecting the actual occurrence of diseases in the field. There is a need to continue developing and improving monitoring methods for soybean bacterial blight disease. Thus, the aim of this study was to explore a method for real-time monitoring of the occurrence of soybean bacterial blight disease using drone-based multispectral images and to provide a theoretical basis for the prediction and forecasting of large-scale occurrences of soybean bacterial spot disease for disease control in Northeast China.

2. Materials and Methods

The workflow of this study is presented in Figure 2, and includes three main parts: (1) UAV image and field dataset collection and preprocessing; (2) polynomial regression and random forest regression (RFR) analyses of the disease grades of soybean bacterial blight, the green normalized difference vegetation index (GNDVI), the chlorophyll content index (CCI) and soybean yield; and (3) screening methods for soybean yield estimation.

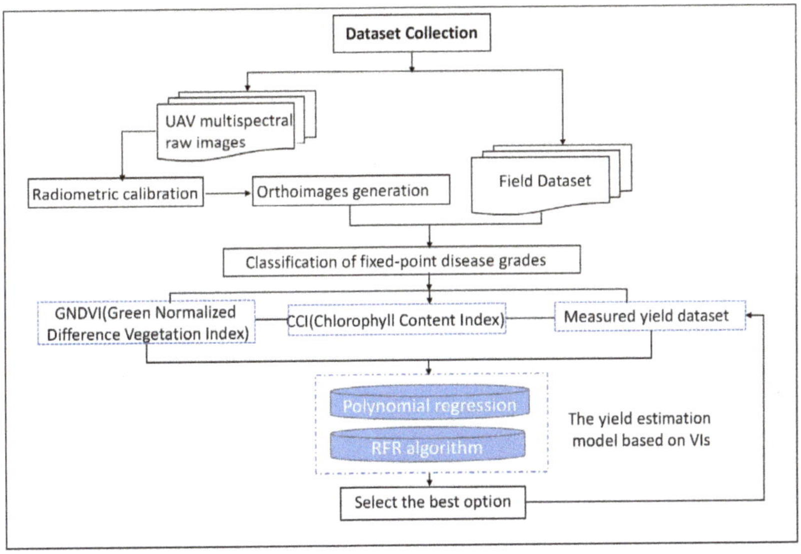

Figure 2. Workflow diagram for this study.

2.1. Overview of the Test Area

The test area is located in the Teaching and Research Experimental Base of Jilin Agricultural University (43°49′14″ N, 125°23′52″ E), Changchun City, Jilin Province, China, at an altitude of about 222.0 m (Figure 3). The geographic coordinate is GCS_WGS_ 1984 (EPSG: 4326). It is in the temperate continental semi-humid monsoon climate zone, with an average annual temperature of 4.6 °C, annual precipitation of 400–600 mm, and average annual sunshine time of about 2600 h. The terrain is flat, easily irrigated, and drained, and the soil is mainly black soil that is rich in organic substances and suitable for soybean growth. The experimental area was 946.4 m^2, which was divided into 25 plots. Each plot consisted of 6 rows; each row was 20 m long and 0.65 m wide. The soybean variety 'Jiyu47' that was planted is susceptible to soybean bacterial blight disease [9,10].

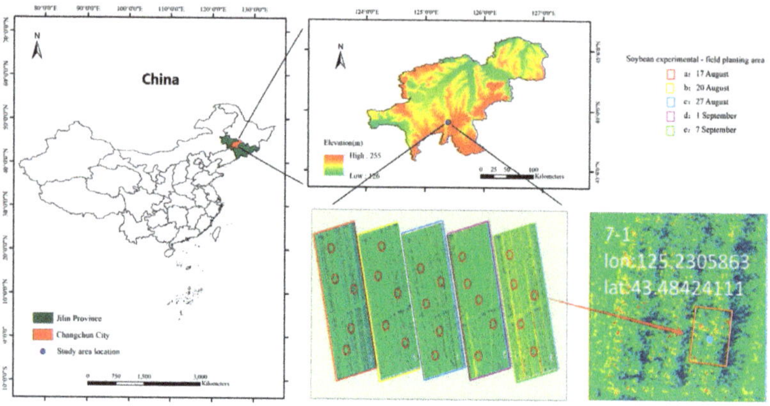

Figure 3. Study area diagram.

2.2. *Establishment of Sampling Points*

The sampling points were located in two adjacent rows that were 0.65 m long and 1.5 m wide along the furrow, and contained a total area of about 2 m^2 of plants; the soybean seedling densities were approximately the same between sampling points. Two adjacent sampling points were marked with a tag inserted into the soil between the two points (Table S1). In order to achieve real-time monitoring, manual periodic tracking was conducted and the information on disease grade was supplemented and updated in a timely manner based on disease markers in the field prior to each UAV flight.

2.3. *Pathogen Culture and Inoculation*

2.3.1. Preparation of Bacterial Suspensions

A sample of the bacterial strain that causes soybean bacterial blight disease was taken from the test area of this study and then isolated, purified, and cultured in the lab in NBY medium (liquid), which consists of 3 g of beef extract, 5 g of peptone, 2.5 g of sucrose (fructose), 2 g of yeast extract, 2 g of K_2HPO_4, 0.5 g of KH_2PO_4, 7.5 g of $MgSO_4$, 0.5 g of KH_2PO_4, 1.5 g of $MgSO_4{\cdot}7H_2O$, and 1000 mL of distilled water. A 20 μL sample of the bacterial suspension was transferred to 30 mL of NBY liquid medium using a pipette gun in an ultra-clean bench. The bacterial suspension (concentration: 1×10^8 CFU/mL) was shaken at 177 rpm for 24 h at 28 °C in a gas bath incubator. Then, 160 mL of the bacterial suspension was added to 4.5 L of sterile water and used immediately to inoculate in the field.

2.3.2. Inoculations of Pathogens in Field

The field inoculation was carried out twice on 1 August and 15 August 2022. The inoculation area was the 25 plots, which had fewer than three replicates of the different canopy leaf disease grades to ensure that the disease occurrence in the field showed all the disease grades. To ensure the effectiveness of the inoculation, we avoided the midday hours, when the sun is strong, and periods before rainfall. The prepared bacterial suspension was placed in a Drexel hand sprayer and sprayed on the soybean canopy leaves in a uniform manner so that each plant was evenly wetted. In order to achieve better inoculation efficiency, the bacterial suspension was sprayed onto the back of the soybean leaves from the bottom to the top to increase the contact area between the bacterial solution and the leaf surface. This method effectively reduces the risk of burns to the leaves caused by volatilization of the bacterial solution and the concentration of light, and it helps the bacteria invade the plant through the stomata on the back of the leaves during respiration, thus improving the inoculation effect.

2.4. Ground Data Acquisition

The ground data mainly included the CCI of the soybean leaves and the incidence of soybean bacterial blight disease [31]. The CCI values were measured 24–36 h prior to each UAV spectral acquisition using a CCM-200 handheld chlorophyll meter (Opti-Sciences, Hudson, NH, USA). At each sampling point, onset leaves were identified based on the presence of a large area of spots that did not contain cavities or other damage. Since only the metal probe part of the meter was used, the measurement of the affected leaves was performed at leaf spots, at spots without disease, and at the intersection between disease spots and health tissues. Several groups (10–20 samples) of samples were selected for averaging, and the equipment was calibrated after every 3–5 measurements to avoid measurement errors. The data obtained from the drone image acquisition were the soybean canopy leaf physical condition and disease level (Table S2).

2.5. UAV Image Acquisition and Preprocessing

2.5.1. UAV Multispectral Remote Sensing Platform

The UAV multispectral remote sensing platform mainly consisted of a UAV flight platform and an airborne imaging spectroscopy system [32]. The platform uses DJJ Wizard 4 (Shenzhen DJI Innovation Technology Co., Ltd., Shenzhen, China), a multispectral version (maximum takeoff weight: 1.482 kg; maximum single sortie duration: 27 min), and a multispectral camera (Figure 4). The multispectral parameters of Genie 4 have a spectral range of 460–950 nm, 6 sensors, 5 bands, 208 $\times$ 10^4 pixels, and a spectral imaging speed of 10^{-2}–10^{-4} s (website: https://www.dji.com/cn/support/product/p4-multispectral, accessed on 7 April 2025).

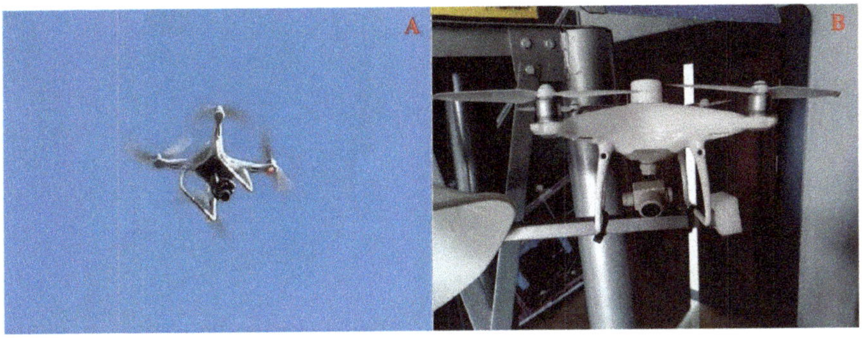

Figure 4. UAV equipment diagram. (**A**) Flying UAV in the air. (**B**) UAV in the lab. (Photographed by Xiaoshuang Li and Weishi Meng).

2.5.2. Monitoring Method

Monitoring Time

In order to obtain high-quality multispectral images, the time period in which the observation data were collected was from 10:00 to 14:00 to avoid the influence of factors such as uneven light and unstable wind speeds. The remote sensing image acquisition by the UAV was performed under windy and sunny weather conditions, with wind speeds less than 5.4 m/s. The spectral camera was whiteboard corrected before the UAV took off, and the flight was completed within 40–50 min in order to avoid errors caused by equipment factors. The flight altitudes of the UAV were 30 m and 50 m to adapt to different flight requirements. After considering the above factors, the UAV remote sensing images were taken on 17 August (podding stage to the beginning of the grain stage), 20 August (beginning of the grain stage to the full grain stage), 27 August (the full grain stage to the first maturity stage), 1 September (the first maturity stage to the full grain stage) and 7 September (the full grain stage), 2022 [6,33]. The output of each photograph produced a high resolution TIFF image, which was used to extract spectral information and calculate the GNDVI by using ArcGIS 10.2.2, which was used in the next step of constructing a relational model of the impact on yield.

Image Spectral Bands

The parameters for each band in the UAV multispectral images are shown in Table 1. Red light is strongly absorbed by plant leaves, which reflect and transmit very low amounts of red light. Red edge (RE) is invisible light with a wavelength range of 700–780 nm and can indicate vegetation nutrition, growth, moisture, leaf area, etc. Near-infrared (NIR) is also invisible light, with a wavelength range of 780–1350 nm. When soybean vegetation is lush, with vigorous growth and high pigmentation, the red edge of the reflected light spectrum will shift towards near-infrared, and when the vegetation is affected by a variety of insect pests and grasses, environmental pollution, leaf aging, or other factors, the red edge will shift towards the red light [34].

Table 1. Band parameters for the multispectral images.

Band	Wavelength Range (nm)
Blue	450 ± 16
Green	560 ± 16
Red	650 ± 16
Red Edge	730 ± 16
Near-Infrared	840 ± 26

Vegetation Index

The GNDVI (green normalized difference vegetation index) is a modification of the NDVI and is used to monitor the degree of denseness of the canopy at the maturity stage of soybean plants. It is calculated as the near-infrared band minus the green light band, divided by the near-infrared band plus the green light band [35]. Because soybean leaf photosynthetic pigments, especially chlorophyll, absorb red and blue light and reflect green light, the GNDVI replaces the red light band in the NDVI with the green light band, which is more suitable for disease development monitoring. The vegetation index formulas used in this study are shown in Table 2.

Table 2. Vegetation index equations.

Vegetation Index	Equation
Normalized Red Light (R)	$R/(R+G+B)$
Normalized Green Light (G)	$G/(R+G+B)$
Normalized Blue Light (B)	$B/(R+G+B)$
Green Normalized Vegetation Index (GNDVI)	$(NIR-G)/(NIR+G)$

2.6. Soybean Yield Estimation

Before the soybeans reached maturity, from 18 to 25 September, the location of each sampling point and the occurrence of disease were recorded. All soybean samples within the delineated area were harvested, and the harvested samples were fully threshed and counted in the laboratory to calculate the soybean yield data for each sampling point. The soybean variety used (Jiyu 47) has a theoretical weight of 20 g per 100 grains. The yield was calculated using the following formula: Yield = (Total number of grains from sampling point/100) × 20. The yield per 2 m^2 was calculated. The final yield per acre was then determined by scaling up the data using the conversion factor of 666 m^2 = 1 acre. To ensure accuracy, three sets of replicate sampling point yield data were collected for all disease classes.

2.7. Statistical Analysis

The linear regression between CCI and disease grade, the calculation of the GNDVI range for each disease grade, and the construction of a regression model for GNDVI and CCI and for GNDVI and yield using the polynomial regression equation and random forest method were performed using Python 3.10.11 (Python software, Python Software Foundation, Wilmington, DE, USA). One-way ANOVA of the CCI at different disease grades and at each soybean vegetative stage and one-way ANOVA of soybean yield was performed using SPSS 26.0 (IBM Inc., Chicago, IL, USA).

3. Results

3.1. Correlation Analysis Between Soybean CCI and Disease Grades of Soybean Bacterial Blight

The soybean CCI was significantly and negatively correlated with the disease grade of soybean bacterial blight. From 17 August to 7 September, the soybean plants developed from the pod stage to the filling and maturity stages, and the leaves gradually changed from green to yellow. The CCI at the different observation dates and disease grades are shown in Table 3, and the results of the regression fitting are shown in Figure 5.

Table 3. Soybean CCI at different disease grades on various observation dates.

Disease Grade	17 August (Podding Stage to Beginning of Grain Stage)	20 August (Beginning of Grain Stage to Full Grain Stage)	27 August (Full Grain Stage to First Maturity Stage)	1 September (First Maturity Stage to Full Grain Stage)	7 September (Full Grain Stage)
0	43.28 ± 0.72 [a]	42.18 ± 0.82 [a]	35.74 ± 2.06 [a]	30.57 ± 1.04 [a]	29.80 ± 1.10 [a]
1	39.04 ± 0.45 [b]	37.84 ± 0.59 [b]	31.12 ± 1.32 [b]	28.57 ± 0.93 [b]	25.55 ± 1.15 [b]
2	33.94 ± 0.53 [c]	33.26 ± 0.29 [c]	27.88 ± 1.26 [c]	25.90 ± 0.36 [c]	23.00 ± 1.40 [c]
3	30.04 ± 0.43 [d]	30.02 ± 0.36 [d]	24.50 ± 0.80 [d]	23.97 ± 0.47 [d]	20.65 ± 0.85 [d]
4	27.18 ± 0.24 [e]	27.80 ± 0.23 [e]	21.70 ± 0.74 [e]	21.00 ± 0.80 [e]	18.30 ± 0.60 [e]
5	23.80 ± 0.34 [f]	23.74 ± 0.48 [f]	19.28 ± 0.73 [f]	18.90 ± 0.78 [f]	15.20 ± 0.90 [f]
6	19.84 ± 0.18 [g]	19.52 ± 0.32 [g]	17.18 ± 0.88 [g]	16.73 ± 0.58 [g]	12.70 ± 0.60 [g]
7	15.76 ± 0.43 [h]	15.90 ± 0.45 [h]	14.70 ± 1.17 [h]	13.77 ± 1.07 [h]	10.55 ± 0.35 [h]
8	12.10 ± 0.45 [i]	12.06 ± 0.43 [i]	12.64 ± 1.15 [i]	12.13 ± 0.72 [i]	9.00 ± 0.10 [i]
9	7.74 ± 0.70 [j]	7.94 ± 0.76 [j]	10.36 ± 1.17 [j]	8.77 ± 0.29 [j]	8.10 ± 0.20 [j]
10	3.38 ± 0.68 [k]	3.00 ± 0.42 [k]	8.52 ± 1.43 [k]	3.17 ± 0.35 [k]	6.10 ± 0.80 [k]

Note: Data are shown as the mean ± S.E. Lower case letters indicate differences between disease grades on the same survey date.

With an increase in disease grade, the CCI gradually decreased. This indicates that the higher the incidence, the more pronounced the decrease in the soybean CCI (Figure 6). The coefficients of determination (R^2) of the regression models on different observation dates were all greater than 0.98, the F-values of the dates were high, and the p-values were much less than 0.05, indicating that the models had high significance (Table 4). There was a tendency for the CCI to decrease over time for the same disease grade. The coefficient of determination (R^2) of the regression model for the different observation dates was greater than 0.98, the F-value for each date was high, and the p-value was much less than 0.05. This indicates that the model is highly significant (Table 4).

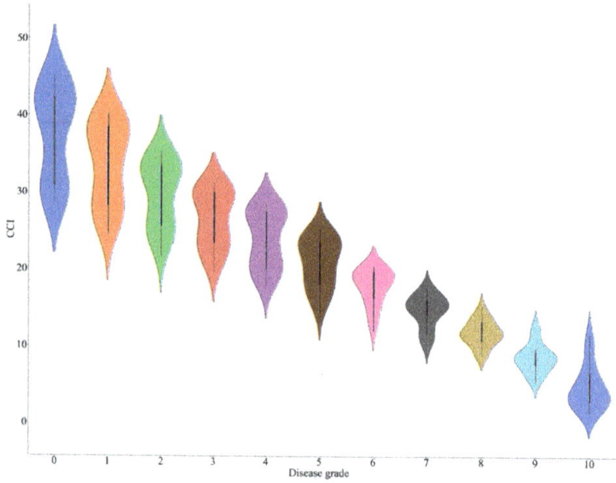

Figure 5. Soybean CCI at different disease grades. Different colors indicate different disease grades.

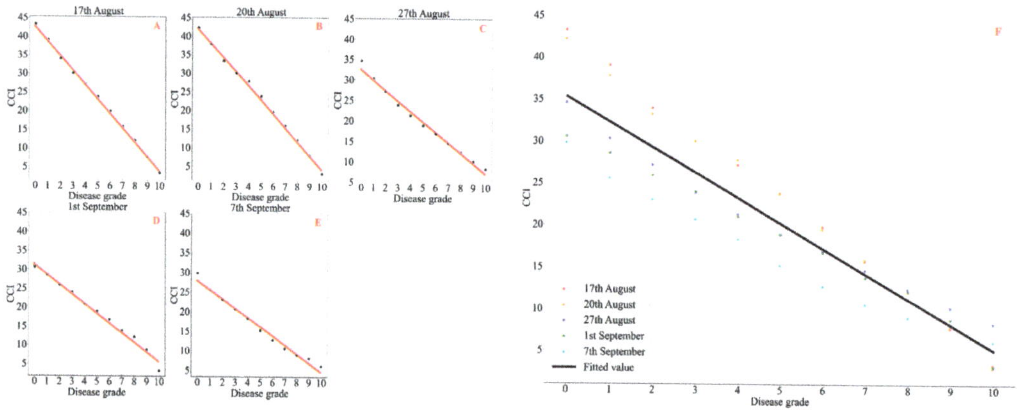

Figure 6. Regression fitting of soybean CCI and disease grade. (**A**) 17 August; (**B**) 20 August; (**C**) 27 August; (**D**) 1 September; (**E**) 1 September; (**F**) Regression fitting of five observations.

Table 4. Regression fitting of soybean CCI and disease grade.

Observation Date	Fitting Model	R^2	F	p	mae	mse	rmse
17 August (podding stage to beginning of grain stage)	$y = -3.75x + 45.71$	0.998	3801.137	3.918×10^{-13}	0.460	0.305	0.552
20 August (beginning of grain stage to full grain stage)	$y = -3.78x + 41.92$	0.997	2901.179	1.318×10^{-12}	0.561	0.443	0.665
27 August (full grain stage to first maturity stage)	$y = -2.68x + 35.26$	0.988	716.605	6.854×10^{-10}	0.741	0.852	0.923
1 September (first maturity stage to full grain stage)	$y = -2.87x + 37.51$	0.982	484.379	3.898×10^{-9}	0.691	0.916	0.957
7 September (full grain stage)	$y = -2.57x + 31.65$	0.980	435.187	6.258×10^{-9}	0.819	1.015	1.000

Note: mae—mean absolute error; mse—mean square error; rmse—root mean square error. These abbreviations are also used below.

3.2. Correlation Analysis of Soybean CCI and GNDVI

The multispectral camera carried by the UAV was used to capture images of the soybean test field five times, and the GNDVI subplot of the multispectral images was calculated according to the disease grade in the plot and the selected sampling points (Figure 7). The GNDVI in all the sampling points showed a clear decreasing trend over time and with increasing disease severity (Table 5). When the CCI exceeded 30, the GNDVI no longer showed a linear growth trend and showed a flat trend (Figure 8), suggesting that soybeans tend to mature over time and that the CCI trend tends to flatten under disease stress. A regression model was established with GNDVI as the dependent variable and CCI as the independent variable (Table 6), and the fitted equation was $y = -0.000250086570x^2 + 0.0240266388x + 0.387753418$, with a coefficient of determination R^2 of 0.849, an F-value of 298.264, and a p-value (1.110×10^{-16}) less than 0.05, indicating that the variables are correlated (Figure 6). The random forest regression models had a higher explanatory accuracy than the regression models (Figure 9). Therefore, the disease grade of soybean plants can be indirectly determined using the GNDVI, which can be used to monitor soybean bacterial blight disease.

Table 5. GNDVI at different disease grades on different observation dates.

Disease Grade	GNDVI					GNDVI_Sd	GNDVI_Range
	17 August (Podding Stage to Beginning of Grain Stage)	20 August (Beginning of Grain Stage to Full Grain Stage)	27 August (Full Grain Stage to First Maturity Stage)	1 September (First Maturity Stage to Full Grain Stage)	7 September (Full Grain Stage)	17th August to 7th September (Podding Stage to Full Grain Stage)	
0	0.9287	0.9621	0.9408	0.9420	0.9160	0.938 ± 0.017	0.9621~0.9160
1	0.9067	0.9525	0.9305	0.9243	0.8843	0.920 ± 0.026	0.9525~0.8843
2	0.8222	0.8596	0.8471	0.7874	0.7916	0.822 ± 0.032	0.8596~0.7874
3	0.7786	0.8418	0.8128	0.7578	0.7523	0.789 ± 0.038	0.8418~0.7523
4	0.7311	0.8256	0.7904	0.7480	0.7163	0.762 ± 0.045	0.8256~0.7163
5	0.6390	0.7953	0.7764	0.7146	0.6726	0.720 ± 0.066	0.7953~0.6390
6	0.5560	0.7530	0.7465	0.6788	0.6382	0.675 ± 0.082	0.7465~0.5560
7	0.5057	0.6754	0.6806	0.5965	0.5721	0.606 ± 0.074	0.6806~0.5057
8	0.4689	0.6535	0.6586	0.5592	0.4717	0.562 ± 0.093	0.6586~0.4689
9	0.4041	0.5961	0.5903	0.5299	0.4278	0.510 ± 0.090	0.5961~0.4041
10	0.3621	0.5270	0.5335	0.4569	0.3459	0.445 ± 0.089	0.5335~0.3459

Table 6. Polynomial regression and random forest regression between GNDVI and soybean CCI on different observation dates.

Observation Date	Polynomial Regression				Random Forest Regression			
	Fitting Model	R^2	F	p	mae	mse	rmse	R^2
17 August (podding stage to beginning of grain stage)	$y = -0.00011x^2 + 0.02x + 0.313$	0.981	465.967	<0.001	0.011	0.001	0.013	0.995
20 August (beginning of grain stage to full grain stage)	$y = -0.00007x^2 + 0.013x + 0.553$	0.985	587.210	<0.001	0.009	0.001	0.011	0.991
27 August (full grain stage to first maturity stage)	$y = -0.00033x^2 + 0.027x + 0.403$	0.987	672.916	<0.001	0.009	0.001	0.012	0.989
1 September (first maturity stage to full grain stage)	$y = 0.016x + 0.439$	0.965	249.450	<0.001	0.012	0.001	0.015	0.988
7 September (full grain stage)	$y = -0.00100x^2 + 0.044x + 0.205$	0.972	316.937	<0.001	0.015	0.001	0.016	0.989
17 August to 7 September (podding stage to full grain stage)	$y = -0.00025x^2 + 0.02403x + 0.38775$	0.849	298.264	<0.001	0.024	0.001	0.031	0.957

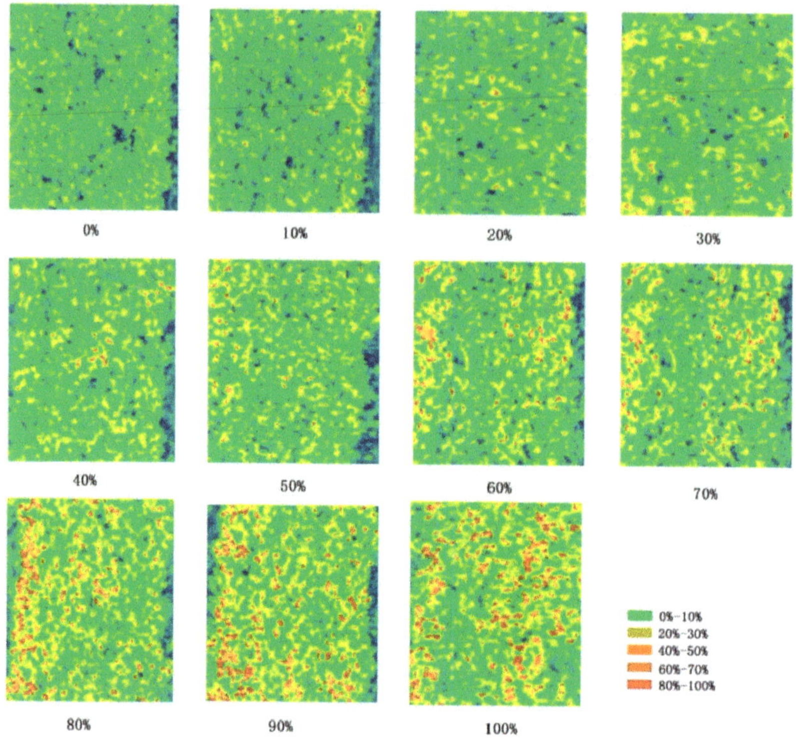

Figure 7. GNDVI at different disease grades.

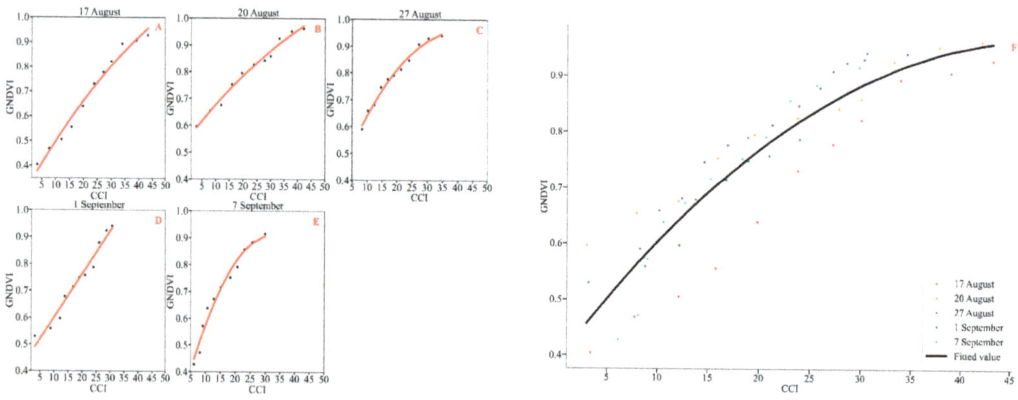

Figure 8. Polynomial regression analysis between GNDVI and soybean CCI on different observation dates. (**A**) 17 August; (**B**) 20 August; (**C**) 27 August; (**D**) 1 September; (**E**) 1 September; (**F**) Polynomial regression of five observations.

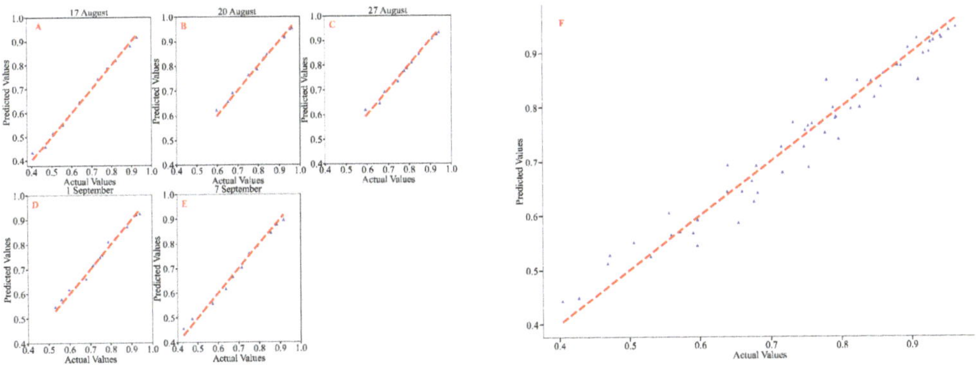

Figure 9. Random forest regression analysis between GNDVI and soybean CCI on different observation dates. (**A**) 17 August; (**B**) 20 August; (**C**) 27 August; (**D**) 1 September; (**E**) 1 September; (**F**) Random forest regression of five observations. Blue triangles represent the actual values.

3.3. Estimation of Soybean Yields

The one-way analysis of variance (ANOVA) between soybean bacterial blight disease grade and yield showed that the most significant yield loss of 8.6% was observed when the disease reached grade 4, which showed significant differences (Figure 10). Since the results of the above analyses concluded that there was a significant positive correlation between the GNDVI and disease grade, a model for estimating the GNDVI and soybean yield was constructed using polynomial regression (Figure 11) and random forest regression (Figure 12). The results showed that both models had the highest accuracy on 7 September (polynomial regression: R^2 = 0.983; random forest regression: R^2 = 0.978). On the other observation dates, the random forest model had a higher accuracy than the polynomial regression model. Thus, the random forest model can better explain the performance on 17 August, the models are comparable for 27 August, and the polynomial regression model was more suitable for the performance on 7 September (Table 7). Therefore, the GNDVI and random forest model could be used to estimate soybean yields. By optimizing the yield estimation model, we can obtain a more reliable method for soybean yield prediction based on multispectral technology, which can aid in decision-making in preventative disease control.

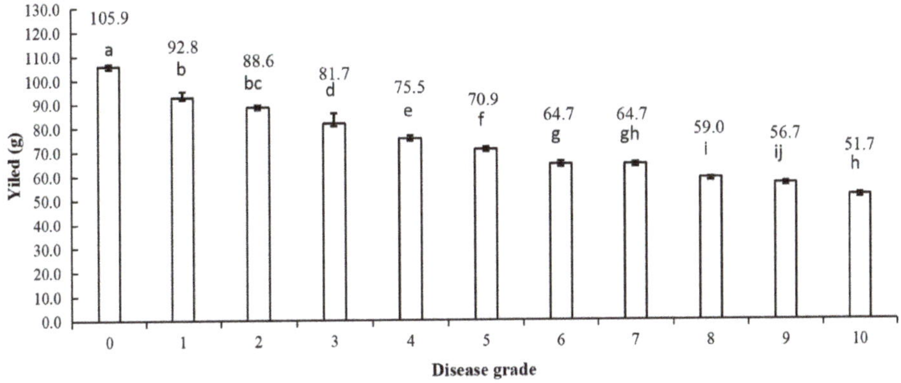

Figure 10. Soybean yields at different disease grades. Lowercase letters indicate differences among disease grades.

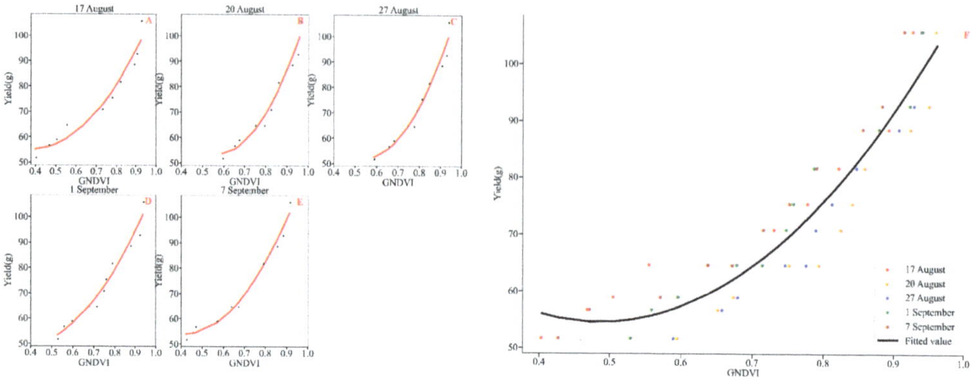

Figure 11. Polynomial regression analysis between GNDVI and soybean yield on different observation dates. (**A**) 17 August; (**B**) 20 August; (**C**) 27 August; (**D**) 1 September; (**E**) 1 September; (**F**) Polynomial regression of five observations.

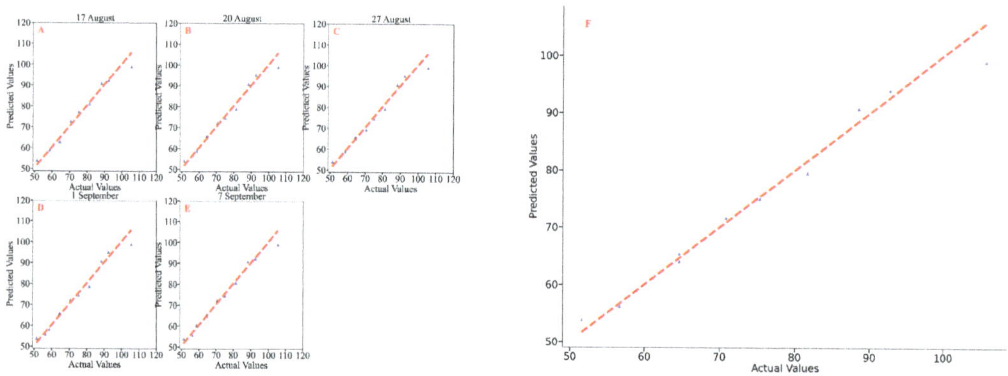

Figure 12. Random forest regression analysis between GNDVI and soybean yield on different observation dates. (**A**) 17 August; (**B**) 20 August; (**C**) 27 August; (**D**) 1 September; (**E**) 1 September; (**F**) Random forest regression of five observations. Blue triangles represent the actual values.

Table 7. Polynomial regression and random forest regression between GNDVI and soybean yield on different observation dates.

Observation Date	Polynomial Regression				Random Forest Regression			
	Fitting Model	R^2	F	p	mae	mse	rmse	R^2
17 August (podding stage to beginning of grain stage)	$y = 145.981x^2 - 112.79x + 77.039$	0.953	182.185	<0.001	1.681	6.100	2.470	0.976
20 August (beginning of grain stage to full grain stage)	$y = 315.913x^2 - 366.992x + 160.348$	0.961	223.983	<0.001	1.868	6.799	2.607	0.974
27 August (full grain stage to first maturity stage)	$y = 303.712x^2 - 330.855x + 142.121$	0.967	260.446	<0.001	1.859	6.666	2.582	0.974
1 September (first maturity stage to full grain stage)	$y = 146.966x^2 - 100.863x + 65.566$	0.964	242.715	<0.001	2.019	6.962	2.638	0.973
7 September (full grain stage)	$y = 195.218x^2 - 164.304x + 88.513$	0.983	523.781	<0.001	1.602	5.708	2.389	0.978

4. Discussion

In this study, UAV-borne multispectral imaging technology and ground-based data collection were combined to monitor soybean bacterial blight disease. This is a new method for monitoring soybean bacterial blight disease, and this study provides a more feasible solution for the application of this technology in actual soybean production. While previous studies have used spectral technology or ground surveys alone [27,28], by combining multispectral images of soybean canopy leaves using a UAV and ground surveys collecting data such as the CCI values of soybean leaves and the extent of disease incidence, we can obtain a comprehensive understanding of soybean growth and disease at both the macro- and microscopic levels. This multi-technology integration provides richer data support for accurate disease monitoring and analysis and enables more timely and precise disease detection. In selecting the research area and sampling points, the actual planting environment and disease occurrence pattern were fully considered, and the disease occurrence in the field, manipulated through artificial inoculation of pathogenic bacteria, reflected similar conditions to those observed in actual production.

The number of sampling points was insufficient to cover each disease level adequately and as a result, in-depth causal analysis and related vegetation index modeling could not be conducted. Further experiments are required in the future to explore these aspects. To reach the desired disease incidence levels (disease grades), especially during the less satisfactory mid stage of soybean spot development, leaves with no or a low disease incidence were intensively and manually inoculated with a specific dose of the bacterial solution. This was performed to increase the area of soybean canopy leaf spots. During the subsequent field recording and remote sensing monitoring, this treatment yielded favorable experimental results and achieved the expected goals. After a thorough and comprehensive analysis, it was found that this method has a certain degree of feasibility, and the experimental results provide a certain theoretical foundation and reference value. To ensure that the experimental process was close to that of real soybean production, the natural progression of disease development was utilized to allow the test plots to develop soybean bacterial blight disease.

UAV technology has been widely used in crop phenotypic research because of its flexibility and efficiency [32]. In this study, to initially establish the monitoring and early warning model, image data collected at a height of 30 m were used. However, the cost of acquiring UAV images at a height of 30 m is high, making it unsuitable for large-scale monitoring. For large areas of land, remote sensing images taken at higher altitudes can be used for broad-area exploration. Suspected disease areas can be segmented, and then more accurate image acquisition can be carried out at 30 m or other appropriate heights. Based on the results of this study, to obtain more accurate data for guiding agricultural production and developing disease prediction models, we need to continuously address the deficiencies in the experiment and conduct multiple repetitions for verification to obtain a more accurate and efficient prediction model.

The GNDVI is an improvement of the traditional NDVI, which replaces the red light band of the NDVI with a green light band for the absorption and reflection of light by photosynthetic pigments in soybean leaves and is therefore more effective in monitoring the density of the soybean canopy and the development of diseases at the maturity stage. This targeted application of the GNDVI provides new perspectives and methods for soybean disease monitoring [35,36]. The GNDVI was found to be significantly correlated with wheat yield; the GNDVI had better discriminating efficiency, allowing for better predictions of yield when recorded at early vegetative stages, and showed better results compared to the NDVI [37]. The GNDVI was also found to predict the aboveground biomass yield of maize better than the NDVI [38]. This was corroborated by the results of this study.

After the disease outbreak, there were relatively few completely healthy, disease grade 0 sampling points. Thus, it was challenging to establish a large disease-free area and select a sufficient number of grade 0 sampling points for the control group. Due to the lack of field inoculation for soybean bacterial mottle disease and suboptimal field conditions for disease development, there were fewer samples with higher disease grades (7 to 9). This made it difficult to select a control for severely diseased samples. In contrast, the distribution of disease samples at grades 1-6 was more extensive, with grade 1 disease samples being the most prevalent. For the less severe grades 1 to 4, during image color rendering and considering the gradient differences among the different disease grades, choosing the green to brownish-red color band gradient was more effective. However, most of the images of disease grades 1 to 4 consisted of light green and yellow, making image recognition more difficult compared to the other disease grades. The images of disease grades 5 to 6 had a medium level of image recognition difficulty. Although a small number of sampling points were inaccurately recognized, they could be accurately identified after calibration. Future research can optimize the monitoring and synchronous calibration technology, which will significantly improve the monitoring and UAV accuracy. The final image was presented in a vertical perspective, facilitating the data processing by the researchers.

5. Conclusions

This study provides a basis for field trials using drone-mounted spectrometers to estimate soybean disease grades and yields using random forest regression and polynomial regression. The most significant yield loss of 8.6% was observed when the disease reached grade 4. The random forest regression model was found to be more accurate in explaining the yield on the different observation dates, especially on 7 September. The monitoring method applied in this study is effective in visualizing the damage due to soybean bacterial blight disease, but further improvements are required in the evaluation of intermediate damage and the generalization of the evaluation procedure.

Supplementary Materials: The following supporting information can be downloaded at: https://www.mdpi.com/10.3390/agronomy15040921/s1; Table S1: Field markers for disease locations; Table S2: Grading criteria for foliar conditions of soybean bacterial blight disease.

Author Contributions: Conceptualization, J.Z. (Jiahuan Zhang); methodology, J.Z. (Jiahuan Zhang); software, W.M. and X.L.; investigation, X.L., W.M. and J.Z. (Jing Zhang); data curation, T.P. and X.L.; writing—original draft preparation, X.L., W.M., J.Z. (Jing Zhang) and T.P.; writing—review and editing, W.M. and J.Z. (Jiahuan Zhang); funding acquisition, J.Z. (Jiahuan Zhang). All authors have read and agreed to the published version of the manuscript.

Funding: This research was supported by the Capital Construction Funds within the Provincial Budget in 2020 (Innovation Capacity Building) of Jilin Province, Jilin Provincial Development and Reform Commission (Grant No. 2020C019-5) and the Earmarked Fund for the China Agriculture Research System of MOF and MARA (Grant No. CARS04).

Data Availability Statement: The original contributions presented in the study are included in the article. Further inquiries can be directed to the corresponding author.

Acknowledgments: We thank Shaozhong Song from Jilin College of Engineering and Technology, Huanjun Liu and Yion Wang from the Northeast Institute of Geography and Agroecology, Chinese Academy of Sciences, for their guidance and assistance in performing the remote sensing. We thank the anonymous reviewers for reviewing this paper and providing constructive comments.

Conflicts of Interest: The authors declare no conflicts of interest.

References

1. Bandara, A.Y.; Weerasooriya, D.K.; Bradley, C.A.; Allen, T.W.; Esker, P.D. Dissecting the economic impact of soybean diseases in the United States over two decades. *PLoS ONE* **2020**, *15*, e0231141. [CrossRef] [PubMed]
2. Tarakanov, R.; Shagdarova, B.; Lyalina, T.; Zhuikova, Y.; Il'ina, A.; Dzhalilov, F.; Varlamov, V. Protective properties of copper-loaded chitosan nanoparticles against soybean pathogens *Pseudomonas savastanoi* pv. *glycinea* and *Curtobacterium flaccumfaciens* pv. *flaccumfaciens*. *Polymers* **2023**, *15*, 1100. [CrossRef]
3. El-Esawi, M.A.; Ali, H.M.; Hatamleh, A.A.; Al-Dosary, M.A.; El-Ballat, E.M. Multi-functional PGPR Serratia liquefaciens confers enhanced resistance to lead stress and bacterial blight in soybean (*Glycine max* L.). *Curr. Plant Biol.* **2024**, *40*, 100403. [CrossRef]
4. Feng, J. Recent advances in taxonomy of plant pathogenic bacteria. *J. Integr. Agric.* **2017**, *50*, 2305–2314. (In Chinese with English Abstract) [CrossRef]
5. Wang, C.; Linderholm, H.W.; Song, Y.; Wang, F.; Liu, Y.; Tian, J.; Xu, J.; Song, Y.; Ren, G. Impacts of drought on maize and soybean production in Northeast China during the past five decades. *Int. J. Environ. Res. Public Health* **2020**, *17*, 2459. [CrossRef]
6. Ye, W.W.; Liu, W.C.; Wang, Y.C. Occurrence status and whole-process green control technologies for soybean diseases and pests in China. *J. Plant Prot.* **2023**, *50*, 265–273. (In Chinese with English Abstract) [CrossRef]
7. Surbhi, K.; Singh, K.; Aravind, T.; Bhatt, P.; Jeena, H.; Rakhonde, G. GIS-based survey and molecular detection of bacterial blight of soybean in Sub-Himalayan Ranges of Uttarakhand, India. *Trop. Plant Pathol.* **2023**, *48*, 332–346. [CrossRef]
8. Zhao, F.Z.; Wang, Y.A.; Cheng, W.; Antwi-Boasiako, A.; Yan, W.K.; Zhang, C.T.; Gao, X.W.; Kong, J.J.; Liu, W.S.; Zhao, T.J. Genome-wide association study of bacterial blight resistance in soybean. *Plant Dis.* **2025**, *109*, 341–351. [CrossRef]
9. Zhang, J.H.; Gao, J.; Yuan, M.L.; Li, Y. Studies on the distribution of physiological races of *Pseudomonas syringae* pv. *glycinea*. *J. Jilin Agric. Univ.* **2003**, *25*, 24–26. (In Chinese with English Abstract) [CrossRef]
10. Zhang, J.H.; Gao, J.; Yuan, M.L. Identification of physiological races of *Pseudomonas syringae* pv. *glycinea* in Jilin province. *J. Jilin Agric. Univ.* **1993**, *15*, 24–27+105. (In Chinese with English Abstract) [CrossRef]
11. Jagtap, P.G.; Dhopte, S.B.; Dey, U. Evaluation of different antibacterial antibiotics against bacterial blight of soybean caused by *Pseudomonas syringae* pv. *glycinea* under field conditions. *Ind. Phytopathol.* **2013**, *66*, 411–412.
12. Wang, X.Y.; Li, Y.; Pan, T.; Zhang, Z.G.; Qu, J.Q.; Sun, Y.Z.; Zhao, H.J.; Miao, L.X.; Hu, Z.B.; Zhao, Z.M.; et al. Diagnosis of soybean bacterial blight progress stage based on deep learning in the context of data-deficient. *Comput. Electron. Agric.* **2023**, *212*, 108170. [CrossRef]
13. Sotelo, J.P.; Rovey, M.F.P.; María, E.C.; Moliva, M.V.; Oliva, M.D.L.M. Characterization of *Pseudomonas syringae* strains associated with soybean bacterial blight and in vitro inhibitory effect of oregano and thyme essential oils. *Physiol. Mol. Plant Pathol.* **2023**, *128*, 102133. [CrossRef]
14. Rahman, M.H.; Sejan, M.A.S.; Aziz, M.A.; Tabassum, R.; Baik, J.-I.; Song, H.-K. A comprehensive survey of unmanned aerial vehicles detection and classification using machine learning approach: Challenges, solutions, and future directions. *Remote Sens.* **2024**, *16*, 879. [CrossRef]
15. Mukherjee, A.; Misra, S.; Raghuwanshi, N.S. A survey of unmanned aerial sensing solutions in precision agriculture. *J. Netw. Comput. Appl.* **2019**, *148*, 102461. [CrossRef]
16. Huang, Y.B.; Thomson, S.J.; Brand, H.J.; Reddy, K.N. Development and evaluation of low-altitude remote sensing systems for crop production management. *Int. J. Agric. Biol. Eng.* **2016**, *9*, 1–11. [CrossRef]
17. Yuan, J.; Zhang, Y.; Zheng, Z.; Yao, W.; Wang, W.; Guo, L. Grain crop yield prediction using machine learning based on UAV remote sensing: A systematic literature review. *Drones* **2024**, *8*, 559. [CrossRef]
18. Zahra, A.; Qureshi, R.; Sajjad, M.; Sadak, F.; Nawaz, M.; Khan, H.A.; Uzair, M. Current advances in imaging spectroscopy and its state-of-the-art applications. *Expert Syst. Appl.* **2024**, *238*, 122172. [CrossRef]
19. Hou, Y.; Bao, H.; Rimi, T.I.; Zhang, S.; Han, B.; Wang, Y.; Yu, Z.; Chen, J.; Gao, H.; Zhao, Z.; et al. Rice quality and yield prediction based on multi-source indicators at different periods. *Plants* **2025**, *14*, 424. [CrossRef]
20. Yu, Y.; Li, C.; Shen, W.; Yan, L.; Zheng, X.; Yao, Z.; Cui, S.; Cui, C.; Hu, Y.; Yang, M. Correlation study between canopy temperature (CT) and wheat yield and quality based on infrared imaging camera. *Plants* **2025**, *14*, 411. [CrossRef]
21. Sukhova, E.; Zolin, Y.; Popova, A.; Grebneva, K.; Yudina, L.; Sukhov, V. Broadband normalized difference reflectance indices and the normalized red–green index as a measure of drought in wheat and pea plants. *Plants* **2025**, *14*, 71. [CrossRef] [PubMed]
22. Shrivastava, S.; Singh, K.S.; Hooda, S.D. Color sensing and image processing-based automatic soybean plant foliar disease severity detection and estimation. *Multimed. Tools Appl.* **2015**, *74*, 11467–11484. [CrossRef]
23. Marston, Z.P.D.; Cira, T.M.; Hodgson, E.W.; Knight, J.F.; MacRae, I.V.; Koch, R.L. Detection of stress induced by soybean aphid (Hemiptera: Aphididae) using multispectral imagery from unmanned aerial vehicles. *J. Econ. Entomol.* **2020**, *113*, 779–786. [CrossRef]
24. Mignoni, M.E.; Honorato, A.; Kunst, R.; Righi, R.; Massuquetti, A. Soybean images dataset for caterpillar and *Diabrotica speciosa* pest detection and classification. *Data Brief* **2022**, *40*, 107756. [CrossRef] [PubMed]

25. Yamamoto, S.; Nomoto, S.; Hashimoto, N.; Maki, M.; Hongo, C.; Shiraiwa, T.; Homma, K. Monitoring spatial and time-series variations in red crown rot damage of soybean in farmer fields based on UAV remote sensing. *Plant. Prod. Sci.* **2023**, *26*, 36–47. [CrossRef]
26. Tetila, E.C.; Machado, B.B.; de Souza Belete, N.A.; Guimaraes, D.A.; Pistori, H. Identification of soybean foliar diseases using unmanned aerial vehicle images. *IEEE Geosci. Remote Sens. Lett.* **2017**, *14*, 2190–2194. [CrossRef]
27. Nagasubramanian, K.; Jones, S.; Singh, A.K.; Sarkar, S.; Singh, A.; Ganapathysubramanian, B. Plant disease identification using explainable 3D deep learning on hyperspectral images. *Plant Methods* **2019**, *15*, 98. [CrossRef]
28. Liu, S.; Yu, H.-Y.; Sui, Y.-Y.; Kong, L.-J.; Yu, Z.-D.; Guo, J.-J.; Qiao, J. Hyperspectral data analysis for classification of soybean leaf diseases. *Spectrosc. Spectr. Anal.* **2023**, *43*, 1550–1555. [CrossRef]
29. Afolabi, C.G.; Salihu, S.; Shokalu, O. Screening of soybean [*Glycine max* (L.) merrill] lines for reaction to natural field infection and resistant against bacteria foliar diseases. *Arch. Phytopathol. Plant Protect.* **2023**, *56*, 269–283. [CrossRef]
30. Brown, M.T.; Mueller, D.S.; Kandel, Y.R.; Telenko, D.E.P. Influence of integrated management strategies on soybean sudden death syndrome (SDS) root infection, foliar symptoms, yield and net returns. *Pathogens* **2023**, *12*, 913. [CrossRef]
31. Ravelombola, W.S.; Qin, J.; Shi, A.; Nice, L.; Bao, Y.; Lorenz, A.; Orf, J.H.; Young, N.D.; Chen, S. Genome-wide association study and genomic selection for soybean chlorophyll content associated with soybean cyst nematode tolerance. *BMC Genom.* **2019**, *20*, 904. [CrossRef] [PubMed]
32. Yang, C.; Yang, G.; Wang, H.; Li, S.; Zhang, J.; Pan, D.; Ren, P.; Feng, H.; Li, H. Identifying key traits for screening high-yield soybean varieties by combining UAV-based and field phenotyping. *Remote Sens.* **2025**, *17*, 690. [CrossRef]
33. Maimaitijiang, M.; Sagan, V.; Sidike, P.; Hartling, S.; Esposito, F.; Fritschi, F.B. Soybean yield prediction from UAV using multimodal data fusion and deep learning. *Remote Sens. Environ.* **2020**, *237*, 111599. [CrossRef]
34. Huang, W.J.; Huang, M.Y.; Liu, L.Y.; Wang, J.H.; Zhao, C.J.; Wang, J.D. Inversion of the severity of winter wheat yellow rust using proper hyper spectral index. *Trans. Chin. Soc. Agric. Eng.* **2005**, *21*, 97–103.
35. Shi, W.; Li, Y.; Zhang, W.; Yu, C.; Zhao, C.; Qiu, J. Monitoring and zoning soybean maturity using UAV remote sensing. *Ind. Crops Prod.* **2024**, *222*, 119470. [CrossRef]
36. Jia, S.; Cui, M.; Chen, L.; Guo, S.; Zhang, H.; Bai, Z.; Li, Y.; Deng, L.; Li, F.; Zhang, W. Soybean water monitoring and water demand prediction in arid region based on UAV multispectral data. *Agronomy* **2025**, *15*, 88. [CrossRef]
37. Kyratzis, A.C.; Skarlatos, D.P.; Menexes, G.C.; Vamvakousis, V.F.; Katsiotis, A. Assessment of vegetation indices derived by UAV imagery for durum wheat phenotyping under a water limited and heat stressed mediterranean environment. *Front. Plant Sci.* **2017**, *8*, 1114. [CrossRef]
38. Tiruneh, G.A.; Meshesha, D.T.; Adgo, E.; Tsunekawa, A.; Haregeweyn, N.; Fenta, A.A.; Reichert, J.M. A leaf reflectance-based crop yield modeling in Northwest Ethiopia. *PLoS ONE* **2022**, *17*, e0269791. [CrossRef]

Disclaimer/Publisher's Note: The statements, opinions and data contained in all publications are solely those of the individual author(s) and contributor(s) and not of MDPI and/or the editor(s). MDPI and/or the editor(s) disclaim responsibility for any injury to people or property resulting from any ideas, methods, instructions or products referred to in the content.

agronomy

MDPI

Article

Preparation of Wheat-Straw-Fiber-Based Degradable Mulch Film for Sustained Release of Carbendazim and Its Application for Soybean Root Rot Control

Shuang Liu [1,2], Zhe Jin [1,2], Pengfei Zhou [1,2], Huimin Shang [3], Haiyan Yang [4], Longhai Li [1,2], Rui Li [1,2], Ying Zhang [1,2] and Haitao Chen [1,5,*]

1 College of Engineering, Northeast Agricultural University, Harbin 150030, China; liushuang@neau.edu.cn (S.L.); jinzhe0311@163.com (Z.J.); zpf18800428412@163.com (P.Z.); lilonghai@neau.edu.cn (L.L.); ruili2018@aliyun.com (R.L.); zhangying83@neau.edu.cn (Y.Z.)
2 Heilongjiang Provincial Engineering Research Center for Mechanization and Materialization of Major Crops Production, Harbin 150030, China
3 School of Mechanical Engineering, Dalian University of Technology, Dalian 116000, China; shm2023@mail.dlut.edu.cn
4 College of Arts and Sciences, Northeast Agricultural University, Harbin 150030, China; haiyanyang@neau.edu.cn
5 College of Mechanical and Electronic Engineering, East University of Heilongjiang, Harbin 150066, China
* Correspondence: htchen@neau.edu.cn

Abstract: In order to sustain control over soybean root rot, wheat-straw-fiber-based mulch film (WFM) coated with carbendazim (C) and chitosan (CS) mixture (C-CS-WFM) were prepared through bar coating technology. The Box–Behnken design method was employed to investigate the effects of chitosan concentration, wet film thickness, and carbendazim loading on the dry tensile strength (DTS), wet tensile strength (WTS), and air permeance (AP) of C-CS-WFM. Eventually, the optimization process parameters were determined as follows: a chitosan concentration of 1.83–2.39%, a wet film thickness of 18–24 μm, and a carbendazim loading of 0.05–0.12 g/m^2. These parameters achieved the desired physical properties of C-CS-WFM, i.e., the DTS is not less than 3.5 kN/m, the WTS is not less than 0.8 kN/m, and the AP does not exceed 2.1 μm/(Pa·s). The results showed that after the introduction of the C-CS coating, the DTS and WTS of C-CS-WFM were enhanced by 11.4% and 14.9%, respectively. In contrast, the AP was reduced by 15.6%. FT-IR analysis indicated that carbendazim was embedded in the C-CS composite material without any chemical interaction. Through SEM and sustained-release kinetic analysis, it was found that the sustained-release mechanism of C-CS-WFM conformed to the Ritger–Peppas kinetic model, and its release mechanism was the physical diffusion and matrix erosion. The results of the in vitro antifungal test and pot experiment demonstrated that C-CS-WFM could effectively inhibit the growth of *Fusarium solani* and promote the growth of plants. This study provided new ideas for the comprehensive prevention and control of soybean root rot.

Keywords: mulch film; soybean root rot; wheat straw; chitosan; carbendazim; sustained release

1. Introduction

Soybean root rot is a pervasive and detrimental soil-borne disease in global soybean cultivation, primarily caused by a range of pathogenic fungi, including *Phytophthora sojae*, *Fusarium species*, and *Rhizoctonia solani* [1]. These fungi can persist in the soil as mycorrhizae, mycelium, or in the form of thick-walled spores or oospores, which are dormant for extended periods, leading to widespread distribution, significant crop damage, and challenging control measures [2–4]. Current control strategies encompass agricultural management, chemical treatments, biological control, and the selection of resistant varieties [5]. Chemical agents, despite their rapid action and ease of use, often suffer from quick degradation and short-lived effectiveness [6–8], necessitating the development of new, cost-effective, and enduring application methods.

Mulching is a widely adopted agricultural practice that enhances soil temperature and moisture retention [9–11], thereby promoting crop growth and mitigating soil-borne diseases to some extent [12]. However, the non-biodegradable nature of traditional plastic mulch films results in environment contamination and microplastic pollution, posing risks to ecosystems and the food chain [13,14]. Consequently, the development of biodegradable mulch films infused with fungicides presents a promising solution for both disease management and environment sustainability. Biodegradable mulch films are typically made from plant fiber or synthetic biodegradable materials such as polylactic acid (PLA) [15,16] chitosan [17], starch [18,19] and polyhydroxyalkanoates (PHAs) [20]. These films provide an eco-friendly alternative to conventional plastic mulch and can be combined with fungicides for enhanced agricultural efficiency. Fungicide incorporation methods include direct blending, where fungicides are mixed with film-forming materials during fabrication, ensuring uniform distribution but requiring careful control of processing conditions to maintain fungicide activity [21]. Encapsulation involves embedding fungicides in nanoparticles or microcapsules, improving stability and enabling controlled release [22,23]. Another approach, layered structures, uses multilayer films with a dedicated fungicide-releasing layer to precisely control release rates and improve disease management [24]. Biodegradable films integrated with fungicides create an effective controlled release system, reducing application frequency and minimizing environmental contamination. Release mechanisms depend on factors such as temperature, soil moisture, and pH, with degradation synchronized to the crop's growth cycle, ensuring targeted and efficient disease control [24].

Although researchers have been developing biocidal biodegradable films based on biopolymers such as carboxymethyl chitosan/poly(vinyl alcohol) [25], polyhydroxybutyrate [26], chitosan/hydroxypropyl methylcellulose [27], and poly(vinyl alcohol)/starch [19], However, these materials often face limitations due to high production costs and complex manufacturing processes. In contrast, straw-fiber-based films, derived from agricultural waste, offer a cost-effective and readily available alternative [28–30]. Their unique porous structure and biodegradability make them a promising material for agricultural applications, although their mechanical and barrier properties require enhancement to meet practical needs [31].

This study introduces a novel strategy for modifying straw-fiber-based mulch film by coating them with chitosan, a natural polymer known for its film-forming properties and biocompatibility. The coating process incorporates the broad-spectrum fungicide carbendazim, enhancing the mechanical and barrier properties of the mulch film while also improving the stability and longevity of the fungicide [32–35]. The films were prepared using a classical bar coating method, offering a balance between biodegradability and soil-borne disease control. The study systematically investigated the effects of chitosan concentration, wet film thickness, and carbendazim loading on the dry tensile strength (DTS), wet tensile strength (WTS), and air permeance (AP) of the mulch film. The microstructure of the films was analyzed using scanning electron microscopy (SEM), and the sustained-release mechanism of carbendazim was explored to assess the durability of the film's efficacy. Additionally, in vitro antifungal property tests and pot experiments were conducted using *Fusarium solani* (*F. solani*) as a representative pathogenic fungus to evaluate the film's performance.

2. Materials and Methods

2.1. Materials

Chitosan (CS) (molecular weight 200 kDa; deacetylation degree 95%) and carbendazim (≥99.5%) were purchased from Shanghai Yuanye Biotechnology Co. (Shanghai, China). All other chemical reagents were analytically pure.

2.2. Preparation of WFM

Wheat straw of Dongnong Winter Wheat 2 was used as raw material to prepare wheat-straw-fiber-based mulch film (WFM) according to a previous work [28]. In detail, wheat straw was processed into semi-finished products using a kneading and cutting machine (kneading and cutting machine feed capacity of 1000 kg/h, spindle speed of 1800 r/min). The semi-finished product was soaked and washed to remove impurities and increase the moisture content of the raw material. Then, the semi-finished products were used to produce wheat straw fiber by extrusion blasting with D200 fiber making machine. Then, the wheat straw fiber and the hardwood unbleached kraft pulp cardboard torn into pieces were separately soaked in water for more than 4 h. Then, according to the standard of GB/T24325-2009 "Pulp Valley (valley) pulper method" [36], the prepared wheat straw fibers and kraft pulp were put into the valley pulper for 30 min of defibering process, and then pulped to the required pulping degree (55 ± 5° SR for wheat straw pulp, and 45 ± 2° SR for kraft pulp) for spare. After that, according to the specific needs of the experiment, the wheat straw pulp and kraft pulp were mixed according to the ratio of 65:35 (both measured in terms of absolute dry pulp), and 1.0% wet strength agent and 1.4% neutral sizing agent were added (both measured in terms of absolute dry pulp). Finally, the WFM was prepared by adjusting the parameters of the paper machine for film pressing and drying.

2.3. Preparation of C-CS-WFM

The C-CS-WFM sample preparation procedure is shown in Figure 1. According to the requirements of the test protocol shown in Table 1, different concentrations of chitosan solution and carbendazim solution were first prepared by dissolving CS and carbendazim in aqueous acetic acid solution (2%, v/v), respectively, and the solutions were magnetically stirred at 500 rpm for 1 h at room temperature (25 ± 1 °C temperature and 50 ± 2% relative humidity) to ensure that the solids were completely dissolved, and the solutions were left to stand for 1 h to remove air bubbles. Then, the chitosan solution and carbendazim solution were mixed to prepare C-CS mixtures with different concentrations and drug loadings. Finally, the above mixture was coated on a 20 cm × 35 cm WFM with an OSP coating bar with 15 μm, 20 μm and 25 μm wet film thicknesses, respectively, using a laboratory coating machine (MS-RL 320, Ruilin machinery technology co., Ltd., Xianyou, Fujian, China) to obtain the C-CS-WFM samples. All samples were dried at room temperature for 24 h before testing.

Table 1. Factor-level code.

Factors	Levels Used, Actual (Coded Factor)		
Independent Variables	**Low (−1)**	**Medium (0)**	**High (+1)**
X_1 = Chitosan concentration (%)	1.5	2.0	2.5
X_2 = Wet film thickness (μm)	15	20	25
X_3= Carbendazim loading (g/m^2)	0.05	0.10	0.15

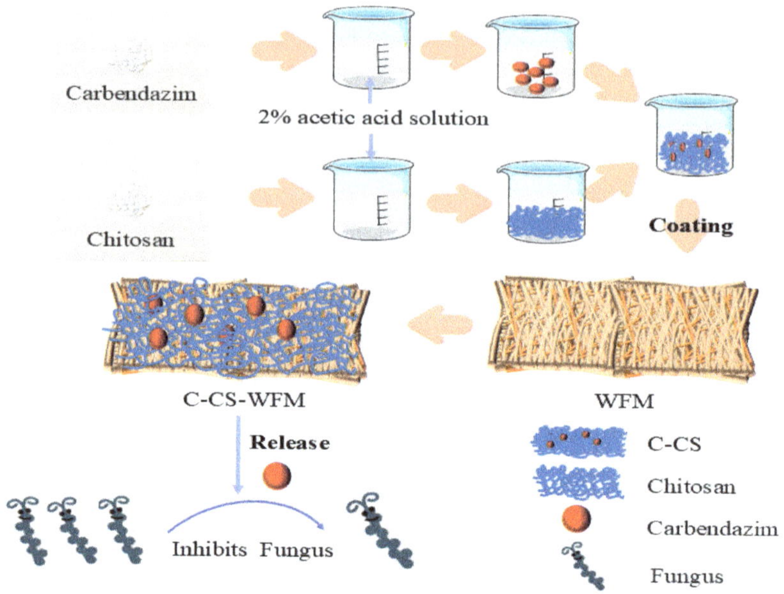

Figure 1. Flow chart of C-CS-WFM preparation.

2.4. Experimental Design and Statistical Analysis

In this study, a three-factor, three-level Box–Behnken design (BBD) was used to optimize the mechanical and barrier properties of C-CS-WFM. Chitosan concentration (X_1, %), wet film thickness (X_2, μm), and carbendazim loading (X_3, g/m^2) were used as independent variables, and dry tensile strength (Y_1, kN/m), wet tensile strength (Y_2, kN/m), and air permeance (Y_3, μm/(Pa·s)) were used as dependent variables, and the coding of the test factor levels is shown in Table 1.

Design Expert 6.0 (Stat-Ease Inc., Minneapolis, MN, USA) was used to design the corresponding BBD tests and statistically analyze the test results.

2.5. Performance Measurement and Characterization

2.5.1. Measurement of Mechanical Properties

According to Equation (1), to calculate the tensile strength S, the unit is expressed in kN/m, with reference to GB/T12914-2018, "Paper and board—Determination of tensile properties—Constant rate of elongation method (20 mm/min)" [37], and GB/T465.2-2008 "Paper and board—Determination of tensile strength after immersion in water" [38]. The film specimen was made into 120 mm × 15 mm, and the dry and wet tensile strength was determined by the Pendulum-type Paper Tensile Strength Measuring Instrument (Jinan Ruidek Instrument Co., Ltd., Jinan, China), and then converted into the dry and wet tensile strength.

$$S = F/W \quad (1)$$

where S is the tensile strength (kN/m), F is the average tensile strength (N), and W is the initial width of the sample (mm).

2.5.2. Measurement of Air Permeance

According to Equation (2) to calculate the Schopper permeance (Ps), the unit is expressed in μm/(Pa·s), with reference to GB/T458-2008 "Paper and board-Determination of air permeance" [39] for the determination of the film specimen made of 100 mm × 60 mm, through the HK-TQD01 Schopper permeance tester (Jinan Drake Instrument Co., Ltd., Jinan, China) to determine the air passing through the specimen within 15 s. The volume of air passing through the specimen in 15 s was measured by the HK-TQD01 Schopper-type air permeance tester (Jinan Derrick Instrument Co., Ltd., Jinan, China) and then converted into air permeance.

$$P_s = V/(\nabla P \cdot t) \quad (2)$$

where P_s is the air permeance (μm/(Pa·s)), V is the volume of air passing through the specimen (mL) during the measurement time, ∇P is the pressure difference between the two sides of the specimen (kPa), and t is the measurement time (s).

2.5.3. SEM

The C-CS-WFM specimens were made into 5 mm × 2 mm, patched and then sprayed with gold, and the microstructures of C-CS-WFM before and after immersion were observed comparatively by SEM (SU8010 field emission scanning electron microscope, Hitachi, Chiyoda, Japan).

2.5.4. FT-IR

Freeze-dried powders of carbendazim, chitosan, and C-CS were prepared as samples using KBr compression technique and analyzed by FT-IR spectroscopy in the range of 400–4000 cm^{-1} using Nicolet iS50 FT-IR spectrometer (Thermo Fisher, Waltham, MA, USA). The test results were used to analyze the interactions between the components of the samples through peak variations.

2.6. Drug Release Kinetics

The carbendazim release test was carried out on C-CS-WFM and wheat-straw-fiber-based mulch film directly coated with the same amount of aqueous carbendazim-loaded solution (C-WFM), in which the C-WFM and the optimized C-CS-WFM were immersed in 120 mL of distilled water at room temperature. At fixed intervals, 10 mL of leachate (containing released contents) was removed from it for determination and the same volume of distilled water was added to the original solution to ensure that its total volume remained constant. Measurements were made at 1 h intervals during 0 to 12 h of immersion and at 12 h intervals during 12 h to 240 h of immersion. A UV-visible spectrophotometer (TU-1810, Beijing Purkinje General Instrument co., LTD, Beijing, China) was used to determine the concentration of carbendazim in the leachate corresponding to different time points at 282 nm according to the Chinese standard GB/T5009.188–2003 [40].

The commonly used first-order kinetic model (Equation (3)), Higuchi kinetic model (Equation (4)) and Ritger–Peppas kinetic model (Equation (5)) were fitted to the experimental data by Origin2024. The R^2 values were compared to explore the release mechanism of carbendazim in water. The equations for each model are given below:

$$ln(1 - M_t/M_\infty) = -kt. \quad (3)$$

$$M_t/M_\infty = kt^{\frac{1}{2}}. \quad (4)$$

$$M_t/M_\infty = kt^n. \quad (5)$$

where M_t is the cumulative release of the drug at time t (μg), M_∞ is the total amount of carbendazim added (μg), M_t/M_∞ is the percentage of cumulative release of the drug (%), t is the controlled release time (h), k is the rate constant of the drug release process, and n is the parameter, characterizing the release mechanism.

2.7. In Vitro Antifungal Properties

The *F. solani* used in this study was provided by JiaMuSi Branch of Heilongjiang Academy of Agricultural Science. It was routinely cultured on potato dextrose agar (PDA) plates at 25 °C in the dark and then cooled to 4 °C to be used as a fungal source in the in vitro activities of film and pot experiment.

In vitro antifungal activity of C-CS-WFM was evaluated by the method where the C-CS-WFM disc (diameter of 6 mm, sterilized under ultraviolet light for 20 min) was placed in the center of PDA plates, and 4 mycelium plugs of *F. solani* were placed symmetrically around it, and then was cultured at 25 °C in the dark. The blank WFM without carbendazim was as control. Three repeats were tested for each group of experiments.

2.8. Pot Experiment

The pot experiment used Heinong 84 soybean seeds, which were sterilized with 10% sodium hypochlorite. Flower cultivate soil which included native soil, substrate, peat soil and perlite, sterilized at 120 °C for 2 h, filled the pots. The soil surface was covered with WFM and C-CS-WFM. A perforator created 5 holes in the mulch. Subsequently, the soybeans were carefully sowed into the holes. A pot without mulch served as the control group. Three repeats were tested, with 5 plants included in each repeat.

After the soybeans sprouted, a certain amount of diluted spore suspension was accurately aspirate using a pipette. The spore suspension was slowly dripped into the soil around the roots of the soybean plants carefully, making every effort to evenly distribute the spore suspension around the root system (10 milliliters of root irrigation per soybean plant). After the operation was completed, the spore suspension was gently covered with a thin layer of soil to prevent it from being exposed to the air and evaporating or being splashed out. The spore suspension was produced as follows: Take the strains cultivated on PDA for 7 days in the 2.8 bacterial inhibition test. Transfer 5 mL of sterile water to the medium plate with a pipette gun, then scrape the plate colonies with a sterilized inoculation spatula in a sterile centrifugal tube and replenish the tube with 10 mL of sterile water, screw the cap on the tube, shake and mix well, and then vibrate well in a shaker to make sure that the spores are uniform and have no obvious stratification. Then, filter it through gauze to make the spore suspension. This was configured to 1×10^8/mL with saline. After 20 d of infection, we started to investigate the incidence of the disease by measuring the root dry weight of soybeans, the stem length of soybean soybeans, the stem fresh weight of soybeans, grading the disease index of root rot with reference to the standards GB/T17980.88-2004 [41] and calculating the disease index and incidence of soybean root rot as shown in Equations (6) and (7).

$$y = \sum[(n_i \times a_i)/(n \times a)] \times 100 \quad (6)$$

$$x = (n_j/n) \times 100\% \quad (7)$$

where y is the soybean root rot disease index, n_i is the number of diseased plants at all levels of soybean, a_i is the value of soybean root rot disease level, n is the total number of plants investigated, a is the highest value of soybean root rot disease level, x is the incidence rate of soybean root rot disease (%), and n_j is the number of diseased plants in soybean.

The test results were also statistically analyzed using Design Expert 6.0 (Stat-Ease Inc., Minneapolis, MN, USA).

3. Results

3.1. Optimization of C-CS-WFM Performance

3.1.1. Experimental Results and Regression Model

The BBD experimental design and results are shown in Table 2.

Table 2. Experimental design and results.

Run	Independent Variables			Dependent Variables		
	Chitosan Concentration (%)	Wet Film Thickness (μm)	Carbendazim Loading (g/m^2)	Dry Tensile Strength (kN/m)	Wet Tensile Strength (kN/m)	Air Permeance (μm/(Pa·s))
1	1.50	15.00	0.10	3.50	0.78	2.91
2	2.50	15.00	0.10	3.61	0.81	2.39
3	1.50	25.00	0.10	3.51	0.77	2.30
4	2.50	25.00	0.10	3.70	0.78	1.98
5	1.50	20.00	0.05	3.51	0.76	2.99
6	2.50	20.00	0.05	3.67	0.85	2.38
7	1.50	20.00	0.15	3.49	0.79	2.68
8	2.50	20.00	0.15	3.59	0.75	2.38
9	2.00	15.00	0.05	3.47	0.85	2.36
10	2.00	25.00	0.05	3.52	0.81	1.71
11	2.00	15.00	0.15	3.36	0.79	2.23
12	2.00	25.00	0.15	3.49	0.79	1.60
13	2.00	20.00	0.10	3.67	0.80	2.00
14	2.00	20.00	0.10	3.65	0.81	2.02
15	2.00	20.00	0.10	3.66	0.81	2.02
16	2.00	20.00	0.10	3.64	0.82	2.04
17	2.00	20.00	0.10	3.66	0.80	2.05

The results of the experiments were analyzed, and the quadratic equation models of Y_1 (DTS), Y_2 (WTS), and Y_3 (AP) were significant ($p < 0.0001$). F-tests were performed at a confidence level of 0.05, and the non-significant terms were excluded to obtain the regression models of each objective function as shown in Equations (8)–(10).

$$Y_1 = 1.92737 - 0.02\,X_1 + 0.12658\,X_2 + 6.17895\,X_3 - 0.00358947\,X_2^2 - 41.89474\,X_3^2 + 0.008\,X_1\,X_2 + 0.08\,X_2\,X_3. \quad (8)$$

$$Y_2 = 0.29917 + 0.51472\,X_1 - 0.006\,X_2 + 1.425\,X_3 - 0.090556\,X_1^2 - 1.3\,X_1\,X_3 + 0.04\,X_2\,X_3. \quad (9)$$

$$Y_3 = 11.6995 - 9.0955\,X_1 + 0.1197\,X_2 - 13.715\,X_3 + 1.987\,X_1^2 - 0.00543\,X_2^2 + 30.7\,X_3^2 + 0.02\,X_1\,X_2 + 3.1\,X_1\,X_3. \quad (10)$$

where X_1 is the chitosan concentration (%), X_2 is the wet film thickness (μm), and X_3 is the carbendazim loading (g/m^2).

Analysis of variance (ANOVA) was used to ensure the accuracy of the model, and the results are shown in Table 3. As can be seen from Table 3, the p value of the regression term for each indicator was <0.05, indicating that the regression equation was highly significant; the p value of the fitted term was >0.05, meaning that the model was significant.

Table 3. ANOVA of regression model.

Source		Sum of Square	df	Mean Square	F Value	*p* Value
Dry tensile strength	Model	0.15	9	0.02	66.10	<0.0001
	Residual	1.72×10^{-3}	7	2.4×10^{-4}		
	Lack of Fit	1.20×10^{-3}	3	4.00×10^{-4}	3.08	0.1530
	Pure Error	5.20×10^{-4}	4	1.30×10^{-4}		
	Cor Total	0.15	16			
Wet tensile strength	Model	0.01	9	1.28×10^{-3}	29.43	<0.0001
	Residual	3.05×10^{-4}	7	4.36×10^{-5}		
	Lack of Fit	2.50×10^{-5}	3	8.33×10^{-6}	0.12	0.9442
	Pure Error	2.80×10^{-4}	4	7.00×10^{-5}		
	Cor Total	0.01	16			
Air permeance	Model	2.25	9	0.25	145.60	<0.0001
	Residual	0.012	7	1.71×10^{-3}		
	Lack of Fit	9.67×10^{-3}	3	3.23×10^{-3}	5.56	0.0654
	Pure Error	2.32×10^{-3}	4	5.80×10^{-4}		
	Cor Total	2.26	16			

3.1.2. Influence of Test Factors on DTS

Figure 2a illustrates the influence of wet film thickness and chitosan concentration on the DTS of the film. At a carbendazim loading of 0.1 g/m^2, DTS is directly proportional to chitosan concentration, with higher concentrations leading to greater DTS. The -NH_2 groups in chitosan form hydrogen bonds with the phenolic -OH groups in pulp fibers effectively enhancing the dry tensile strength. The higher the chitosan concentration, the more hydrogen bonds are formed, resulting in greater DTS [42]. The impact of wet film thickness on DTS is such that when the thickness is less than 20 μm, the DTS increases with the increase in wet film thickness. However, when the wet film thickness exceeds 20 μm, the DTS decreases as the thickness increases further. An increase in wet film thickness can increase the amount of chitosan applied to the film surface, thereby increasing the interaction between chitosan and fibers. However, an excessively thick wet film may lead to an uneven film layer after drying, causing the DTS to decrease after reaching a certain level [43]. The influence of chitosan concentration on DTS is greater than that of wet film thickness, with the maximum value occurring at a chitosan concentration of 2.5% and a wet film thickness of 20 μm.

Figure 2b shows the effect of wet film thickness and carbendazim loading on DTS. At a chitosan concentration of 2%, the influence of wet film thickness on DTS first increases and then decreases. The effect of carbendazim loading on DTS also follows an increasing then decreasing trend. When the carbendazim loading is below 0.1 g/m^2, there is a positive correlation between fungicide loading and DTS, with the DTS increasing as the carbendazim loading increases. When the fungicide loading is above 0.1 g/m^2, there is a negative correlation, with the DTS decreasing as the carbendazim loading increases. Below a loading of 0.1 g/m^2, the mixture of fungicide and chitosan can fill the gaps between fibers, slightly enhancing the DTS of the film. Above a loading of 0.1 g/m^2, an excess of carbendazim can lead to stress concentration, which in turn reduces the DTS of C-CS-WFM [44]. The influence of loading on DTS is greater than that of wet film thickness, with the maximum value occurring at a loading of 0.1 g/m^2 and a wet film thickness of 20 μm.

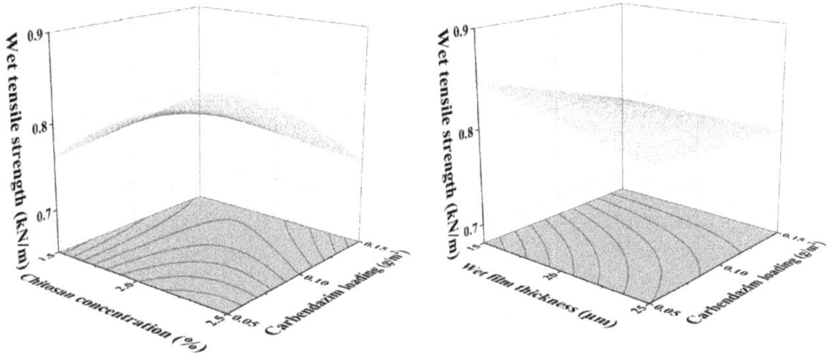

(a) (b)

Figure 2. Three-dimensional (3D) and two-dimensional (2D) response plots of DTS: (**a**) Influence of chitosan concentration and wet film thickness with 0.1 g/m^2 loading on DTS; (**b**) influence of wet film thickness and carbendazim loading with 2% chitosan concentration on DTS.

3.1.3. Influence of Test Factors on WTS

Figure 3a displays the impact of chitosan concentration and carbendazim loading on the WTS of the film. At a constant wet film thickness of 20 μm, it is observed that WTS is directly proportional to chitosan concentration, with higher concentrations resulting in greater WTS [32]. Conversely, WTS is inversely proportional to the amount of carbendazim loading, with higher loadings leading to lower WTS. Chitosan, when applied to the film surface, forms a dense hydrophobic layer that enhances the water resistance of the straw fiber film. Additionally, the chitosan molecules, containing polar groups such as amino and hydroxyl, interact with the cellulose molecules in the straw fiber film through hydrogen bonding. This molecular interaction strengthens the binding between fibers, thereby improving the WTS of the straw fiber film. The effect of fungicide on WTS is similar to its effect on DTS. The influence of chitosan concentration on wet tensile strength is greater than that of carbendazim loading. The maximum WTS is achieved at a chitosan concentration of 2.5% and a carbendazim loading of 0.05 g/m^2.

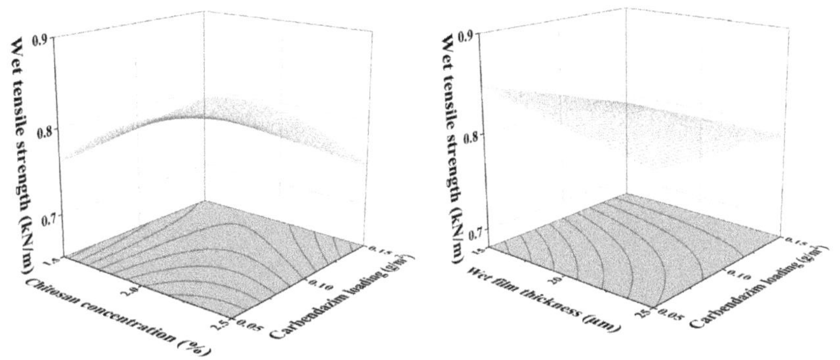

(a) (b)

Figure 3. Three-dimensional (3D) and two-dimensional (2D) response plots of WTS: (**a**) Influence of chitosan concentration and carbendazim loading with 20 μm wet film thickness on WTS; (**b**) influence of wet film thickness and carbendazim loading with 2% chitosan concentration on WTS.

Figure 3b illustrates the effects of carbendazim loading and wet film thickness on the WTS of the film. At a chitosan concentration of 2%, it is evident that WTS is negatively correlated with both carbendazim loading and wet film thickness: higher carbendazim loading and thicker wet film both result in lower WTS. The influence of wet film thickness on WTS is similar to its impact on DTS. The effect of carbendazim loading on WTS is more significant than that of wet film thickness. The maximum WTS is achieved when the wet film thickness is 10 μm and the carbendazim loading is 0.05 g/m^2.

3.1.4. Influence of Test Factors on AP

Figure 4a shows the impact of wet film thickness and chitosan concentration on the AP of the film. At a constant carbendazim loading of 0.1 g/m^2, AP decreases as the wet film thickness increases, indicating a negative correlation. Regarding chitosan concentration, AP is negatively correlated when the concentration is below 2%, meaning AP decreases with increasing chitosan concentration. Beyond 2% chitosan concentration, AP increases with concentration, showing a positive correlation. Below 2% chitosan, the increased concentration fills the gaps between wheat straw fibers, forming a dense layer that reduces AP [45,46]. Above 2%, the higher concentration leads to a less fluid composite film, causing uneven coating and increasing AP. The influence of chitosan concentration on AP is greater than that of wet film thickness, with the minimum AP occurring at a chitosan concentration of 2.0% and a wet film thickness of 25 μm.

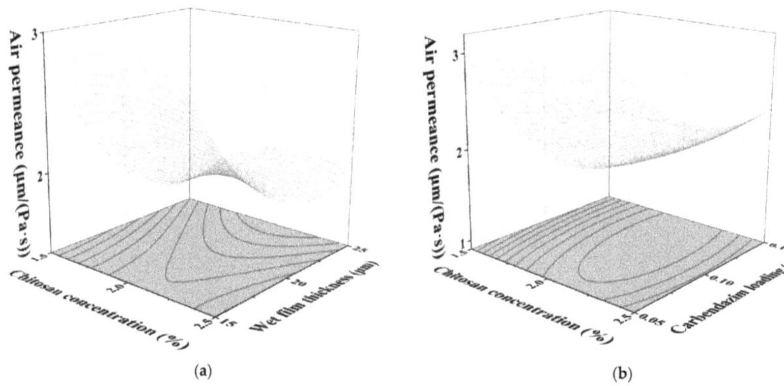

Figure 4. Three-dimensional (3D) and two-dimensional (2D) response plots of AP: (**a**) Influence of chitosan concentration and wet film thickness with 0.1 g/m^2 loading on AP; (**b**) influence of chitosan concentration and carbendazim loading with 20 μm wet film thickness on AP.

Figure 4b examines the effects of chitosan concentration and carbendazim loading on AP. At a wet film thickness of 20 μm, AP is negatively correlated with chitosan concentration below 2%, decreasing as chitosan concentration increases. Above 2% chitosan concentration, AP increases with concentration. Carbendazim loading also negatively correlates with AP. In C-CS-WFM, as carbendazim is added to the chitosan matrix as filler, its particles can block the pores in the chitosan film, which increases the gas diffusion path length [47]. Consequently, an increase in carbendazim loading amount raises the particle density, causing a gradual reduction in AP. The impact of chitosan concentration on AP is significantly greater than that of carbendazim loading. The minimum AP is observed at a chitosan concentration of 2% and a carbendazim loading of 1.5 g/m^2.

3.1.5. Optimization and Model Verification

According to the requirements for the use of C-CS-WFM, the process parameters are optimized to ensure that the DTS is not less than 3.5 kN/m, the WTS is not less than 0.8 kN/m, and the AP does not exceed 2.1 μm/(Pa·s). The optimization results show that the best combination of process parameters is as follows: a chitosan concentration of 1.83–2.39%, a wet film thickness of 18–24 μm, and a carbendazim loading of 0.05–0.12 g/m^2. The optimization analysis results are shown in Figure 5.

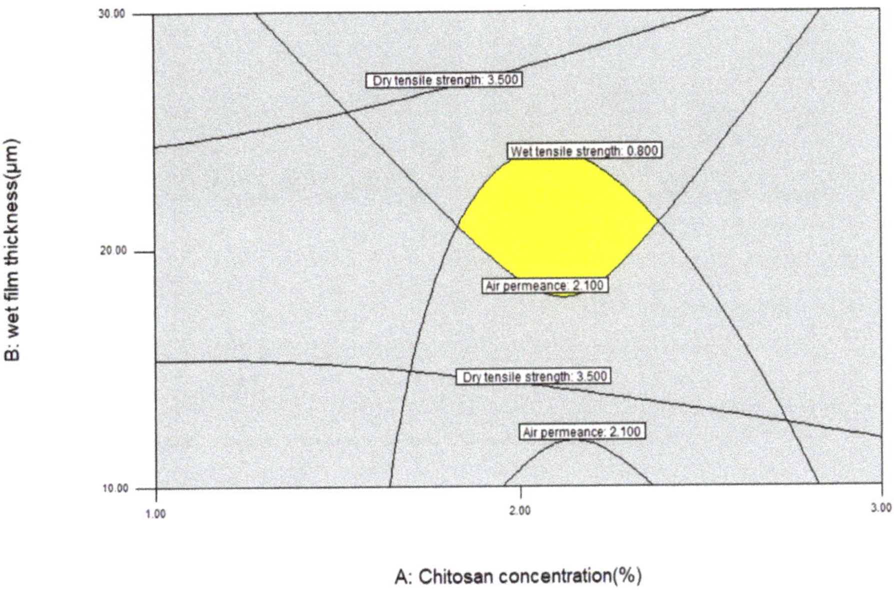

Figure 5. Optimization results.

To validate the regression models' accuracy, we prepared C-CS-WFM samples using an optimized set of parameters: chitosan concentration of 2%, wet film thickness of 20 μm, and carbendazim loading of 0.1 g/m^2. The resulting DTS, WTS, and AP of the C-CS-WFM samples were tested and are detailed in Table 4. The data indicate that the measured values for DTS, WTS, and AP closely matched the regression model estimates. Additionally, the same properties were tested for WFM samples and are also presented in Table 4. The results demonstrate that the optimized C-CS-WFM outperformed WFM, with a 11.4% increase in DTS, a 14.9% increase in WTS, and a 15.6% decrease in AP.

Table 4. Mechanical properties and air permeance of different film samples.

Item		DTS (kN/m)	WTS (kN/m)	AP (μm/(Pa·s))
WFM		3.27 ± 0.13	0.70 ± 0.04	2.40 ± 0.32
C-CS-WFM	predicted value	3.66	0.81	2.04
	measured value	3.64 ± 0.01	0.80 ± 0.01	1.99 ± 0.02

SEM analysis was conducted to examine the microstructures of WFM and the optimal C-CS-WFM, with results depicted in Figure 6. Figure 6a shows that the wheat straw fibers exhibit a obvious fibrillation and under the influence of additives, the fibers are tightly bound, creating a network structure with numerous pores. The morphology of the optimal C-CS-WFM is presented in Figure 6b, where the chitosan/carbendazim mixture not only fills the interfiber pores but also coats the fiber surfaces, resulting in a smoother surface for C-CS-WFM compared to WFM.

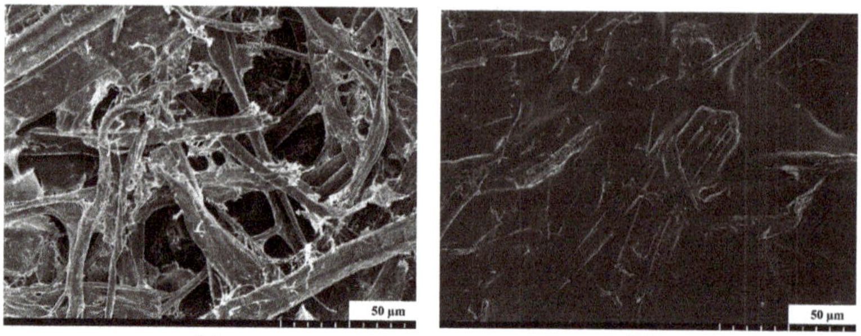

(a) (b)

Figure 6. SEM image: (**a**) WFM; (**b**) C-CS-WFM.

3.2. FT-IR Spectroscopy

FT-IR spectra of chitosan, carbendazim, and the carbendazim/chitosan composite coating are shown in Figure 7. Chitosan, due to the abundance of O-H groups in its molecular structure, exhibits a broad absorption band from 3600 to 3200 cm^{-1}, which overlaps with the weak N-H stretching vibration peaks of the primary amine group. Distinct stretching vibration absorption peaks are observed at 2932 cm^{-1} and 2875 cm^{-1}, corresponding to the C-H stretching vibrations in chitosan. Additionally, a bending vibration absorption peak is observed at 1597 cm^{-1}, which is attributed to the -NH_2 group [48].

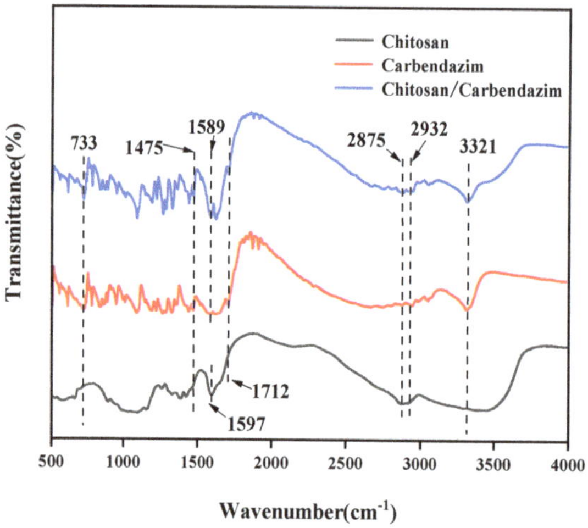

Figure 7. FT-IR spectra.

In contrast, the carbendazim molecule, which lacks hydroxyl groups, shows a relatively narrower N-H absorption band in the 3200–3400 cm^{-1} range compared to chitosan. A moderately strong and sharp peak at 3321 cm^{-1} is attributed to the N-H stretching vibration of the secondary amine group in carbendazim. The peaks at 1475 cm^{-1} and 1589 cm^{-1} are skeletal vibration peaks of the benzene ring, where the 1475 cm^{-1} peak is absent in chitosan. The 1589 cm^{-1} peak is similar to the bending vibration absorption peak of chitosan at 1597 cm^{-1}, further highlighting the differences between the two materials. Furthermore, the C=O stretching vibration peak at 1712 cm^{-1} and peak at 733 cm^{-1} corresponding to the absorption of ortho-disubstituted benzene are distinctive features of carbendazim that clearly differentiates it from chitosan [49].

The spectrum of the carbendazim/chitosan composite coating is essentially the result of the overlapping peaks from both carbendazim and chitosan, with no new characteristic peaks appearing. Moreover, the characteristic peaks of each component do not show significant shifts. For example, the broad absorption band in the 3600 to 3200 cm^{-1} range, which results from the overlap of O–H and primary amine N–H stretching vibrations, is observed, similar to chitosan. Additionally, the N–H stretching vibration peak of the secondary amine group in carbendazim remains at 3321 cm^{-1}. These observations indicate that carbendazim has been successfully incorporated into chitosan without any chemical interaction, reinforcing that the two components retain their individual characteristics in the composite coating.

3.3. Drug Release Kinetics

The in vitro release profiles of carbendazim from C-WFM and C-CS-WFM are depicted in Figure 8. For C-WFM, the initial burst release within the first 0–3 h accounted for a 72% cumulative release rate, primarily due to carbendazim loaded on the surface of C-WFM, which quickly dissolved upon immersion in water. In the subsequent 3–24 h, only an additional 6% was released, attributed to carbendazim residing in the fiber pores of the film, with its diffusion slowed by the fibers. Within 72 h, C-WFM achieved a total of 78% release; the unreleased portion may be adsorbed by the micropores on the wheat straw fiber surface, as the fibrous medium itself has strong adsorption capacity within its micropores, which is not easily desorbed, and complete release might require the degradation of the wheat straw fibers [50]. In contrast to C-WFM, the release profile of C-CS-WFM showed a significant change. During the initial burst release phase within 3 h, the cumulative release was only 26%, a 64% decrease compared to the absence of chitosan. Over the next 3–48 h, carbendazim release followed an approximate linear relationship with time, indicating that the addition of chitosan effectively controlled the release of carbendazim. The carbendazim released during the burst phase was adsorbed and retained near the C-CS surface. Between 3 and 48 h, a constant release rate of 1.1 mg/(m^2·h) was observed, which reflected the sustained release associated with the erosion of chitosan. C-CS-WFM achieved an 84% cumulative release within 72 h, surpassing C-WFM's 78%, suggesting that the surface coating method on the film can enhance the effective utilization rate of the fungicide.

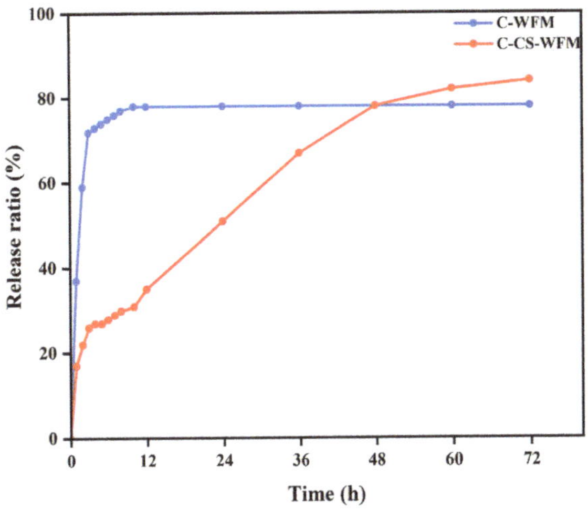

Figure 8. Carbendazim sustained-release curve.

To explore the carbendazim sustained-release mechanism of C-CS-WFM, we applied the first-order kinetic model, Higuchi kinetic model, and Ritger–Peppas kinetic model to fit the carbendazim release data, with the results summarized in Table 5. The determination coefficients (R^2) for these models were 0.9119, 0.9734, and 0.9831, respectively, with the Ritger–Peppas kinetic model showing the highest R^2 value. In this model, the exponent n characterizes the mechanism of drug release. The n value of 0.46 obtained in this study, which is greater than 0.45, indicates that the carbendazim sustained release is controlled by a combination of carbendazim diffusion and carbendazim/chitosan coating erosion [51].

Table 5. Different models to fit the data.

Model	R^2	Modify R^2
The first-order kinetic model	0.9119	0.9056
Higuchi kinetic model	0.9734	0.9735
Ritger–Peppas kinetic model	0.9831	0.9819

To further confirm the sustained-release mechanism of carbendazim in C-CS-WFM, the microstructure of C-CS-WFM after 10 days of immersion in 120 mL of distilled water at room temperature was observed. As shown in Figure 9, the surface of the film exhibited noticeable grooves and gaps compared to the primary C-CS-WFM. This may be attributed to the C-CS film undergoing erosion in the aqueous medium, disrupting the continuous film structure on the surface of C-CS-WFM. The erosion of the C-CS film increased its contact area with water, facilitating the release of carbendazim. Additionally, the rate at which carbendazim dissolves from the film is related to the number of hydrophilic groups in the film. Chitosan, which contains a significant amount of hydrophilic groups (amino and hydroxyl groups), allows for faster water penetration. Under the drive of carbendazim concentration gradient, the embedded carbendazim molecules dissolve and are released into the water more quickly, which also contributes to the sustained release of carbendazim. Therefore, SEM analysis of C-CS-WFM also demonstrates that the release of carbendazim from C-CS-WFM is the result of the combined effects of erosion of the C-CS film and physical diffusion.

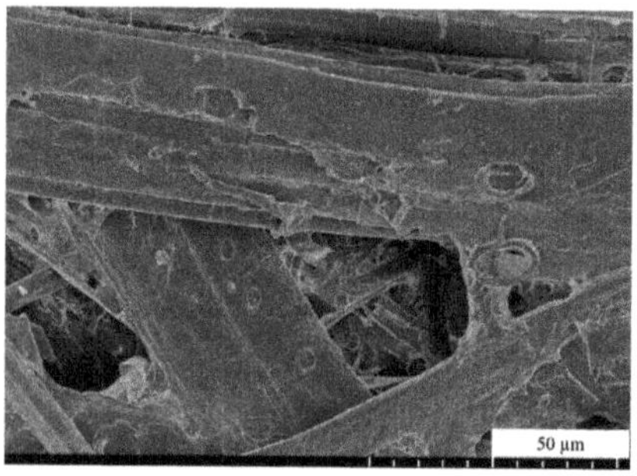

Figure 9. SEM image of C-CS-WFM after 10 days of immersion.

3.4. Antifungal Properties and Pot Experiment

As shown in Figure 10, the blank WFM had no antifungal activity against *F. solani*, while the inhibition zones were formed around the C-CS-WFM. Therefore, the C-CS-WFM had antifungal activity because of the release of carbendazim.

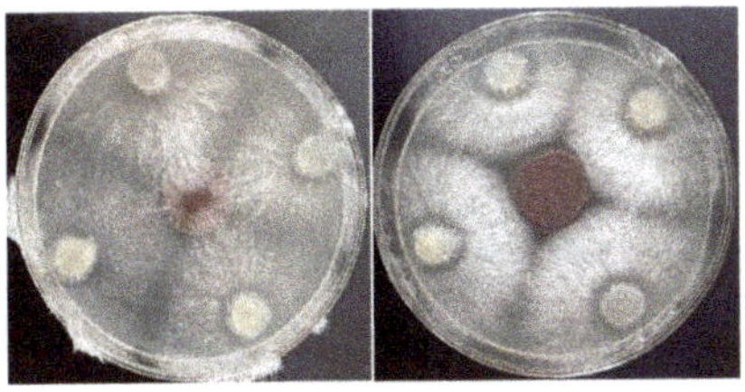

(a) (b)

Figure 10. The antifungal activity: (**a**) WFM; (**b**) C-CS-WFM.

The potential application of C-CS-WFM was further evaluated by a pot experiment. Figure 11a presents the morphology of soybean plants and their roots after 20 days of fungal treatment for C-CS-WFM, WFM, and CK (no mulch film) treatments. The C-CS-WFM treatment group showed significantly better plant morphology, root length, and number of root hairs compared to the other two groups, with WFM outperforming CK. CK exhibited clear symptoms of root rot, with curved, weak plant growth, brown spots on the roots, and fewer root hairs. As shown in Figure 11b–d, both C-CS-WFM and WFM treatments significantly increased the stem length, stem fresh weight, and root dry weight of soybeans, primarily because the film coverage provided better growing conditions that promoted root growth. Notably, the root dry weight of soybeans treated with C-CS-WFM was significantly higher than that of the WFM group, indicating that the growth of seedlings after fungal infection was significantly impeded. The C-CS-WFM effectively suppressed the occurrence of root rot through the sustained release of carbendazim, thereby promoting the growth of soybean plants. Figure 11e,f shows the root rot incidence of soybeans treated with three different methods over 20 days in the presence of *F. solani*. The CK group had the highest disease incidence at 100%, while WFM and C-CS-WFM groups had rates of 89% and 39%, respectively. The disease indices were 64 for CK, 59 for WFM, and 38 for C-CS-WFM, indicating that the antimicrobial film significantly reduced the probability of soybean plants getting sick through the sustained release of carbendazim.

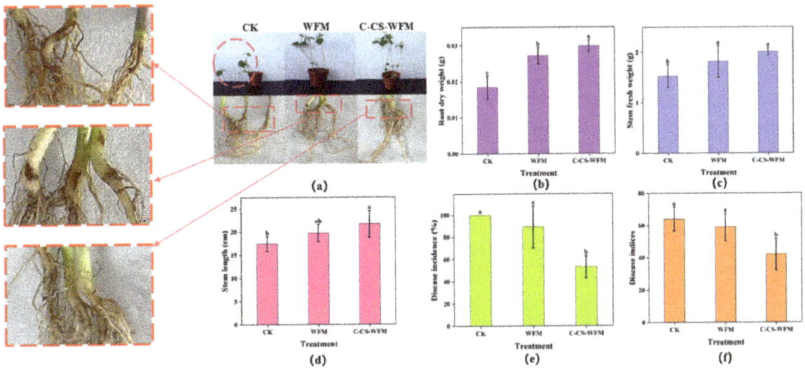

Figure 11. Pot experiment of control soybean root rot: (**a**) Photograph of the incidence of soybean after 20 d treatments with CK: sterile flower cultivate soil and no mulch film; WFM: *F. solani* containing flower cultivate soil and covered WFM; C-CS-WFM: *F. solani* containing flower cultivate soil and covered C-CS-WFM. The growth conditions of soybean: (**b**) Root dry weight; (**c**) Stem fresh weight; (**d**) Stem length. (**e**) Soybean root rot disease incidence and (**f**) Soybean root rot disease indices. Graph represents mean ± SE followed by same letter is not significantly different at $p \leq 0.05$.

4. Discussion

Due to the unique fiber composition and porous structure of plant fiber mulches (also known as paper mulches), their design must consider both the performance requirements during application and the functionality after being laid on the ground [52]. Thus, designing straw-fiber-based mulches is a complex, integrated decision-making process.

4.1. Physical Properties of C-CS-WFM

The core physical properties of straw-fiber-based mulches—DTS, WTS, and AP—directly affect their suitability and functional performance in agricultural applications. DTS reflects the mulch's resistance to tearing under dry conditions and is a critical metric for evaluating mechanical performance, ensuring smooth application and durability in the field. WTS, which describes the mulch's tensile resistance in wet conditions, is crucial for maintaining integrity and functionality under rain or irrigation. This property is a key consideration in determining whether biodegradable mulches can replace plastic mulches [53]. AP, representing the ability of the mulch to allow gas exchange, is closely related to its fiber structure and porosity. It influences soil–air interactions, affecting root respiration and soil moisture evaporation [54]. The application of chitosan coating enhances fiber bonding, forms hydrogen bonds, and improves the fiber network structure, thereby increasing both DTS and WTS while reducing AP. As a natural biodegradable material, chitosan does not compromise the eco-friendly nature of straw-fiber-based mulches [46].

4.2. Drug Release Performance of C-CS-WFM

Functionalizing mulches is an effective way to reduce agricultural costs, such as by incorporating fungicides onto the mulch surface. Compared to conventional fungicide delivery methods, chitosan coating technology effectively controls fungicide release rates while improving utilization efficiency. C-CS-WFM reduced the initial burst release to 26% within the first 3 h, a 64% decrease compared to the 72% observed for C-WFM. C-CS-WFM maintained a sustained-release rate of 1.1 mg/(m^2·h) over 3–48 h. In this study, the cumulative release of C-CS-WFM reached 84% by 72 h, surpassing the 78% of C-WFM, indicating improved fungicide utilization efficiency.

4.3. Cost and Environmental Impact

Compared to traditional polyethylene (PE) mulches, the cost of biodegradable mulches, including those made from PLA, polybutylene adipate terephthalate (PBAT), or plant fibers, remains a critical barrier to widespread adoption. For example, Marí et al. reported that increasing subsidy rates by up to 50.1% of the market price could make biodegradable films and paper mulches economically viable alternatives to PE [55]. Thus, reducing costs and enhancing functionality are essential for advancing biodegradable mulch development.

Raw material costs constitute the primary component of mulch production expenses. In this study, wheat straw was used as the main raw material, costing approximately 50~100 USD/ton in China (market price) or 185 USD/ton in Canada [56], significantly lower than the cost of PLA or PBAT (approximately 2000 USD/ton) [57,58]. Moreover, utilizing agricultural waste offers both economic and environmental benefits.

As a natural biodegradable material, the addition of chitosan does not compromise the environmental performance of straw-fiber-based mulches. Applying a chitosan–fungicide mixture to the mulch surface not only reduces drift losses from spraying and leaching losses from soil drenching but also ensures sustained fungicide release. This prolongs fungicide effectiveness, reduces application frequency, and minimizes overall usage, mitigating the adverse environmental impact of chemical fungicides.

5. Conclusions

Chitosan and carbendazim, as a layer, were applied to WFM via a bar coating method to create C-CS-WFM. This novel film not only excels in mechanical and barrier properties but also exhibits drug-controlled release characteristics.

A three-factor Box–Behnken design was employed to systematically investigate the effects of three key factors—chitosan concentration, wet film thickness, and carbendazim loading—on the performance of the WFM film. The optimization process identified the ideal parameter ranges as follows: a chitosan concentration of 1.83–2.39%, a wet film thickness of 18–24 μm, and a carbendazim loading of 0.05–0.12 g/m^2. Under these optimized conditions, the resulting C-CS-WFM film exhibited desirable performance metrics, including a DTS of not less than 3.5 kN/m, a WTS of at least 0.8 kN/, and an AP not exceeding 2.1 μm/(Pa·s). This demonstrates the film's robustness and suitability for practical applications.

A chitosan/carbendazim coating layer was added on top of the optimized WTS using the bar coating method. FT-IR analysis confirmed carbendazim and chitosan did not undergo any chemical interactions in the chitosan/carbendazim coating. The in vitro release profiles revealed that the incorporation of chitosan into C-CS-WFM significantly improved the controlled release and utilization efficiency of carbendazim compared to C-WFM. C-CS-WFM reduced the initial burst release to 26% within the first 3 h, a 64% decrease compared to C-WFM's 72%. Furthermore, C-CS-WFM maintained a sustained-release rate of 1.1 mg/(m^2·h) over 3–48 h, driven by chitosan erosion, while C-WFM exhibited a much slower additional release of only 6% during the same period. By 72 h, C-CS-WFM achieved an 84% cumulative release, surpassing C-WFM's 78%, indicating an improvement in fungicide utilization efficiency. The carbendazim release data fit the Ritger–Peppas kinetic model better than the first-order and Higuchi kinetic models. An exponent n value of 0.46 from the model suggests that the sustained release is governed by a combination of carbendazim diffusion and chitosan/carbendazim coating erosion. This finding was corroborated by SEM analysis, which showed that after 10 days of immersion in 120 mL of distilled water at room temperature, the C-CS film underwent erosion, disrupting the continuous film structure on the surface of the C-CS-WFM.

The antifungal test showed that C-CS-WFM effectively suppressed *F. solani* infections, as indicated by clear inhibition zones, unlike blank WFM, which lacked antifungal activity. This confirmed its antifungal properties and ability to promote soybean growth through carbendazim release. In pot experiments, C-CS-WFM significantly improved soybean growth compared to WFM and CK, with healthier plants, longer roots, and denser root hairs. C-CS-WFM achieved higher stem length, stem fresh weight, and root dry weight, demonstrating enhanced fungal resistance and better growth conditions.

In summary, this study highlights the successful integration of chitosan and a broad-spectrum fungicide to develop a high-performance film, offering an eco-friendly and sustainable approach to effectively control soybean root rot while promoting green agricultural practices.

Author Contributions: Conceptualization, S.L., Z.J. and H.C.; methodology, L.L.; software, Y.Z.; validation, Z.J., P.Z. and H.S.; formal analysis, S.L. and R.L.; investigation, Z.J., P.Z. and H.S.; resources, L.L. and Y.Z.; data curation, P.Z. and H.S.; writing—original draft preparation, S.L. and Z.J.; writing—review and editing, L.L., H.Y., R.L. and Y.Z.; visualization, Z.J.; supervision, S.L. and H.C.; project administration, H.C.; funding acquisition, S.L. All authors have read and agreed to the published version of the manuscript.

Funding: This research was funded by the Natural Science Foundation of Heilongjiang Province, China, grant number LH2021C030, the postdoctoral scientific research developmental fund of Heilongjiang Province, grant number LBH-Q21066, and University Nursing Program for Young Scholars with Creative Talents in Heilongjiang Province, grant number UNPYSCT-2016130.

Data Availability Statement: Data is contained within the article.

Conflicts of Interest: The authors declare no conflicts of interest.

References

1. Williamson-Benavides, B.A.; Dhingra, A. Understanding Root Rot Disease in Agricultural Crops. *Horticulturae* **2021**, *7*, 33. [CrossRef]
2. Liu, Y.; Wei, X.; Chang, F.; Yu, N.; Guo, C.; Cai, H. Distribution and Pathogenicity of Fusarium Species Associated with Soybean Root Rot in Northeast China. *Plant Pathol. J.* **2023**, *39*, 575–583. [CrossRef] [PubMed]
3. Martins, P.M.M.; Merfa, M.V.; Takita, M.A.; De Souza, A.A. Persistence in Phytopathogenic Bacteria: Do We Know Enough? *Front. Microbiol.* **2018**, *9*, 1099. [CrossRef] [PubMed]
4. Xi, X.; Fan, J.; Yang, X.; Liang, Y.; Zhao, X.; Wu, Y. Evaluation of the Anti-Oomycete Bioactivity of Rhizosphere Soil-Borne Isolates and the Biocontrol of Soybean Root Rot Caused by Phytophthora Sojae. *Biol. Control* **2022**, *166*, 104818. [CrossRef]
5. Rodriguez, M.C.; Sautua, F.; Scandiani, M.; Carmona, M.; Asurmendi, S. Current Recommendations and Novel Strategies for Sustainable Management of Soybean Sudden Death Syndrome. *Pest Manag. Sci.* **2021**, *77*, 4238–4248. [CrossRef]
6. Yang, Q.; Yang, X.; Huang, X.; Ye, W.; Wang, T.; Cheng, Z.; Shi, J.; Li, Y.; Xu, J.; He, Y. Seed Coating with Fungicide Causes a Beneficial Shift in Root-associated Microbiomes of Mature Soybean. *Soil Sci. Soc. Am. J.* **2023**, *87*, 43–62. [CrossRef]
7. Li, G.-B.; Wang, J.; Kong, X.-P. Coprecipitation-Based Synchronous Pesticide Encapsulation with Chitosan for Controlled Spinosad Release. *Carbohydr. Polym.* **2020**, *249*, 116865. [CrossRef] [PubMed]
8. Luo, J.; Gao, Y.; Liu, Y.; Huang, X.; Zhang, D.; Cao, H.; Jing, T.; Liu, F.; Li, B. Self-Assembled Degradable Nanogels Provide Foliar Affinity and Pinning for Pesticide Delivery by Flexibility and Adhesiveness Adjustment. *ACS Nano* **2021**, *15*, 14598–14609. [CrossRef] [PubMed]
9. Yang, F.; He, B.; Dong, B.; Zhang, G. Film-Straw Dual Mulching Improves Soil Fertility and Maize Yield in Dryland Farming by Increasing Straw-Degrading Bacterial Abundance and Their Positive Cooperation. *Agric. Ecosyst. Environ.* **2024**, *367*, 108997. [CrossRef]
10. Saberi Riseh, R. Advancing Agriculture through Bioresource Technology: The Role of Cellulose-Based Biodegradable Mulches. *Int. J. Biol. Macromol.* **2024**, *255*, 128006. [CrossRef] [PubMed]
11. Mansoor, Z.; Tchuenbou-Magaia, F.; Kowalczuk, M.; Adamus, G.; Manning, G.; Parati, M.; Radecka, I.; Khan, H. Polymers Use as Mulch Films in Agriculture—A Review of History, Problems and Current Trends. *Polymers* **2022**, *14*, 5062. [CrossRef]
12. Wen, S.; Dang, P.; Li, D.; Qin, X.; Siddique, K.H.M. Effects of Semi-Film and Full-Film Mulching on Soybean Growth, Biological Nitrogen Fixation and Grain Yield. *J. Agron. Crop Sci.* **2024**, *210*, e12724. [CrossRef]
13. Merino, D.; Simonutti, R.; Perotto, G.; Athanassiou, A. Direct Transformation of Industrial Vegetable Waste into Bioplastic Composites Intended for Agricultural Mulch Films. *Green Chem.* **2021**, *23*, 5956–5971. [CrossRef]
14. Wang, Q.; Duan, Y.; Huang, Y.; Teng, Y.; Li, C.; Tao, Y.; Lu, J.; Du, J.; Wang, H. Multifunctional Soybean Protein Isolate-Graft-Carboxymethyl Cellulose Composite as All-Biodegradable and Mechanically Robust Mulch Film for "Green" Agriculture. *Carbohydr. Polym.* **2024**, *323*, 121410. [CrossRef] [PubMed]
15. Sukwijit, C.; Seubsai, A.; Charoenchaitrakool, M.; Sudsakorn, K.; Niamnuy, C.; Roddecha, S.; Prapainainar, P. Production of PLA/Cellulose Derived from Pineapple Leaves as Bio-Degradable Mulch Film. *Int. J. Biol. Macromol.* **2024**, *270*, 132299. [CrossRef] [PubMed]
16. Parida, M.; Jena, T.; Mohanty, S.; Nayak, S.K. Advancing Sustainable Agriculture: Evaluation of Poly (Lactic Acid) (PLA) Based Mulch Films and Identification of Biodegrading Microorganisms among Soil Microbiota. *Int. J. Biol. Macromol.* **2024**, *269*, 132085. [CrossRef] [PubMed]
17. Kochkina, N.E.; Lukin, N.D. Structure and Properties of Biodegradable Maize Starch/Chitosan Composite Films as Affected by PVA Additions. *Int. J. Biol. Macromol.* **2020**, *157*, 377–384. [CrossRef] [PubMed]
18. Uyarcan, M.; Güngör, S.C. Improving Functional Properties of Starch-Based Films by Ultraviolet (UV-C) Technology: Characterization and Application on Minced Meat Packaging. *Int. J. Biol. Macromol.* **2024**, *282*, 137085. [CrossRef]
19. Xie, Z.; Xiong, Q.; Fang, Y.; Zhang, Q.; Liang, W.; Cheng, J.; Shang, W.; Zhao, W.; Zhao, J. Novel Biodegradable Composite Mulch Film Embedded with Temperature-Responsive Pesticide Microcapsules for Durable Control of Phytophthora Root Rot on Soybean. *ACS Sustain. Chem. Eng.* **2023**, *11*, 9868–9879. [CrossRef]
20. Othman, N.A.F.; Selambakkannu, S.; Seko, N. Biodegradable Dual-Layer Polyhydroxyalkanoate (Pha)/Polycaprolactone (Pcl) Mulch Film for Agriculture: Preparation and Characterization. *Energy Nexus* **2022**, *8*, 100137. [CrossRef]

21. Lv, Z.; Meng, X.; Liang, Q.; Jiang, T.; Sun, S.; Tan, Y.; Feng, J. A Biodegradable Oxidized Starch/Carboxymethyl Chitosan Film Coated with Pesticide-Loaded ZIF-8 for Tomato Fusarium Wilt Control. *Int. J. Biol. Macromol.* **2024**, *259*, 129249. [CrossRef] [PubMed]
22. Rehman, A.; Feng, J.; Qunyi, T.; Korma, S.A.; Assadpour, E.; Usman, M.; Han, W.; Jafari, S.M. Pesticide-Loaded Colloidal Nanodelivery Systems; Preparation, Characterization, and Applications. *Adv. Colloid Interface Sci.* **2021**, *298*, 102552. [CrossRef]
23. Saberi Riseh, R.; Vatankhah, M.; Hassanisaadi, M.; Kennedy, J.F. Macromolecules-Based Encapsulation of Pesticides with Carriers: A Promising Approach for Safe and Effective Delivery. *Int. J. Biol. Macromol.* **2024**, *269*, 132079. [CrossRef] [PubMed]
24. Zhou, L.; Wang, H.; Sun, F. Environmentally Responsive Mulch Films for Sustainable Pest Control. *Adv. Agric. Technol.* **2022**, *7*, 145–160.
25. Lv, Z.; Meng, X.; Sun, S.; Jiang, T.; Zhang, S.; Feng, J. Biodegradable Carboxymethyl Chitosan/Polyvinyl Alcohol Hymexazol-Loaded Mulch Film for Soybean Root Rot Control. *Agronomy* **2023**, *13*, 2205. [CrossRef]
26. Chen, G.; Cao, L.; Cao, C.; Zhao, P.; Li, F.; Xu, B.; Huang, Q. Effective and Sustained Control of Soil-Borne Plant Diseases by Biodegradable Polyhydroxybutyrate Mulch Films Embedded with Fungicide of Prothioconazole. *Molecules* **2021**, *26*, 762. [CrossRef] [PubMed]
27. Liang, W.; Zhao, Y.; Xiao, D.; Cheng, J.; Zhao, J. A Biodegradable Water-Triggered Chitosan/Hydroxypropyl Methylcellulose Pesticide Mulch Film for Sustained Control of Phytophthora Sojae in Soybean (*Glycine max* L. Merr.). *J. Clean. Prod.* **2020**, *245*, 118943. [CrossRef]
28. Zhao, C.S.; Zuo, P.; Wang, X.; He, Y.-Z.; Chen, H.-T.; Zhang, Y.; Li, L.-H. Parameter Optimization of a Biodegradable Agricultural Film Manufactured with Wheat Straw Fiber. *BioResources* **2022**, *17*, 2331–2346. [CrossRef]
29. Gumber, S.; Kanwar, S.; Mazumder, K. Properties and Antimicrobial Activity of Wheat-Straw Nanocellulose-Arabinoxylan Acetate Composite Films Incorporated with Silver Nanoparticles. *Int. J. Biol. Macromol.* **2023**, *246*, 125480. [CrossRef] [PubMed]
30. Li, R.; Wang, Z.; Dong, H.; Yang, M.; Sun, X.; Zong, Q.; Xu, Z. Lattice Boltzmann Modeling of the Effective Thermal Conductivity in Plant Fiber Porous Media Generated by Quartet Structure Generation Set. *Mater. Des.* **2023**, *234*, 112303. [CrossRef]
31. Li, Y.; Liu, C.; Gai, X.; Deng, C.; Wei, H.; Liu, Y.; Xiao, H. Mulch from Lignocellulose as Agricultural Plastic Alternative for Sustained-Release of Photosensitive Pesticide. *Int. J. Biol. Macromol.* **2024**, *255*, 128347. [CrossRef]
32. Francolini, I.; Galantini, L.; Rea, F.; Di Cosimo, C.; Di Cosimo, P. Polymeric Wet-Strength Agents in the Paper Industry: An Overview of Mechanisms and Current Challenges. *Int. J. Mol. Sci.* **2023**, *24*, 9268. [CrossRef] [PubMed]
33. Wang, S.; Jing, Y. Effects of a Chitosan Coating Layer on the Surface Properties and Barrier Properties of Kraft Paper. *BioResources* **2016**, *11*, 1868–1881. [CrossRef]
34. Zhou, Y.; Wu, J.; Zhou, J.; Lin, S.; Cheng, D. pH-Responsive Release and Washout Resistance of Chitosan-Based Nano-Pesticides for Sustainable Control of Plumeria Rust. *Int. J. Biol. Macromol.* **2022**, *222*, 188–197. [CrossRef]
35. Bakshi, P.S.; Selvakumar, D.; Kadirvelu, K.; Kumar, N.S. Chitosan as an Environment Friendly Biomaterial—A Review on Recent Modifications and Applications. *Int. J. Biol. Macromol.* **2020**, *150*, 1072–1083. [CrossRef]
36. *GB/T 24325-2009*; Pulps--Laboratory Beating-Valley Beater Method. Standardization Administration of China: Beijing, China, 2009.
37. *GB/T 12914-2018*; Paper and Board-Determination of Tensile Properties Constant Rate of Elongation Method (20 Mm/Min). Standardization Administration of China: Beijing, China, 2018.
38. *GB/T 465.2-2008*; Paper and Board-Determination of Tensilestrength After Immersion in Water. Standardization Administration of China: Beijing, China, 2008.
39. *GB/T 458-2008*; Paper and Board-Determination of Air Permeance. Standardization Administration of China: Beijing, China, 2008.
40. *GB/T 5009.188-2003*; Determination of Thiophanate-Methyl, Carbendazimin Vegetables and Fruits. Standardization Administration of China: Beijing, China, 2003.
41. *GB/T 17980.88-2004*; Pesticide—Guidelines for the Field Efficacy Trials (II)—Part 88: Fungicides Against Root Rot of Soybean. Standardization Administration of China: Beijing, China, 2004.
42. Tanpichai, S.; Witayakran, S.; Wootthikanokkhan, J.; Srimarut, Y.; Woraprayote, W.; Malila, Y. Mechanical and Antibacterial Properties of the Chitosan Coated Cellulose Paper for Packaging Applications: Effects of Molecular Weight Types and Concentrations of Chitosan. *Int. J. Biol. Macromol.* **2020**, *155*, 1510–1519. [CrossRef]
43. Tanpichai, S.; Srimarut, Y.; Woraprayote, W.; Malila, Y. Chitosan Coating for the Preparation of Multilayer Coated Paper for Food-Contact Packaging: Wettability, Mechanical Properties, and Overall Migration. *Int. J. Biol. Macromol.* **2022**, *213*, 534–545. [CrossRef] [PubMed]
44. Zang, Y.H.; Aspler, J.S. The Effect of Surface Binder Content on Print Density and Ink Receptivity of Coated Paper. *J. PULP Pap. Sci.* **1998**, *24*, 141–145.
45. Wang, S.; Jing, Y. Effects of Formation and Penetration Properties of Biodegradable Montmorillonite/Chitosan Nanocomposite Film on the Barrier of Package Paper. *Appl. Clay Sci.* **2017**, *138*, 74–80. [CrossRef]

46. Reis, A.B.; Yoshida, C.M.P.; Reis, A.P.C.; Franco, T.T. Application of Chitosan Emulsion As A Coating on Kraft Paper. *Polym. Int.* **2011**, *60*, 963–969. [CrossRef]
47. Quyen, D.T.M.; Adisak, J.; Rachtanapun, P. Relationship between Solubility, Moisture Sorption Isotherms and Morphology of Chitosan/Methylcellulose Films with Different Carbendazim Content. *J. Agric. Sci.* **2012**, *4*, p187. [CrossRef]
48. Sandhya; Kumar, S.; Kumar, D.; Dilbaghi, N. Preparation, Characterization, and Bio-Efficacy Evaluation of Controlled Release Carbendazim-Loaded Polymeric Nanoparticles. *Environ. Sci. Pollut. Res. Int.* **2016**, *24*, 926–937. [CrossRef]
49. Negi, G.; Pankaj; Srivastava, A.; Sharma, A. In Situ Biodegradation of Endosulfan, Imidacloprid, and Carbendazim Using Indigenous Bacterial Cultures of Agriculture Fields of Uttarakhand, India. *World Acad. Sci. Eng. Technol. Int. J. Biotechnol. Bioeng.* **2014**, *1*, 973–981.
50. Wang, R.; Liu, Y.; Lu, Y.; Liang, S.; Zhang, Y.; Zhang, J.; Shi, R.; Yin, W. Fabrication of a Corn Stalk Derived Cellulose-Based Bio-Adsorbent to Remove Congo Red from Wastewater: Investigation on Its Ultra-High Adsorption Performance and Mechanism. *Int. J. Biol. Macromol.* **2023**, *241*, 124545. [CrossRef] [PubMed]
51. Siepmann, J.; Peppas, N.A. Modeling of Drug Release from Delivery Systems Based on Hydroxypropyl Methylcellulose (HPMC). *Adv. Drug Deliv. Rev.* **2001**, *48*, 139–157. [CrossRef]
52. Haapala, T.; Palonen, P.; Korpela, A.; Ahokas, J. Feasibility of Paper Mulches in Crop Production —A Review. *Agric. Food Sci.* **2014**, *23*, 60–79. [CrossRef]
53. Li, A.; Zhang, J.; Ren, S.; Zhang, Y.; Zhang, F. Research Progress on Preparation and Field Application of Paper Mulch. *Environ. Technol. Innov.* **2021**, *24*, 101949. [CrossRef]
54. Saglam, M.; Sintim, H.Y.; Bary, A.I.; Miles, C.A.; Ghimire, S.; Inglis, D.A.; Flury, M. Modeling the Effect of Biodegradable Paper and Plastic Mulch on Soil Moisture Dynamics. *Agric. Water Manag.* **2017**, *193*, 240–250. [CrossRef]
55. Marí, A.I.; Pardo, G.; Cirujeda, A.; Martínez, Y. Economic Evaluation of Biodegradable Plastic Films and Paper Mulches Used in Open-Air Grown Pepper (*Capsicum annum* L.) Crop. *Agronomy* **2019**, *9*, 36. [CrossRef]
56. McKnight, C.; Qiu, F.; Luckert, M.; Hauer, G. Prices for a Second-generation Biofuel Industry in Canada: Market Linkages between Canadian Wheat and US Energy and Agricultural Commodities. *Can. J. Agric. Econ. Can. Agroecon.* **2021**, *69*, 337–351. [CrossRef]
57. Teixeira, L.V.; Bomtempo, J.V.; Oroski, F.D.A.; Coutinho, P.L.D.A. The Diffusion of Bioplastics: What Can We Learn from Poly(Lactic Acid)? *Sustainability* **2023**, *15*, 4699. [CrossRef]
58. Wang, B.X.; CortesPeña, Y.; Grady, B.P.; Huber, G.W.; Zavala, V.M. Techno-Economic Analysis and Life Cycle Assessment of the Production of Biodegradable Polyaliphatic–Polyaromatic Polyesters. *ACS Sustain. Chem. Eng.* **2024**, *12*, 9156–9167. [CrossRef]

Disclaimer/Publisher's Note: The statements, opinions and data contained in all publications are solely those of the individual author(s) and contributor(s) and not of MDPI and/or the editor(s). MDPI and/or the editor(s) disclaim responsibility for any injury to people or property resulting from any ideas, methods, instructions or products referred to in the content.

agronomy

MDPI

Article

Exploring the Efficacy of Four Essential Oils as Potential Insecticides against *Thrips flavus*

Yulong Niu [1,†], Tianhao Pei [1,†], Yijin Zhao [1,2,†], Changjun Zhou [3], Bing Liu [3], Shusen Shi [1], Meng-Lei Xu [4,*] and Yu Gao [1,*]

1 Key Laboratory of Soybean Disease and Pest Control (Ministry of Agriculture and Rural Affairs), College of Plant Protection, Jilin Agricultural University, Changchun 130118, China
2 Dalian City Investment Asset Management Co., Ltd., Dalian 116021, China
3 Daqing Branch, Heilongjiang Academy of Agricultrual Science, Daqing 163316, China
4 State Key Laboratory of Supramolecular Structure and Materials, College of Food Science and Engineering, Jilin University, Changchun 130062, China
* Correspondence: xumenglei@jlu.edu.cn (M.-L.X.); gaothrips@jlau.edu.cn (Y.G.)
† These authors contributed equally to this work.

Abstract: Plant essential oils are important alternatives in green integrated pest management. This study examined the chemical composition, bioactivity, and control efficacy of four Lamiaceae essential oils (EOs) against *Thrips flavus* Schrank in laboratory conditions with the goal of exploiting plant-derived insecticides to control *Thrips flavus*. The four EOs tested were marjoram oil (*Origanum majorana* L.), clary sage oil (*Salvia sclarea* L.), perilla leaf oil (*Perilla frutescens* (L.) Britt.), and spearmint oil (*Mentha spicata* L.). All these EOs exhibited a certain degree of insecticidal activity against *Thrips flavus*. The median lethal concentration (LC_{50}) was determined after treatment by the leaf-dipping method in laboratory bioassays, and its values were 0.41 mg/mL for marjoram oil, 0.42 mg/mL for clary sage oil, 0.43 mg/mL for perilla leaf oil, and 0.54 mg/mL for spearmint oil. In the pot experiment, the number of dead insects was recorded at 1, 3, and 7 days post-application, and the control efficacy of EOs against *Thrips flavus* was calculated. The concentration of 900.00 g a.i.$\cdot$hm^{-2} of spearmint oil was 100% lethal against *Thrips flavus* after treating potted plants for seven days. The Y-tube olfactometer method was used to test for the attraction or repellent response of EOs against *Thrips flavus*. The spearmint oil significantly attracted female adults in the olfactory test. Furthermore, gas chromatography–mass spectrometry (GC–MS) was used to examine the chemical composition of the EOs. Linalool (24.52%), isopropyl myristate (28.74%), (+)-limonene (32.44%), and (+)-carvone (70.3%) were their primary ingredients. The findings suggest that all four EOs are highly effective against *Thrips flavus* and may be a possible alternative in the management of *Thrips flavus*, especially when considering reducing the use of synthetic pesticides.

Keywords: *Thrips flavus*; Lamiaceae; essential oil; plant-derived insecticides; insecticidal toxicity

1. Introduction

Thrips flavus Schrank (Thysanoptera, Thripidae) is an important pest affecting cash crops, a member of the Thripidae (Thysanoptera), which is found worldwide and can seriously damage crops at various stages of growth [1]. *Thrips flavus* is regarded as a major pest of flowering plants of the Asteraceae (Compositae) and Fabaceae (Leguminosae) families, among others, in Northern China. This pest typically causes premature senescence or deformation of flowers and curling, distortion, and wilting of leaves [2,3]. Various ecological factors have been shown to affect *Thrips flavus* survival and development. Recent research indicates that *Cucumis sativus* L. and *Glycine max* (L.) Merr. were two potentially suitable host plants for *Thrips flavus* [1]. The ambient CO_2 concentrations can accelerate the development of thrips but reduce their survival rate [4]. The survival of thrips gradually decreased with increasing temperature from 19 °C to 31 °C [5]. Currently, chemical insecticides are predominantly used to control *Thrips flavus*, which shows greater sensitivity to imidacloprid, avermectin, and lambda-cyhalothrin emulsifiable concentrate compared to *Frankliniella occidentalis* (Pergande) (Thripidae) in the Yunnan region [6]. However, concerns over the increasing resistance of Thripidae to chemical insecticides and their environmental and ecological impacts have led to a search for safer, environmentally friendly alternatives [7,8]. Essential oils (EOs), secondary plant metabolites, offer such an alternative. These substances possess a range of bioactivities, including oviposition inhibition, avoidance, egg hatching inhibition, larval development suppression, antifeedant and repellent effects, as well as the capacity to knock down, poison, and fumigate phytophagous pests [9,10]. EOs also exhibit attractive behavior to certain pests [11]. They are characterized by their diverse bioactivities, safety for non-target organisms, availability from various sources, environmental friendliness, natural degradation, the ability to delay the development of pest resistance, and the potential to replace chemical insecticides [12–14]. The target pests, thrips, are unlikely to become resistant to these EOs, as products containing EOs are more complex chemically than traditional insecticide products with a single active ingredient [15]. *Thrips flavus* can potentially be controlled using EOs and products derived from EOs.

Lamiaceae, the sixth largest angiosperm family, comprises more than 7000 species across approximately 230 genera [16]. Many of these species are well-known as ornamental and medicinal plants; examples include lavender (*Lavandula angustifolia* Mill.), basil (*Ocimum basilicum* L.), peppermint (*Mentha × piperita*), rosemary (*Rosmarinus officinalis* L.), thyme (*Thymus mongolicus* (Ronniger) Ronniger), etc. The EOs are synthesized and accumulated in the leaves, stems, and epidermal glands of the reproductive structures [17]. Numerous EOs have been widely utilized in the food, cosmetic, pharmaceutical, and crop protection industries [18–20]. Lamiaceae EOs consist of a diverse array of chemical components, including aliphatic and aromatic molecules, with general compounds such as β-caryophyllene, linalool, limonene, β-pinene, 1,8-cineol, α-pinene and thymol [21–23]. For instance, terpenoids and isoprenoids are a class of organic substances found in peppermint oil that are among the most diverse naturally occurring compounds derived from plants [24]. The chemical composition of Lamiaceae EOs determines their mechanism of action and their target applications, and their active ingredients with insecticidal properties show promise for developing plant-derived insecticides [25,26]. Numerous studies have demonstrated the versatile effects of EOs derived from plants of the Lamiaceae family against various agricultural pests. Marjoram oil showed significant antifeedant activity against *Hylobius abietis* (L.) (Curculionidae) and significant toxicity as a fumigant against *Tribolium castaneum* (Herbst) (Tenebrionidae) [27,28]. Rosemary oil contains 1,8-eudesmol, which demonstrated attractant activity against *Frankliniella occidentalis* (Pergande), and lavandula oil exhibited attractant activity against *Drosophila suzukii* (Matsumura) (Drosophilidae) [29,30]. Conversely, thyme oil showed remarkable repellent activity against *Sitophilus zeamais* Motsch (Curculionidae) [31]. *Culex pipiens* L. (Culicidae) oviposition is strongly inhibited by the EOs of mint and basil [32]. However, no studies have yet explored the use of Lamiaceae EOs to control soybean (*Glycine max* (L.) Merr.) pest thrips. This study aimed to assess the insecticidal potential of EOs against *Thrips flavus*, providing insights for the development of botanical insecticides and offering practical guidelines to protect crops from *Thrips flavus* infestations.

2. Materials and Methods

2.1. Insects

The thrips were collected from a soybean field of the Ministry of Agriculture and Rural Development (Jilin) (located at 125°24′19″ E, 43°48′17″ N, approximately 225 m above sea level). No pesticides are applied to this experimental field during the experimental period. The region is characterized by a temperate continental semi-humid monsoon climate. Thrips were brought back to the laboratory and placed in a rearing cage using the sweeping method. Thrips were identified using the key of Y.F. Han (1997) [33] and Mound et al. (2018) [34] and continuously reared three generations. The insect cage was then kept in an incubator with soybean plants for two to three days at 25 ± 1 °C, 70 ± 5% relative humidity, and a 16:8 h light: dark photoperiod. The soybean plants were cultivated for two to three days under a 16:8 h light–dark photoperiod and were watered three to five times weekly [35].

2.2. Essential Oils

The four EOs assessed were perilla leaf oil (*Perilla frutescens* (L.) Britt.), marjoram oil (*Origanum majorana* L.), clary sage oil (*Salvia sclarea* L.), and spearmint oil (*Mentha spicata* L.), which were obtained from Lamiaceae species by Ji'an Zhongxiang Natural Plants Co. Ltd. (Ji'an City, Jiangxi Province, China) The EOs were extracted by steam distillation, and their purity was >98%.

2.3. Chemical Composition Analysis

Gas chromatography–mass spectrometry (GC–MS) (QP2010 plus, Shimadzu, Japan) was utilized to analyze the chemical composition of the four EOs [36–38]. The heating procedure was according to the methodology adopted by Pei et al. [36]. GC–MS software (version 2.53) tools (NIST 147 and NIST 27) were used for compound identification.

2.4. Laboratory Bioassay

The leaf-dipping method was used to determine the toxicity of the four EOs [36]. Fresh soybean leaves of the same size that were undamaged and free of pests and diseases were selected, washed in water, and allowed to dry naturally. Serial dilutions of the four EOs (0.2, 0.4, 0.6, 0.8, 1.0 mg/mL) were prepared in acetone to be tested. The leaves were then immersed in the different EOs, removed after ten seconds, and dried again naturally. One leaf (4 cm × 2 cm) was then placed in a centrifuge tube (50 mL) with moistened filter paper inside. Thirty female thrips (3-day-old) were introduced into the tubes. Subsequently, the tube opening was promptly sealed with Parafilm sealing film. Approximately 70 micro-holes were punched into the film using a $2^{\#}$ insect pin, ensuring an even distribution of the micro-holes. Leaves from the control group were treated without EOs. The 30% thiamethoxam SC was purchased from Hebei Zhongbaolvnong Science & Technology Co., Ltd. (Langfang City, Hebei Province, China) and used as a control group of commonly used insecticides, with serial dilutions of 0.005, 0.008, 0.012, 0.015, 0.018 mg/mL. This procedure was repeated for each EO. Each group was replicated three times. After 24 h of treatment at room temperature, thrips mortality was assessed. After counting the number of dead and surviving insects, the mortality and adjusted mortality rates were calculated [39]. The mortality rate was calculated by Equation (1):

$$M_0 = \frac{M_1}{M_2} \times 100 \tag{1}$$

where M_0 is the mortality rate (%); M_1 is the number of dead thrips; M_2 is the total number of thrips in each treatment.

The adjusted mortality rate was calculated by Equation (2):

$$AM = \frac{N_1 - N_2}{1 - N_2} \times 100 \tag{2}$$

where AM is the adjusted mortality rate (%); N_1 is the mortality rate of the treatment group (%); N_2 is the mortality rate of the control group (%).

2.5. Olfactory Test

The ability of each EO to attract or repel adult *Thrips flavus* was assessed by an olfactory test. A Y-tube olfactometer was used to test the attraction or repellent rate [40]. Prior to the olfactory tests, both female and male adults were removed and cultivated separately.

Lights were positioned in parallel to avoid light interference, and humidified air was used as a control. The EO was diluted to 1.0 mg/mL with acetone and tested with a 1 μL volume dropped on filter paper in the odor chamber. Clean, humid air was passed through each arm of the odor chamber and Y-tube at a flow rate of 300 mL/min. An adult *Thrips flavus* was placed in the front end inside the olfactometer. Its position in the tube was recorded after five minutes. Female and male adults were tested separately, with at least thirty individuals of each sex being treated with each EO. The response criteria were as follows: the adult was considered to have selected that odor source if it crawled more than half-length into either of the tubes and remained there for at least one minute; if the adult did not select or remained motionless after five minutes, it was considered not to have selected the odor source (no response) [36]. The selection rate was used as an informative indicator to evaluate the level of activity.

2.6. Pot Experiment

A pot experiment was conducted to assess the control efficacy of four EOs against *Thrips flavus*. The pot experiment procedure was performed as described by Pei et al. [36]. The schematic diagram of the pot experiment is shown in Figure S1. Soybean plants were planted in pots in batches, and when the 2nd compound leaf was grown, plants with uniform growth were selected, and one soybean seedling free of pests and diseases was kept in each pot. Five concentrations of four EOs were set at 180.00, 360.00, 540.00, 720.00, and 900.00 g a.i.·hm^{-2}. The control treatment was acetone without the EOs. Each concentration and control treatment was replicated three times. After spraying 5 mL evenly per pot with a spray bottle, the potted plants dried naturally in a windless area, and then thirty adult thrips were introduced per pot. The treated pots were arranged in randomized blocks one meter apart. At one, three, and seven days after treatment, the number of insects that died was counted, and the control efficacy was calculated. The control efficacy was calculated using Equation (3):

$$CE = (1 - \frac{P_1 \times P_2}{P_3 \times P_4}) \times 100 \quad (3)$$

where CE is the control efficacy (%); P_1 is the number of thrips in the treatment group after treatment with EOs; P_2 is the number of thrips in the control group before treatment with EOs; P_3 is the number of thrips in the treatment group before treatment with EOs, and P_4 is the number of thrips in the control group after treatment with EOs [36].

2.7. Statistical Analysis

The laboratory bioassay results were used to calculate the 95% confidence intervals, the median lethal concentration (LC_{50}), and the bioactivity regression equation using DPS 9.50 (Hangzhou Ruifeng Information Technology Co., Ltd.) [41]. One-way analysis of variance (ANOVA) was performed using IBM SPSS 20.0, and Tukey's test was used to compare significant differences between treatments [42]. The olfactory behavior response test results were processed using IBM SPSS 20.0, and the chi-square test was used to determine the main differences between treatments. The results were plotted using Origin 2021.

3. Results

3.1. Laboratory Bioassay

In the toxicity test, although the LC_{50} value of marjoram oil was slightly lower than those of the other three EOs (perilla leaf oil, spearmint oil, and clary sage oil), the 95% confidence intervals overlapped, and, therefore, there was no difference in the toxicity of these four essential oils (Table 1). The four EOs exhibited similar bioactivity for *Thrips flavus*. The LC_{50} of thiamethoxam used in this experiment was only 0.0077 mg/mL, significantly lower than those of the four EOs above. This indicates that the toxicity of these EOs is significantly lower than 30% thiamethoxam.

Table 1. The toxicity of four essential oils to adult *Thrips flavus*.

Essential Oils	95% Confidence Interval	LC_{50} (mg/mL)	Regression Equation	Correlation Coefficient	χ^2	*df*
Perilla leaf oil	0.35~0.51	0.43	$y = 6.2134 + 3.3226x$	0.89	7.03	3
Marjoram oil	0.17~0.60	0.41	$y = 6.5178 + 3.9312x$	0.88	17.87	3
Clary sage oil	0.35~0.48	0.42	$y = 6.1932 + 4.1952x$	0.99	0.94	3
Spearmint oil	0.47~0.62	0.54	$y = 6.1934 + 4.4029x$	0.94	7.41	3
30% thiamethoxam	0.0053~0.0095	0.0077	$y = 10.4813 + 2.5941x$	0.92	3.1347	3

Note: LC_{50} = concentration to kill 50% of thrips.

3.2. Pot Experiment

After one day of treatment, there were no significant differences in the control efficacy against *Thrips flavus* between the 180.00 g a.i.·hm^{-2} ($F = 2.818$, $p = 0.107$), 360.00 g a.i.·hm^{-2} ($F = 1.699$, $p = 0.244$), and 540.00 g a.i.·hm^{-2} ($F = 3.792$, $p = 0.058$) application rates. At 720.00 g a.i.·hm^{-2} ($F = 5.684$, $p = 0.022$), the control efficacy of spearmint oil was not significantly different from that of the other two EOs and was significantly higher than that of clary sage oil. In addition to this, marjoram oil at 900.00 g a.i.·hm^{-2} ($F = 4.605$, $p = 0.037$) concentration was significantly more effective than clary sage oil against thrips, and there was no significant difference between the former and the other two EOs (Figure 1).

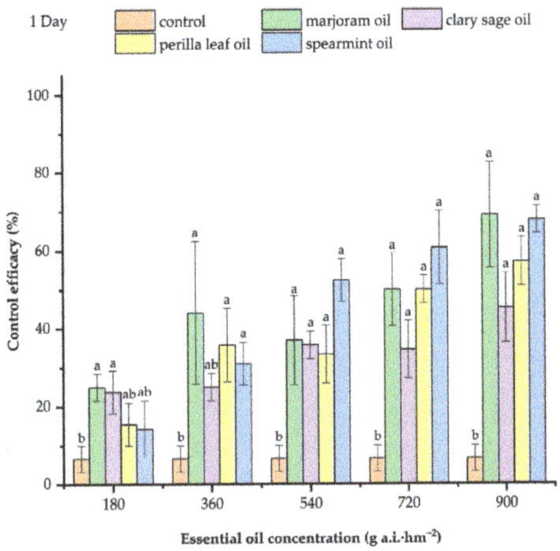

Figure 1. The control efficacy of four essential oils against *Thrips flavus* after one day of application in pot experiment. Different lowercase letters indicated significant differences ($p < 0.05$) among the control and essential oil treatments at the same concentration.

The control efficacy of perilla leaf oil at 900.00 g a.i.·hm^{-2} did not differ significantly from the other two EOs three days after application, although it was significantly more effective against thrips than clary sage oil ($F = 6.624$, $p = 0.015$). The control efficacy of the four EOs against thrips did not differ significantly at concentrations of 180.00 g a.i.·hm^{-2} ($F = 3.841$, $p = 0.057$), 360.00 g a.i.·hm^{-2} ($F = 0.406$, $p = 0.753$), 540.00 g a.i.·hm^{-2} ($F = 0.649$, $p = 0.605$), and 720.00 g a.i.·hm^{-2} ($F = 1.230$, $p = 0.361$) (Figure 2).

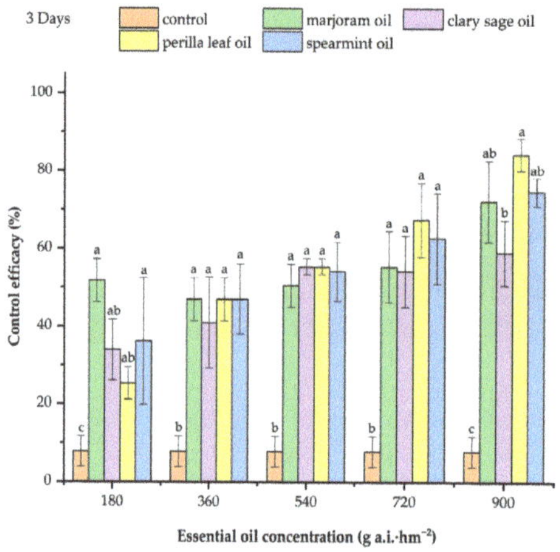

Figure 2. The control efficacy of four essential oils against *Thrips flavus* after three days of application in pot experiment. Different lowercase letters indicated significant differences ($p < 0.05$) among the control and essential oil treatments at the same concentration.

After seven days of treatment at 180.00 g a.i.·hm^{-2} ($F = 3.054$, $p = 0.092$), 720.00 g a.i.·hm^{-2} ($F = 1.203$, $p = 0.369$), and 900.00 g a.i.·hm^{-2} ($F = 1.562$, $p = 0.273$), there was no significant difference in the control efficacy against *Thrips flavus*. In contrast, both marjoram oil and spearmint oil had significantly higher control efficacy compared to perilla leaf oil at 360.00 g a.i.·hm^{-2} ($F = 5.870$, $p = 0.020$). Moreover, spearmint oil and clary sage oil were significantly higher than perilla leaf oil at 540.00 g a.i.·hm^{-2} ($F = 5.428$, $p = 0.025$), and marjoram oil was not significantly different from the first two EOs (Figure 3).

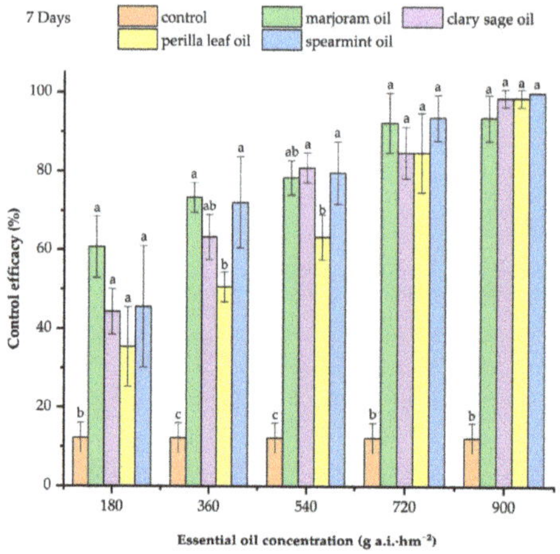

Figure 3. The control efficacy of four essential oils against *Thrips flavus* after seven days of application in pot experiment. Different lowercase letters indicated significant differences ($p < 0.05$) among the control and essential oil treatments at the same concentration.

The control of thrips by the four EOs, marjoram oil, clary sage oil, perilla leaf oil, and spearmint oil, was significantly improved with longer application times and higher concentrations. Spearmint and marjoram oil had a higher efficacy at all concentrations, one day, three days, and seven days after application. Perilla leaf oil had a better control efficacy at one day and three days after application. However, seven days after application, perilla leaf oil was inferior to the other EOs at low concentrations. One day and three days after application, perilla leaf oil applied at a low concentration had a significantly lower control efficacy than the other oils, but at seven days, at the higher concentrations, it was not significantly different compared to the other oils. Clary sage oil showed better efficacy seven days after application and was not significantly different from spearmint and marjoram oil. Clary sage oil applied at a high concentration was significantly less effective than the other EOs after one day of application and three days after application. In contrast, the low concentration of clary sage oil did not differ in efficacy from the other EOs. The four EOs reached the highest efficacy seven days after application, with no significant differences observed between the different concentrations.

3.3. Olfactory Test

Female adults of *Thrips flavus* showed different olfactory behavioral responses to the EOs of different Lamiaceae species. Female adult thrips were significantly attracted to spearmint oil (χ^2 = 4.948, p = 0.026) with an attraction rate of 71.05%. There were no significant differences in attraction between clary sage oil (χ^2 = 0.178, p = 0.673), marjoram oil (χ^2 = 0.530, p = 0.467), and perilla leaf oil (χ^2 = 0.166, p = 0.684). *Thrips flavus* showed 67.86%, 84.09%, 56.45%, and 81.25% selectivity for the four EOs (Figure 4). Among the four EOs tested, spearmint oil had a significant attractant effect on adult females.

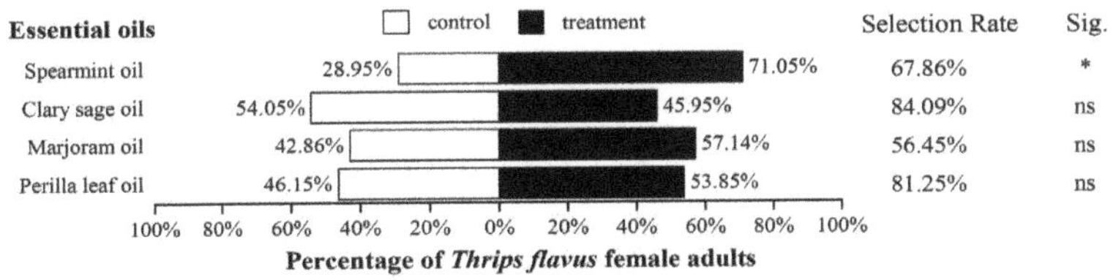

Figure 4. Olfactory behavioral response of female adult *Thrips flavus* to four essential oils. The symbol "ns" indicated 'no significance' ($p > 0.05$), while an asterisk "*" indicated significance ($p < 0.05$).

Male adults of *Thrips flavus* show no significant attraction response to spearmint oil (χ^2 = 0.000, p = 1.000), clary sage oil (χ^2 = 2.381, p = 0.123), marjoram oil (χ^2 = 0.617, p = 0.432), and perilla leaf oil (χ^2 = 0.617, p = 0.432). *Thrips flavus* exhibited a selectivity of 65.22%, 80.00%, 65.96%, and 63.27% for the four EOs (Figure 5).

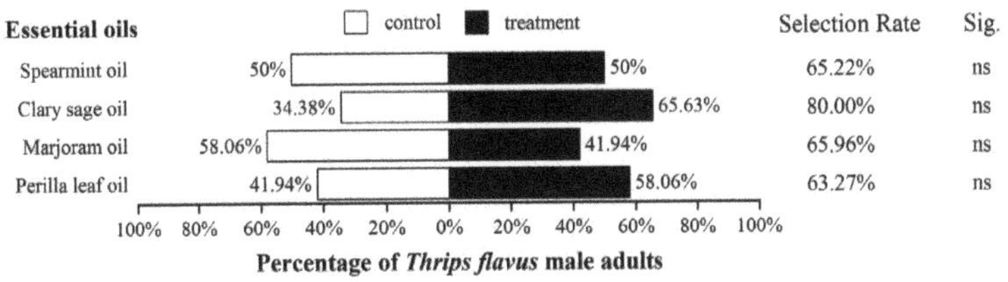

Figure 5. Olfactory behavioral response of male adult *Thrips flavus* to four essential oils. The symbol "ns" indicated 'no significance' ($p > 0.05$).

3.4. Chemical Analysis of Essential Oils

Marjoram oil consisted of seventeen major compounds, ranging in concentration from 24.52% to 0.26%, the most abundant being linalool (24.52%), followed by benzyl acetate (16.42%) and α-hexylcinnamaldehyde (14.28%), and the least abundant leaf alcohol (0.26%). Of these, five belong to esters, representing 29.27% of the constituents. They were followed by three types of terpene alcohols, accounting for 25.35% of the constituents. There were also aldehydes (14.28%), alcohols (8.87%), phenols (3.99%), amides (2.95%), and ketones (0.96%) (Table 2).

Table 2. Chemical constituents of marjoram oil (*Origanum majorana* L.).

No.	Retention Time (min)	Retention Index	Compounds	Relative Percentage (%)
1	8.14	1085	linalool	24.52
2	3.308	843	leaf alcohol	0.26
3	7.637	1069	methyl benzoate	2.95
4	9.765	1145	benzyl acetate	16.42
5	11.575	1215	ethyl phenylacetate	4.31
6	11.96	1240	nerol	0.57
7	13.012	1306	methyl aminobenzoate	2.95
8	13.135	1317	2-tert-butylcyclohexanol	3.09
9	13.307	1332	phenol, 2-methoxy-3-(2-propenyl)	1.64
10	13.579	1355	4-tert-butylcyclohexanol	5.78
11	13.758	1370	2,6,6-trimethyl-2,4-cycloheptadien-1-one	0.29
12	15.226	1494	butylated hydroxytoluene	2.35
13	16.93	1618	methyl dihydrojasmonate	4.65
14	17.368	1645	cis-3-hexenyl salicylate	0.94
15	18.666	1720	α-hexylcinnamaldehyde	14.28
16	19.198	1747	1-phenyl-1-nonen-3-one	0.67

Clary sage oil consisted of nineteen major chemical constituents, ranging in concentration from 28.74% to 0.34%. The most abundant component of clary sage oil was isopropyl myristate (28.74%), followed by linalyl acetate (20.07%), and the least abundant was linalyl anthranilate (0.34%). Of these, six belong to the ester compound group, representing 62.14% of the content. They were followed by two types of terpene alcohols, accounting for 22.49% of the total constituents. There were also alcohols (4.74%), ketones (4.13%), terpenes (3.07%), phenols (2.23%), and aldehydes (1.2%) (Table 3).

Table 3. Chemical constituents of clary sage oil (*Salvia sclarea* L.).

Number	Retention Time (min)	Retention Index	Compounds	Relative Percentage (%)
1	14.557	1438	α-guaiene	0.44
2	15.348	1504	α-bulnesene	0.57
3	17.518	1654	patchouli alcohol	1.10
4	6.142	1013	*p*-cymene	2.23
5	4.554	933	α-pinene	1.24
6	5.261	973	β-pinene	0.37
7	14.357	1420	bicyclo[5.2.0]nonane, 4-ethenyl-4,8,8-trimethyl-2-methylene-	0.45
8	6.336	1021	1,8-cineole	7.31
9	10.482	1171	terpinen-4-ol	1.12
10	10.731	1179	α-terpineol	2.52
11	12.056	1246	linalyl acetate	20.07
12	13.476	1346	geranyl acetate	0.52
13	8.132	1085	linalool	15.18
14	9.286	1126	DL-camphor	4.13
15	11.479	1208	7-methoxy-3,7-dimethyloctanal	1.20
16	13.382	1338	linalyl anthranilate	0.34
17	13.687	1364	neryl acetate	0.6
18	18.819	1728	benzyl benzoate	11.87
19	20.475	1813	isopropyl myristate	28.74

Perilla leaf oil consisted of nine major chemical constituents ranging in concentration from 32.44% to 0.59%. The most abundant was (+)-limonene (32.44%), followed by γ-terpinene (23.92%), and the least abundant was cis-linalool oxide (0.59%). Of these, three belong to the terpene group, representing 56.97% of the total constituents. Aldehydes followed with 15.61% of the total. There are also phenols (13.25%), terpene alcohols (10.53%), olefins (2.34%), ketones (0.70%), and alcohols (0.59%) (Table 4).

Table 4. Chemical constituents of perilla leaf oil (*Perilla frutescens* (L.) Britt.).

Number	Retention Time (min)	RI	Compounds	Relative Percentage (%)
1	6.139	1013	*p*-cymene	13.25
2	6.375	1023	(+)-limonene	32.44
3	7.097	1051	γ-terpinene	23.92
4	7.956	1079	terpinolene	10.53
5	6.783	1039	1,3,6-octatriene, 3,7-dimethyl-, (z)-	2.34
6	7.2	1054	3-carene	0.61
7	7.33	1059	cis-linalool oxide	0.59
8	7.608	1068	(-)-fenchone	0.70
9	11.87	1234	cinnamaldehyde	15.61

Spearmint oil consists of five major chemical constituents ranging in concentration from 70.34% to 1.05%. The most abundant is (+)-carvone (70.34%), followed by (+)-limonene (26.22%), and the least abundant is α-pinene (1.05%). Of these, three belong to the terpene group, representing 28.64% of the total constituents. There are also ketones (70.34%) and alcohols (1.20%) (Table 5).

Table 5. Chemical constituents of spearmint oil (*Mentha spicata* L.).

Number.	Retention Time (min)	RI	Compounds	Relative Percentage (%)
1	4.557	934	α-pinene	1.05
2	6.373	1023	(+)-limonene	26.22
3	5.263	973	β-pinene	1.19
4	10.382	1167	3-p-menthol	1.20
5	11.614	1217	(+)-carvone	70.34

4. Discussion

EOs have several advantages when used as biocontrol agents for pest control. Firstly, they are made from natural plant components and can be broken down in the environment, making them environmentally friendly. Secondly, the extraction of EOs can effectively use excess peelings, twigs, and leaves from agricultural production, thus minimizing the waste of resources. Thirdly, EOs have specific insecticidal activity against certain pests and are safer against non-target species [43,44]. This study investigated the toxicity against *Thrips flavus* of four EOs: marjoram oil (LC_{50} = 0.41 mg/mL); clary sage oil (LC_{50} = 0.42 mg/mL); perilla leaf oil (LC_{50} = 0.43 mg/mL); and spearmint oil (LC_{50} = 0.54 mg/mL) under laboratory conditions. All four EOs exhibited some degree of toxicity. The use of Lamiaceae EOs for pest control has been studied extensively. Spearmint oil was shown to be toxic as a fumigant to *Reticulitermes dabieshanensis* Wang and Li (Rhinotermitidae) with an LC_{50} of 0.194 μL/L [45]. At a concentration of 5 mg, spearmint oil and basil oil could control 100% of male *Blattella germanica* L. (Blattidae) [46]. Similar results in the Lamiaceae reported that pennyroyal oil (*Mentha pulegium* L.) and *Thymus mastichina* L. essential oil exhibited fumigant effect against *Frankliniella occidentalis* with LC_{50} of 3.1 mg/L and 3.6 mg/L [47]. *Mentha pulegium* essential oil treatment also showed a significant fumigant effect on *Thrips tabaci* Lindeman [48] and was the most toxic fumigant to *Thrips palmi* Karny (Thripidae) [49]. The EOs of *Syzygium aromaticum* Merr. and L.M. Perry, *Cinnamomum bejolghota* (Buch.–Ham.) Sweet and *Cymbopogon citratus* (Dc.ex.Nees) showed high fumigant toxicity against *Frankliniella schultzei* (Trybom) (Thripidae) [50]. Whether the four EOs in this study have fumigant activity against *Thrips flavus* will be further investigated by fumigation toxicology assays. To further confirm the insecticidal activity of the four EOs, the pot experiment was conducted using live soybean potted plants. Spearmint oil showed 100% lethality against *Thrips flavus* at a concentration of 900.00 g a.i.·hm^{-2} after seven days of treatment. At the same time, after seven days of treatment, the other EOs showed more than 90% lethality against *Thrips flavus* at a concentration of 900.00 g a.i.·hm^{-2}. These results suggest that the exploitation and development of Lamiaceae EOs into plant-derived insecticides holds great potential. Because the insecticidal mechanism of these EOs is not yet clear, further research will be conducted exploratively.

EOs can be found in large quantities and with various chemical compositions. Several factors influence the biological activity, content, and composition of EOs. These include the plant growth stage [51], the extraction site and technique [52–54], and the geographical environment [55]. The use of innovative technologies to increase extraction efficiency can also improve the quality of EOs [56,57]. The chemical composition of the four EOs used in this study varied significantly based on GC–MS analysis. The highest concentration of (+)-carvone was found in spearmint oil, while isopropyl myristate was most abundant in clary sage oil, limonene in perilla leaf oil, and linalool in marjoram oil. Previous studies have shown that the main constituents of perilla leaf oil are 2-furyl methyl ketone (71.83%), decahydro-1-methyl-2-methylene-naphthalene (10.47%), limonene (5.16%) and caryophyllene (1.66%) [58]. On the other hand, β-terpineol and γ-terpinene were shown to be the main constituents of marjoram oil [27]. Similarly, limonene (+1,8-cineole; 14.3%) and carvone (67.1%) were identified as the major constituents of spearmint oil by GC–MS analysis [59]. When EOs are used as insecticides for pest control, one of the problems to be solved is that the production and quality properties of plant material cannot be standardized, and the control efficacy varies widely. Therefore, further research is needed to study the efficacy of individual chemical constituents in controlling pests and to refine the component with the best control efficacy to be used in the development of plant-derived insecticides.

This study found that of the four EOs tested, spearmint oil was a particularly strong attractant for mature female *Thrips flavus*. Recent studies have shown that some Lamiaceae EOs have a significant attractant effect on certain pests [60]. *Rosmarinus officinialis* L. was attractive to *Frankliniella occidentalis*, which is attracted by the major component 1,8-cineole [30]. Rosemary oil with a certain concentration of β-caryophyllene and limonene was found to be an attractant for *Bemisia tabaci* (Gennadius) [61]. The EO of *Tetradenia riparia* (Hochst.) Codd, of which the primary compounds are fenchone, δ-cadinene, 14-hydroxy-β-caryophyllene, and *tau*-cadinol, is an attractant for *Ceratitis capitata* (Wiedemann) (Tephritidae) [62]. In addition, some pests can be repelled by Lamiaceae EOs. For example, patchouli oil has significantly repelled *Tribolium castaneum* [63]. EO of *Origanum majorana* L. captured 87% fewer *Thrips tabaci* (Thripidae) than that in the control treatment, and this EO is a promising onion thrips repellent [64]. *Cinnamomum verum* Presl showed repellent activity against *Hercinothrips femoralis* (Reuter) (Thripidae) [65]. *Rosmarinus. officinalis* EO showed repellent activity against female adults of *Thrips tabaci* [66]. *Rosmarinus officinialis* EO can also inhibit the oviposition of *Frankliniella occidentalis* (Thripidae), *Frankliniella intonsa* (Thripidae), and *Thrips palmi*. α-Pinene was repellent to *Frankliniella occidentalis* and *Frankliniella intonsa*. Eucalyptol showed significant repellent activity in these three thrips species [67]. Moreover, carvacrol, cinnamaldehyde, and thymol, common constituents of Lamiaceae EOs, can be combined with nanogel technology to repel mosquitoes [68]. The EOs can be utilized as attractants or repellents for pest management by exploiting their ability to attract or repel pests [69]. In the olfactory test, spearmint oil had a different effect on the male and female *Thrips flavus*. This may be due to the chemical composition of the spearmint oil affecting adult male and female thrips differently. This phenomenon was reported in some thrips species [66,67]. The volatile dihydrotagetone alone attracted females *Megalurothrips sjostedti* (Trybom) (Thripidae) but had neither repellent nor attractive activity to males [70]. *Tagetes minuta* (L.) flower oil resulted in different olfactory responses to different sexes of *Ceratitis capitata*, causing attraction to males and avoidance to females. This suggests that the composition of EOs influences their olfactory properties [71]. In addition, their different effects on males and females may be due to other environmental factors. At a 0.01% concentration, the EO derived from celery seeds attracted both male and female adult *Tribolium castaneum*. However, at a 0.1% concentration, the EO had the opposite effect, repelling male adults but attracting female adults. In addition, basil oil at 0.01% repelled adult males but did not affect adult females [72]. This suggests that the olfactory effects of EOs are concentration-dependent and can have different effects on different sexes of insects of the same species. The behavioral responses of *Thrips flavus* to single compounds and different compound blends will be investigated.

It has been shown that changing the application method could effectively improve the control efficacy of EOs against pests [73–75]. The utilization of nanomaterials is an important part of the implementation of novel application methods. When rosemary essential oil was incorporated into biodegradable poly (epsilon-caprolactone) nanoparticles and administered topically to *Drosophila suzukii*, it effectively prolonged and enhanced the plant's resistance to the insect. The results of this study will provide useful inspiration for our subsequent research [76,77]. In addition, nanoemulsion technologies are also being investigated; research is underway to use the EOs of *Ayapana triplinervis* (Vahl) R.M. King and H. Robinson to create more stable and long-lasting nanoemulsions to inhibit the reproduction of *Aedes aegypti* L. (Culicidae) larvae [78]. In addition, the stability of EOs has been improved through the use of microencapsulation technologies. An advantage of this method is that bioactive components' release, solubilization, and protection can be controlled [79]. The application method used in the pot experiment in this study was spraying, which is proven to be almost safe for soybean potted plants (Table S1). The appropriate application method requires further testing and screening. In this study, the main aim is to screen the EOs for their potential insecticidal activities against *Thrips flavus* under laboratory conditions. Further studies will encompass greenhouse or field trials aimed at further assessing the effectiveness of the efficacy of different EOs for thrips control. The proven potential for control of Thysanoptera pests should be fully exploited in management strategies that include combined approaches [80].

5. Conclusions

Four Lamiaceae EOs showed significant control efficacy against thrips. The EOs caused more than 90% mortality at the concentration of 900.00 g a.i.$\cdot$hm^{-2} and higher after 7 days. Spearmint oil caused a significant attraction response in adult females. The other EOs showed no significant attraction or repellent effects toward adult *Thrips flavus* of either sex. Linalool (24.52%), isopropyl myristate (28.74%), (+)-limonene (32.44%), and (+)-carvone (70.3%) were the primary chemical constituents of the EOs. The results of this preliminary study demonstrate that the four Lamiaceae EOs have the potential to develop plant-based insecticides and may be a possible alternative in the management of *Thrips flavus*, especially when considering reducing the use of synthetic pesticides.

Supplementary Materials: The following supporting information can be downloaded at https://www.mdpi.com/article/10.3390/agronomy14061212/s1, Figure S1: Schematic diagram of the structure of a Y-tube olfactometer. Table S1 The phytotoxicity grades of four essential oils to soybean potted plants.

Author Contributions: Conceptualization, Y.G.; methodology, S.S. and Y.G.; investigation, T.P., Y.N., M.-L.X. and Y.Z.; data curation, T.P. and Y.N.; writing—original draft preparation, Y.N., T.P., C.Z., B.L., S.S., Y.G., M.-L.X. and Y.Z.; writing—review and editing, Y.G., Y.N., M.-L.X. and T.P.; funding acquisition, Y.G. All authors have read and agreed to the published version of the manuscript.

Funding: This work was financially supported by the National Key Research and Development Program of China (Grant No. 2023YFD1401000), the Science and Technology Projects of the Education Department of Jilin Province of China (Grant No. JJKH20230408KJ), and the Earmarked Fund for China Agriculture Research System of MOF and MARA (Grant No. CARS–04).

Data Availability Statement: Data are contained within this article or Supplementary Materials.

Acknowledgments: We would like to express our gratitude to our students, Hui Wang, Long Wang, and He-Xin Gao, for the plant management and insect collection.

Conflicts of Interest: Author Yijin Zhao was employed by the company Dalian City Investment Asset Management Co., Ltd. The remaining authors declare that the research was conducted in the absence of any commercial or financial relationships that could be construed as a potential conflict of interest.

References

1. Gao, Y.; Zhao, Y.J.; Wang, D.; Yang, J.; Ding, N.; Shi, S.S. Effect of different plants on the growth and reproduction of *Thrips flavus* (Thysanoptera: Thripidae). *Insects* **2021**, *12*, 502. [CrossRef] [PubMed]
2. Dillon, F.M.; Panagos, C.; Gouveia, G.; Tayyari, F.; Chludil, H.D.; Edison, A.S.; Zavala, J.A. Changes in primary metabolite content may affect thrips feeding preference in soybean crops. *Phytochemistry* **2024**, *220*, 114014. [CrossRef] [PubMed]
3. Adhikari, R.; Seal, D.R.; Schaffer, B.; Liburd, O.E.; Khan, R.A. Within-plant and within-field distribution patterns of asian bean thrips and melon thrips in snap bean. *Insects* **2023**, *14*, 175. [CrossRef] [PubMed]
4. Gu, Z.Y.; Zhang, T.; Long, S.C.; Li, S.; Wang, C.; Chen, Q.C.; Chen, J.; Feng, Z.Y.; Cao, Y. Responses of *Thrips hawaiiensis* and *Thrips flavus* populations to elevated CO_2 concentrations. *J. Econ. Entomol.* **2023**, *116*, 416–425. [CrossRef] [PubMed]
5. Gao, Y.; Ding, N.; Wang, D.; Zhao, Y.J.; Cui, J.; Li, W.B.; Pei, T.H.; Shi, S.S. Effect of temperature on the development and reproduction of *Thrips flavus* (Thysanoptera: Thripidae). *Agric. For. Entomol.* **2022**, *24*, 279–288. [CrossRef]
6. Sun, Y.; Hu, C.X.; Chen, G.H.; Li, X.X.; Liu, J.H.; Xu, Z.W.; Zhou, Y.; Wu, D.H.; Zhang, X.M. Insecticide-mediated changes in the population and toxicity of the thrips species, *Frankliniella occidentalis* (Pergande) and *Thrips flavus* (Schrank) (Thysanoptera: Thripidae). *J. Econ. Entomol.* **2024**, *117*, 293–301. [CrossRef] [PubMed]
7. Shen, X.J.; Chen, J.C.; Cao, L.J.; Ma, Z.Z.; Sun, L.N.; Gao, Y.F.; Ma, L.J.; Wang, J.X.; Ren, Y.J.; Cao, H.Q.; et al. Interspecific and intraspecific variation in susceptibility of two co-occurring pest thrips, *Frankliniella occidentalis* and *Thrips palmi*, to nine insecticides. *Pest Manag. Sci.* **2023**, *79*, 3218–3226. [CrossRef] [PubMed]
8. Reitz, S.R.; Gao, Y.L.; Kirk, W.D.J.; Hoddle, M.S.; Leiss, K.A.; Funderburk, J.E. Invasion biology, ecology and management of western flower thrips. *Annu. Rev. Entomol.* **2020**, *65*, 17–37. [CrossRef] [PubMed]
9. Kilaso, M. Toxicity for control of *Frankliniella schultzei* and *Selenothrips rubrocinctus* (Thysanoptera: Thripidae) of several common synthetic insecticides. *Fla. Entomol.* **2022**, *105*, 155–159. [CrossRef]
10. Monzote, L.; Scull, R.; Cos, P.; Setzer, W.N. Essential oil from piper aduncum: Chemical analysis, antimicrobial assessment, and literature review. *Medicines* **2017**, *4*, 49. [CrossRef]
11. Kheloul, L.; Kellouche, A.; Bréard, D.; Gay, M.; Gadenne, C.; Anton, S. Trade-off between attraction to aggregation pheromones and repellent effects of spike lavender essential oil and its main constituent linalool in the flour beetle *Tribolium confusum*. *Entomol. Exp. Appl.* **2019**, *167*, 826–834. [CrossRef]
12. Li, Y.; Yu, S.; Huang, J.; Wang, Z.Y.; Zeng, Y.F.; Wu, X.M.; Han, K.Z.; Zhou, H.J.; Wang, G.H.; Yu, Z.W. Study of behavioral, electrophysiological response, and the active compounds of the essential oils from six kinds of flowers against *Solenopsis invicta* Buren (Hymenoptera: Formicidae). *Ind. Crop. Prod.* **2022**, *188*, 115603. [CrossRef]
13. Hu, Z.J.; Yang, J.W.; Chen, Z.H.; Chang, C.; Ma, Y.P.; Li, N.; Deng, M.; Mao, G.L.; Bao, Q.; Deng, S.Z.; et al. Exploration of clove bud (*Syzygium aromaticum*) essential oil as a novel attractant against *Bactrocera dorsalis* (Hendel) and its safety evaluation. *Insects* **2022**, *13*, 918. [CrossRef]
14. Ikbal, C.; Pavela, R. Essential oils as active ingredients of botanical insecticides against aphids. *J. Pest Sci.* **2019**, *92*, 971–986. [CrossRef]
15. Nenaah, G.E.; Alasmari, S.; Almadiy, A.A.; Albogami, B.Z.; Shawer, D.M.; Fadl, A.E. Bio-efficacy of *Salvia officinalis* essential oil, nanoemulsion and monoterpene components as eco-friendly green insecticides for controlling the granary weevil. *Ind. Crop. Prod.* **2023**, *204*, 117298. [CrossRef]
16. Zhao, F.; Chen, Y.P.; Salmaki, Y.; Drew, B.T.; Wilson, T.C.; Scheen, A.C.; Celep, F.; Bräuchler, C.; Bendiksby, M.; Wang, Q.; et al. An updated tribal classification of Lamiaceae based on plastome phylogenomics. *BMC Biol.* **2021**, *19*, 2. [CrossRef]
17. Singh, P.; Pandey, A.K. Prospective of essential oils of the Genus *Mentha* as biopesticides: A review. *Front. Plant Sci.* **2018**, *9*, 1295. [CrossRef]
18. Patrignani, F.; Prasad, S.; Novakovic, M.; Marin, P.D.; Bukvicki, D. Lamiaceae in the treatment of cardiovascular diseases. *Front. Biosci.* **2021**, *26*, 612–643. [CrossRef]
19. Ahmed, A.M.A.; Elsayed, A.A.A.; El-Gohary, A.E.; Khalid, K.A. Exogenous l-tyrosine motivates diversities in horsemint essential oil. *J. Essent. Oil Bear. Plants.* **2022**, *25*, 601–610. [CrossRef]
20. Chrysargyris, A.; Tomou, E.M.; Goula, K.; Dimakopoulou, K.; Tzortzakis, N.; Skaltsa, H. *Sideritis* L. essential oils: A systematic review. *Phytochemistry* **2023**, *209*, 113607. [CrossRef]
21. Chen, Y.J.; Luo, J.X.; Zhang, N.; Yu, W.J.; Jiang, J.X.; Dai, G.H. Insecticidal activities of *Salvia hispanica* L. essential oil and combinations of their main compounds against the beet armyworm *Spodoptera exigua*. *Ind. Crop. Prod.* **2021**, *162*, 113271. [CrossRef]
22. Karpinski, T.M. Essential oils of lamiaceae family plants as antifungals. *Biomolecules* **2020**, *10*, 103. [CrossRef]
23. Bedini, S.; Djebbi, T.; Ascrizzi, R.; Farina, P.; Pieracci, Y.; Echeverría, M.C.; Flamini, G.; Trusendi, F.; Ortega, S.; Chiliquinga, A.; et al. Repellence and attractiveness: The hormetic effect of aromatic plant essential oils on insect behavior. *Ind. Crop. Prod.* **2024**, *210*, 118122. [CrossRef]
24. Fuchs, L.K.; Holland, A.H.; Ludlow, R.A.; Coates, R.J.; Armstrong, H.; Pickett, J.A.; Harwood, J.L.; Scofield, S. Genetic manipulation of biosynthetic pathways in mint. *Front. Plant Sci.* **2022**, *13*, 928178. [CrossRef]
25. Ebadollahi, A.; Ziaee, M.; Palla, F. Essential oils extracted from different species of the Lamiaceae plant family as prospective bioagents against several detrimental pests. *Molecules* **2020**, *25*, 1556. [CrossRef]

26. Peschiutta, M.L.; Achimon, F.; Brito, V.D.; Pizzolitto, R.P.; Zygadlo, J.A.; Zunino, M.P. Fumigant toxicity of essential oils against *Sitophilus zeamais* (Motschulsky) (Coleoptera: Curculionidae): A systematic review and meta-analysis. *J. Pest Sci.* **2022**, *95*, 1037–1056. [CrossRef]
27. Teke, M.A.; Mutlu, Ç. Insecticidal and behavioral effects of some plant essential oils against *Sitophilus granarius* L. and *Tribolium castaneum* (Herbst). *J. Plant Dis. Prot.* **2021**, *128*, 109–119. [CrossRef]
28. Azeem, M.; Iqbal, Z.; Emami, S.N.; Nordlander, G.; Nordenhem, H.; Mozuraitis, R.; El-Seedi, H.R.; Borg-Karlson, A.K. Chemical composition and antifeedant activity of some aromatic plants against pine weevil (*Hylobius abietis*). *Ann. Appl. Biol.* **2020**, *177*, 121–131. [CrossRef]
29. Galland, C.D.; Glesner, V.; Verheggen, F. Laboratory and field evaluation of a combination of attractants and repellents to control *Drosophila suzukii*. *Entomol. Gen.* **2020**, *40*, 263–272. [CrossRef]
30. Katerinopoulos, H.E.; Pagona, G.; Afratis, A.; Stratigakis, N.; Roditakis, N. Composition and insect attracting activity of the essential oil of *Rosmarinus officinalis*. *J. Chem. Ecol.* **2005**, *31*, 111–122. [CrossRef]
31. Barros, F.A.P.; Radünz, M.; Scariot, M.A.; Camargo, T.M.; Nunes, C.F.P.; de Souza, R.R.; Gilson, I.K.; Hackbart, H.C.S.; Radünz, L.L.; Oliveira, J.V.; et al. Efficacy of encapsulated and non-encapsulated thyme essential oil (*Thymus vulgaris* L.) in the control of *Sitophilus zeamais* and its effects on the quality of corn grains throughout storage. *Crop Prot.* **2022**, *153*, 105885. [CrossRef]
32. Farag, S.M.; Moustafa, M.A.M.; Fónagy, A.; Kamel, O.; Abdel-Haleem, D.R. Chemical composition of four essential oils and their adulticidal, repellence, and field oviposition deterrence activities against *Culex pipiens* L. (Diptera: Culicidae). *Parasitol. Res.* **2024**, *123*, 110. [CrossRef] [PubMed]
33. Han, Y.F. *Fauna of Economic Insects in China (Thysanoptera)*; Science Press: Beijing, China, 1997.
34. Mound, L.A.; Collins, D.W.; Hastings, A. *Thysanoptera Britannica et Hibernica–Thrips of the British Isles*; Lucidcentral.org, Identic Pty Ltd.: Queensland, Australia, 2018. Available online: https://keys.lucidcentral.org/keys/v3/british_thrips/index.html (accessed on 15 May 2024).
35. Brenner, R.; Prischmann-Voldseth, D.A. Influence of a neonicotinoid seed treatment on a nontarget herbivore of soybean (twospotted spider mite) and diet switching by a co-occurring omnivore (western flower thrips). *Environ. Entomol.* **2020**, *49*, 461–472. [CrossRef] [PubMed]
36. Pei, T.H.; Zhao, Y.J.; Wang, S.Y.; Li, X.F.; Sun, C.Q.; Shi, S.S.; Xu, M.L.; Gao, Y. Preliminary study on insecticidal potential and chemical composition of five Rutaceae essential oils against *Thrips flavus* (Thysanoptera: Thripidae). *Molecules* **2023**, *28*, 2998. [CrossRef] [PubMed]
37. Baviskar, K.P.; Jain, D.V.; Pingale, S.D.; Wagh, S.S.; Gangurde, S.P.; Shardul, S.A.; Dahale, A.R.; Jain, K.S. A review on hyphenated techniques in analytical chemistry. *Curr. Anal. Chem.* **2022**, *18*, 956–976. [CrossRef]
38. Smelcerovic, A.; Djordjevic, A.; Lazarevic, J.; Stojanovic, G. Recent advances in analysis of essential oils. *Curr. Anal. Chem.* **2013**, *9*, 61–70. [CrossRef]
39. Huang, X.; Ge, S.Y.; Liu, J.H.; Wang, Y.; Liang, X.Y.; Yuan, H.B. Chemical composition and bioactivity of the essential oil from *Artemisia lavandulaefolia* (Asteraceae) on *Plutella xylostella* (Lepidoptera: Plutellidae). *Fla. Entomol.* **2018**, *101*, 44–48. [CrossRef]
40. Zhang, Z.Q.; Sun, X.L.; Xin, Z.J.; Luo, Z.X.; Gao, Y.; Bian, L.; Chen, Z.M. Identification and field evaluation of non-host volatiles disturbing host location by the tea geometrid, *Ectropis obliqua*. *J. Chem. Ecol.* **2013**, *39*, 1284–1296. [CrossRef]
41. Tang, Q.Y.; Zhang, C.X. Data Processing System (DPS) software with experimental design, statistical analysis and data mining developed for use in entomological research. *Insect Sci.* **2013**, *20*, 254–260. [CrossRef]
42. Zouirech, O.; El Moussaoui, A.; Saghrouchni, H.; Gaafar, A.R.Z.; Nafidi, H.A.; Bourhia, M.; Khallouki, F.; Lyoussi, B.; Derwich, E. Prefatory *in silico* studies and *in vitro* insecticidal effect of *Nigella sativa* (L.) essential oil and its active compound (carvacrol) against the *Callosobruchus maculatus* adults (Fab), a major pest of chickpea. *Open Chem.* **2023**, *21*, 20230133. [CrossRef]
43. Elumalai, K.; Krishnappa, K.; Pandiyan, J.; Alharbi, N.S.; Kadaikunnan, S.; Khaled, J.M.; Barnard, D.R.; Vijayakumar, N.; Govindarajan, M. Characterization of secondary metabolites from Lamiaceae plant leaf essential oil: A novel perspective to combat medical and agricultural pests. *Physiol. Mol. Plant Pathol.* **2022**, *117*, 101752. [CrossRef]
44. Palazzolo, E.; Laudicina, V.A.; Germanà, M.A. Current and potential use of citrus essential oils. *Curr. Org. Chem.* **2013**, *17*, 3042–3049. [CrossRef]
45. Yang, X.; Han, H.; Li, B.L.; Zhang, D.Y.; Zhang, Z.L.; Xie, Y.J. Fumigant toxicity and physiological effects of spearmint (*Mentha spicata*, Lamiaceae) essential oil and its major constituents against *Reticulitermes dabieshanensis*. *Ind. Crop. Prod.* **2021**, *171*, 113894. [CrossRef]
46. Yeom, H.J.; Lee, H.R.; Lee, S.C.; Lee, J.E.; Seo, S.M.; Park, I.K. Insecticidal activity of lamiaceae plant essential oils and their constituents against *Blattella germanica* L. adult. *J. Econ. Entomol.* **2018**, *111*, 653–661. [CrossRef] [PubMed]
47. Stepanycheva, E.; Petrova, M.; Chermenskaya, T.; Pavela, R. Fumigant effect of essential oils on mortality and fertility of thrips *Frankliniella occidentalis* Perg. *Environ. Sci. Pollut. Res.* **2019**, *26*, 30885–30892. [CrossRef] [PubMed]
48. Topuz, E. Insecticidal activity of *Mentha pulegium* essential oil against *Thrips tabaci*, *Bemisia tabaci* and *Tuta absoluta* adults. *Int. J. Trop. Insect Sci.* **2023**, *43*, 1475–1483. [CrossRef]
49. Yi, C.G.; Choi, B.R.; Park, H.M.; Park, C.G.; Ahn, Y.J. Fumigant toxicity of plant essential oils to *Thrips palmi* (Thysanoptera: Thripidae) and *Orius strigicollis* (Heteroptera: Anthocoridae). *J. Econ. Entomol.* **2006**, *99*, 1733–1738. [CrossRef] [PubMed]
50. Pumnuan, J.; Insung, A. Fumigant toxicity of plant essential oils in controlling thrips, *Frankliniella schultzei* (Thysanoptera: Thripidae) and mealybug, *Pseudococcus jackbeardsleyi* (Hemiptera: Pseudococcidae). *J. Entomol.* **2016**, *40*, 1–10. [CrossRef]

51. Ni, Z.J.; Wang, X.; Shen, Y.; Thakur, K.; Han, J.Z.; Zhang, J.G.; Hu, F.; Wei, Z.J. Recent updates on the chemistry, bioactivities, mode of action, and industrial applications of plant essential oils. *Trends Food Sci. Technol.* **2021**, *110*, 78–89. [CrossRef]
52. Russo, A.; Bruno, M.; Avola, R.; Cardile, V.; Rigano, D. Chamazulene-Rich *Artemisia arborescens* essential oils affect the cell growth of human melanoma cells. *Plants* **2020**, *9*, 1000. [CrossRef]
53. da Silva, W.M.F.; Kringel, D.H.; de Souza, E.J.D.; Zavareze, E.D.; Dias, A.R.G. Basil essential oil: Methods of extraction, chemical composition, biological activities, and food applications. *Food Bioprocess Technol.* **2022**, *15*, 1–27. [CrossRef]
54. Saleh, I.A.; El Gendy, A.N.G.; Afifi, M.A.; El-Seedi, H.R. Microwave extraction of essential oils from *Senecio serpens* GD rowly and comparison with conventional hydro-distillation method. *J. Essent. Oil Bear. Plants.* **2019**, *22*, 955–961. [CrossRef]
55. Passos, B.G.; de Albuquerque, R.; Muñoz-Acevedo, A.; Echeverria, J.; Llaure-Mora, A.M.; Ganoza-Yupanqui, M.L.; Rocha, L. Essential oils from *Ocotea* species: Chemical variety, biological activities and geographic availability. *Fitoterapia* **2022**, *156*, 105065. [CrossRef]
56. Kaur, H.; Bhardwaj, U.; Kaur, R. *Cymbopogon nardus* essential oil: A comprehensive review on its chemistry and bioactivity. *J. Essent. Oil Res.* **2021**, *33*, 205–220. [CrossRef]
57. Ayub, M.A.; Goksen, G.; Fatima, A.; Zubair, M.; Abid, M.A.; Starowicz, M. Comparison of conventional extraction techniques with superheated steam distillation on chemical characterization and biological activities of *Syzygium aromaticum* L. essential oil. *Separations* **2023**, *10*, 27. [CrossRef]
58. You, C.X.; Yang, K.; Wu, Y.; Zhang, W.J.; Wang, Y.; Geng, Z.F.; Chen, H.P.; Jiang, H.Y.; Du, S.S.; Deng, Z.W.; et al. Chemical composition and insecticidal activities of the essential oil of *Perilla frutescens* (L.) Britt. aerial parts against two stored product insects. *Eur. Food Res. Technol.* **2014**, *239*, 481–490. [CrossRef]
59. Eliopoulos, P.A.; Hassiotis, C.N.; Andreadis, S.S.; Porichi, A.E.E. Fumigant toxicity of essential oils from basil and spearmint against two major pyralid pests of stored products. *J. Econ. Entomol.* **2015**, *108*, 805–810. [CrossRef]
60. Shivaramu, S.; Parepally, S.K.; Byregowda, V.Y.; Damodaram, K.J.P.; Bhatnagar, A.; Naga, K.C.; Sharma, S.; Kumar, M.; Kempraj, V. Estragole, a potential attractant of the winged melon aphid *Aphis gossypii*. *Pest Manag. Sci.* **2023**, *79*, 2365–2371. [CrossRef]
61. Sadeh, D.; Nitzan, N.; Shachter, A.; Chaimovitsh, D.; Dudai, N.; Ghanim, M. Whitefly attraction to rosemary (*Rosmarinus officinialis* L.) is associated with volatile composition and quantity. *PLoS ONE* **2017**, *12*, e0177483. [CrossRef]
62. Blythe, E.K.; Tabanca, N.; Demirci, B.; Kendra, P.E. Chemical composition of essential oil from *Tetradenia riparia* and its attractant activity for mediterranean fruit fly, *Ceratitis capitata*. *Nat. Prod. Commun.* **2020**, *15*, 6. [CrossRef]
63. Feng, Y.X.; Wang, Y.; You, C.X.; Guo, S.S.; Du, Y.S.; Du, S.S. Bioactivities of patchoulol and phloroacetophenone from *Pogostemon cablin* essential oil against three insects. *Int. J. Food Prop.* **2019**, *22*, 1365–1374. [CrossRef]
64. van Tol, R.; James, D.E.; de Kogel, W.J.; Teulon, D.A.J. Plant odours with potential for a push-pull strategy to control the onion thrips, *Thrips tabaci*. *Entomol. Exp. Appl.* **2007**, *122*, 69–76. [CrossRef]
65. Zvaríková, M.; Masarovic, R.; Zvarík, M.; Bagová, K.; Procházková, L.; Prokop, P.; Fedor, P. The effect of plant essential oils on the Banded Greenhouse Thrips (*Hercinothrips femoralis* O. M. Reuter 1891) (Thysanoptera: Thripidae: Panchaetothripinae). *J. Plant Dis. Prot.* **2023**, *130*, 747–755. [CrossRef]
66. Koschier, E.H.; Sedy, K.A. Labiate essential oils affecting host selection and acceptance of *Thrips tabaci* lindeman. *Crop Prot.* **2003**, *22*, 929–934. [CrossRef]
67. Li, X.W.; Zhang, Z.J.; Hafeez, M.; Huang, J.; Zhang, J.M.; Wang, L.K.; Lu, Y.B. *Rosmarinus officinialis* L. (Lamiales: Lamiaceae), a promising repellent plant for Thrips Management. *J. Econ. Entomol.* **2021**, *114*, 131–141. [CrossRef]
68. Sanei-Dehkordi, A.; Hatami, S.; Zarenezhad, E.; Montaseri, Z.; Osanloo, M. Efficacy of nanogels containing carvacrol, cinnamaldehyde, thymol, and a mix compared to a standard repellent against *Anopheles stephensi*. *Ind. Crop. Prod.* **2022**, *189*, 115883. [CrossRef]
69. Liu, S.Y.; Zhao, J.; Hamada, C.; Cai, W.L.; Khan, M.; Zou, Y.L.; Hua, H.X. Identification of attractants from plant essential oils for *Cyrtorhinus lividipennis*, an important predator of rice planthoppers. *J. Pest Sci.* **2019**, *92*, 769–780. [CrossRef]
70. Diabate, S.; Martin, T.; Murungi, L.K.; Fiaboe, K.K.M.; Subramanian, S.; Wesonga, J.; Deletre, E. Repellent activity of *Cymbopogon citratus* and *Tagetes minuta* and their specific volatiles against *Megalurothrips sjostedti*. *J. Appl. Entomol.* **2019**, *143*, 855–866. [CrossRef]
71. López, S.B.; López, M.L.; Aragón, L.M.; Tereschuk, M.L.; Slanis, A.C.; Feresin, G.E.; Zygadlo, J.A.; Tapia, A.A. Composition and anti-insect activity of essential oils from *Tagetes* L. Species (Asteraceae, Helenieae) on *Ceratitis capitata* Wiedemann and *Triatoma infestans* Klug. *J. Agric. Food Chem.* **2011**, *59*, 5286–5292. [CrossRef]
72. Dukic, N.; Markovic, T.; Mikic, S.; Cutovic, N. Repellent activity of basil, clary sage and celery essential oils on *Tribolium castaneum* (Herbst). *J. Stored Prod. Res.* **2023**, *103*, 102150. [CrossRef]
73. Jesser, E.; Yeguerman, C.; Stefanazzi, N.; Gomez, R.; Murray, A.P.; Ferrero, A.A.; Werdin-González, J.O. Ecofriendly approach for the control of a common insect pest in the food industry, combining polymeric nanoparticles and post-application temperatures. *J. Agric. Food Chem.* **2020**, *68*, 5951–5958. [CrossRef] [PubMed]
74. Chatzidaki, M.D.; Demisli, S.; Zingkou, E.; Liggri, P.G.V.; Papachristos, D.P.; Balatsos, G.; Karras, V.; Nallet, F.; Michaelakis, A.; Sotiropoulou, G.; et al. Essential oil-in-water microemulsions for topical application: Structural study, cytotoxic effect and insect repelling activity. *Colloid Surf. A-Physicochem. Eng. Asp.* **2022**, *654*, 10. [CrossRef]

75. Kotronia, M.; Kavetsou, E.; Loupassaki, S.; Kikionis, S.; Vouyiouka, S.; Detsi, A. Encapsulation of oregano (*Origanum onites* L.) essential oil in -cyclodextrin (-CD): Synthesis and characterization of the inclusion complexes. *Bioengineering* **2017**, *4*, 74. [CrossRef] [PubMed]
76. Caetano, A.R.S.; Cardoso, M.D.; Haddi, K.; Campolina, G.A.; De Souza, B.M.; Da SilvaLunguinho, A.; De Souza, L.; Nelson, D.L.; De Oliveira, J.E. *Rosmarinus officinalis* essential oil incorporated into nanoparticles as an efficient insecticide against *Drosophila suzukii* (Diptera: Drosophilidae). *Austral Entomol.* **2022**, *61*, 265–272. [CrossRef]
77. Gebaly, A.S.E.; Sofy, A.R.; Hmed, A.A.; Youssef, A.M. Combination of nanoparticles (NPs) and essential oils (EOs) as promising alternatives to non-effective antibacterial, antifungal and antiviral agents: A review. *Biocatal. Agric. Biotechnol.* **2024**, *57*, 103067. [CrossRef]
78. Rodrigues, A.B.L.; Martins, R.L.; Rabelo, É.; Tomazi, R.; Santos, L.L.; Brandao, L.B.; Faustino, C.G.; Farias, A.L.F.; dos Santos, C.B.R.; Cantuária, P.D.; et al. Development of nano-emulsions based on *Ayapana triplinervis* essential oil for the control of *Aedes aegypti* larvae. *PLoS ONE* **2021**, *16*, e0254225. [CrossRef] [PubMed]
79. Indriyani, N.N.; Al Anshori, J.; Permadi, N.; Nurjanah, S.; Julaeha, E. Bioactive components and their activities from different parts of *Citrus aurantifolia* (Christm.) swingle for food development. *Foods* **2023**, *12*, 2036. [CrossRef]
80. Koschier, E.H. Essential oil compounds for thrips control—A review. *Nat. Prod. Commun.* **2008**, *3*, 1171–1182. [CrossRef]

Disclaimer/Publisher's Note: The statements, opinions and data contained in all publications are solely those of the individual author(s) and contributor(s) and not of MDPI and/or the editor(s). MDPI and/or the editor(s) disclaim responsibility for any injury to people or property resulting from any ideas, methods, instructions or products referred to in the content.

agronomy

MDPI

Article

Thermal Tolerance and Host Plant Suitability of *Bemisia tabaci* MED (Gennadius) in Brazilian Legume Crops

Daniel de Lima Alvarez [1], Rafael Hayashida [2,*], Daniel Mariano Santos [1], Felipe Barreto da Silva [3], Cristiane Müller [4], Renate Krause-Sakate [1], William Wyatt Hoback [2] and Regiane Cristina de Oliveira [1]

[1] Crop Protection Department, School of Agronomic Sciences, São Paulo State University "Júlio de Mesquita Filho" (FCA/UNESP), Botucatu 18610-034, Brazil; daniel.alvarez@unesp.br (D.d.L.A.); daniel-mariano.santos@unesp.br (D.M.S.); renate.krause@unesp.br (R.K.-S.); regiane.cristina-oliveira@unesp.br (R.C.d.O.)

[2] Department of Entomology and Plant Pathology, Oklahoma State University, Stillwater, OK 74078, USA; whoback@okstate.edu

[3] Gulf Coast Research and Education Center, University of Florida, Wimauma, FL 33598, USA; barretodasilva.f@ufl.edu

[4] Corteva™ Agrisciences, Mogi Mirim 13814-000, Brazil; cristiane.muller@corteva.com

* Correspondence: rafael.hayashida@okstate.edu; Tel.: +1-405-894-0673

Abstract: The whitefly, *Bemisia tabaci,* is a complex of cryptic species that is a significant pest of different legume hosts that inhabits various regions worldwide with diverse climates and characteristics. Its adaptability is often facilitated by the insect's microbiome, which can contribute to both the metabolism of host plant secondary compounds and insecticide resistance. The most relevant biotypes in Brazil are Middle East-Asia Minor 1 (MEAM1) and Mediterranean (MED), because of their ability to damage different hosts. Although MEAM1 is the prevalent species in Brazil, MED has great potential to spread, and there is little current knowledge about the biology of this biotype in the country. Therefore, the objective of this study was to evaluate the development and viability of MED on two legumes, soybean and common bean, alongside cotton, bell pepper, and tomato, at temperatures of 20 °C, 23 °C, 26 °C, 29 °C, 32 °C, and 35 °C and characterize the composition of its endosymbionts. Temperatures between 23 °C and 32 °C were the most suitable for *B. tabaci* MED development and viability across all tested host plants, whereas 35 °C proved harmful for insects reared on legumes. We observed a temperature threshold (°C) and thermal constant (degree-days) that varied according to the host plant, ranging from 9.81 °C and 384.62 for soybean to 11.17 °C and 333.33 for bell pepper, respectively. The main endosymbionts were in a ratio of 80% *Hamiltonella* and 20% *Cardinium*. These results allow the future mapping of risk for the MED biotype on different host plants in Brazil and elsewhere in South America.

Keywords: whitefly; soybean; common bean; thermal thresholds; thermal constant

1. Introduction

The whitefly, *Bemisia tabaci* (Gennadius) (Hemiptera: Aleyrodidae), is a species complex recognized as a major global pest, with its wide host range and broad geographical distribution. It has been reported on over 700 plant species, including numerous economically important legumes [1,2]. Currently, *B. tabaci* is recognized as a complex of 45 cryptic species that are morphologically indistinguishable and can only be differentiated through molecular analysis [3–8]. Among these, the most prominent biotypes worldwide

https://doi.org/10.3390/agronomy15071622

are Middle East-Asia Minor 1 (MEAM1) and Mediterranean (MED), which are also the most economically significant members of the *B. tabaci* complex [4,9].

Brazil stands as a very important global producer of legumes such as soybean (*Glycine max*) and common bean (*Phaseolus vulgaris*). According to the USDA Foreign Agricultural Service (FAS), the planted area for soybeans in Brazil during the 2024/2025 season was 47.4 million hectares, with an annual production of 169 million tons [10]. The common bean is a staple food in Brazil, playing a crucial role in the country's food security and agricultural economy. In the 2024/25 season, the planted area was 2784.8 thousand hectares, with a total production of 3172 thousand tons [11]. The large areas of legumes planted, combined with climatic conditions and agronomic practices, contribute to the rapid increase in whitefly population sizes and dispersal [12–14].

The *B. tabaci* management strategies in Brazil—particularly for soybean (*G. max*), common bean (*P. vulgaris*), and cotton (*Gossypium hirsutum*)—have largely relied on insecticides as the primary method of control [15–17]. This approach may contribute to the competitive advantage of the MED species over MEAM1 in soybean fields, as MED possesses a natural resistance to several insecticides [18]. Additionally, Brazil's tropical climate supports year-round whitefly populations. The country's agricultural landscape, characterized by the proximity of soybean, common bean, cotton, and vegetable crops, further facilitates the continuous presence and spread of whiteflies under field conditions.

For the whitefly biotypes, temperature also plays an important role in influencing distribution [19–21]. Although MEAM1 remains the dominant biotype in Brazil, the presence of MED has also been reported, including in soybean fields [15,22], and in some regions of the world MED has already become the prevailing biotype [23]. Previous studies have identified MED as an efficient vector of cowpea mild mottle virus (CPMMV) in both soybean and common bean [22]. This highlights potential future challenges in managing MED whiteflies and suggests a likely increase in CPMMV incidence in soybean and common bean crops.

Numerous studies have examined the impact of temperature on the performance of MEAM1 and/or MED whiteflies [19–21,24–28]. Both whitefly biotypes typically complete their development from egg to adult within a temperature range of 15–35 °C, although survival rates are often significantly reduced at temperatures below 20 °C or above 30 °C [27–29]. However, the two biotypes exhibit significant differences, which may have important implications for competitive interactions [30]. The MED biotype generally shows greater tolerance to high temperatures than the MEAM1 biotype, particularly during the adult stage [25]. In areas where both biotypes have invaded, MED has been reported to displace MEAM1, indicating that thermal tolerance may influence inter-biotype competition [30,31].

Recent studies have increasingly emphasized the importance of characterizing bacterial endosymbionts in whitefly populations [32–35]. The primary endosymbiont *Portiera aleyrodidarum* is consistently present across all whitefly species, while secondary endosymbionts—such as *Arsenophonus*, *Cardinium*, *Fritschea*, *Hamiltonella*, *Hemipteriphilus*, *Rickettsia*, and *Wolbachia*—occur in varying combinations [34,35]. Wild-type populations may host over 60 bacterial genera, some of which remain undescribed [36,37]. Because the presence of these endosymbionts can provide adaptive plasticity to whiteflies [35], assessing their influence on thermal susceptibility and host plant suitability is essential for a comprehensive understanding.

Given the limited number of studies in Brazil examining the factors influencing the establishment of *B. tabaci* MED, and the growing significance of this biotype in legume crops, this study aimed to assess the performance of MED on key legumes crops in Brazil—soybean and common bean—under controlled constant temperature regimes, in comparison with three additional host plants, and characterization of the whitefly microbiome. Understanding these dynamics is essential for predicting the potential establishment and spread of MED across Brazil's diverse climatic regions—particularly considering the country's continental scale and substantial thermal variability. Additionally, preliminary assessment of the microbiome was conducted as part of this study as previous research has shown the microbiome to influence insecticide resistance [32], but less is known about its importance with host plant use.

2. Materials and Methods

2.1. *Whitefly Colony*

Samples of *B. tabaci* previously maintained in the laboratory were identified as MED through molecular analysis. MED adults were placed in cages made of polyester mesh (45 × 45 × 55 cm) and reared on bell pepper (*Capsicum annuum*) and common bean (*P. vulgaris*). The rearing was conducted in a climate-controlled environment set at 26 ± 2 °C, with 70 ± 10% relative humidity and a 14 h photoperiod.

2.2. *Identification of the Biotype of B. tabaci and Presence of Endosymbiont Group*

The whitefly biotype was identified by analysis of the mitochondrial gene cytochrome oxidase I (mtCOI). Initially, individual extraction of the DNA of 20 individuals was performed for each sample collected, using the Chelex (Bio-Rad, Richmond, CA, USA) protocol [38]. The samples were submitted to a PCR reaction with the genetic primers C1-J-2195 and TL2-N-3014 [39]. The PCR reaction was performed in a final volume of 50 µL, containing 50 mM $MgCl_2$, 2.5 mM dNTPs, and 1 µM oligonucleotides, using 0.5 units of Taq polymerase. The reaction cycle consisted of an initial denaturation at 94 °C for 5 min, followed by 35 cycles of 94 °C for 30 s, 45 °C for 45 s, and 72 °C for 1 min, with a final extension at 72 °C for 10 min. Next, polymorphism analysis of the amplicons was conducted using Restriction Fragment Length Polymorphism (RFLP) [40]. From each PCR product (880 bp), 5 µL were used and digested with one unit of TaqI at 65 °C for 2 h in a final volume of 15 µL. The digested DNA was evaluated on a 1% agarose gel stained with ethidium bromide.

The same samples were also subjected to the detection of the primary endosymbiont, *Portiera aleyrodidarum*, and six secondary endosymbionts: *Hamiltonella*, *Rickettsia*, *Wolbachia*, *Cardinium*, *Arsenophonus*, and *Fritschea* [21,32,34,41]. Specific primers for these genera were used to amplify regions of the 16S and 23S genes.

2.3. *Determination of Thermal Requirements in Different Hosts*

The period of development of *B. tabaci* MED was evaluated at constant temperatures of 20 °C, 23 °C, 26 °C, 29 °C, 32 °C, and 35 °C on five different hosts: "TAA Dama" common bean *P. vulgaris*, "NEO 580" soybean (*G. max*), "Taurus" bell pepper (*C. annuum*), "FM 985 GLTP" cotton (*Gossypium hirsutum*), and "IPA 6" tomato (*Solanum lycopersicum*), in a BOD (Biochemical Oxygen Demand) incubator chamber (ELETROlab, model EL 202/4, São Paulo, Brazil) with a relative humidity of 70 ± 10% and photophase of 14:10 h L:D. Plant hosts aged 4 to 6 weeks after germination were used in the experiment. The experiment was conducted in September 2022. In this experiment, we tested whiteflies originating from a colony reared on bell pepper. However, to avoid using a pre-adapted host, we reared the whiteflies on common beans for about ten generations before testing bell pepper as the host plant.

For each temperature, the five hosts were tested, with six replications of 20 individuals per sample unit. Approximately 30 adult whiteflies with a maximum age of 72 h were collected from the colony using a Pasteur-type glass pipette and a flexible hose and placed in a 'clip-cage' on their respective hosts. Whiteflies remained in the cage for an oviposition period of 48 h. After this period, the adults were removed from the clip-cages. The leaves were examined using a magnifying glass and 20 eggs were selected for monitoring. The area for observation was marked with non-toxic paint glue. Subsequently, the host plants containing the eggs were placed at their respective pre-established temperatures. The development of individuals was monitored daily until they reached their adult stage.

2.4. Analysis

From the data collected, the temperature thresholds (Tb) and the value of the thermal constant (K) in degree-days (GD) were calculated using the Hyperbole method [42]. The development time for each temperature and host plant was established by Kaplan–Meier curves [43]. For the comparison of curves, the non-parametric analysis was used by the log-rank test at 5% of significance [44].

The number of *B. tabaci* MED individuals that completed egg–adult stage development was counted and viability (%) was calculated. The number of individuals completing development was obtained for each host at each temperature and analyzed using a two-way ANOVA, with post hoc comparisons performed using Tukey's tests. The assumptions of normality and homogeneity of variances were tested before proceeding with the tests. The statistical analyses were performed in the R computing environment, utilizing the "AgroR" (version 1.3.6), "survival" (version 3.8-3), "survminer" (version 0.5.0), "ggsurvplot" (version 0.4.9) and "ggplot2" (version 3.5.2) packages.

3. Results

The analysis of the mitochondrial gene cytochrome oxidase I (mtCOI) revealed that the colony whiteflies were all the MED biotype. The corresponding nucleotide sequence was deposited in GenBank under accession number MK900733.1. The presence of the primary endosymbiont *Portiera aleyrodidarum* was detected in 100% of our population. The composition of secondary endosymbionts in the tested population was 80% *Hamiltonella* and 20% *Cardinium*, while *Rickettsia*, *Wolbachia*, *Arsenophonus*, and *Fristchea* were not detected (Table 1).

Table 1. Infection frequencies (%) of secondary endosymbionts in the *Bemisia tabaci* MED population.

Species	**Endosymbionts (%)**					
	Hamiltonella	***Rickettsia***	***Wolbachia***	***Cardinium***	***Arsenophonus***	***Fristchea***
Bemisia tabaci MED	80	0	0	20	0	0

Significant interactions were observed in the number of viable individuals at the egg-to-adult stage among temperature and host plants tested (F [4,5] = 1.79; $p = 0.026$) (Table 2). The highest viability of individuals on soybean was recorded at temperatures between 23 °C and 32 °C, with mean values ranging from 9.33 ± 2.28 to 13.17 ± 1.89 (mean ± 1 SE). In contrast, the lowest viability occurred at 35 °C, with only 2.00 ± 0.52 individuals surviving.

In common bean, the temperature range supporting viability was broader than in soybean. No significant differences were observed in the number of viable individuals reared at temperatures between 20 °C and 32 °C, ranging between 7.67 ± 2.42 and 12.67 ± 1.65. However, no individuals completed development at 35 °C (Table 2).

Table 2. Number of viable *Bemisia tabaci* MED individuals in the egg–adult stage development (mean ± SE) and viability (%) in different hosts across different temperatures [1].

Treatments	20 °C	%	23 °C	%	26 °C	%	29 °C	%	32 °C	%	35 °C	%
Soybean	4.17 ± 0.95 aBC	20.8	10.83 ± 2.17 aAB	54.2	13.17 ± 1.89 aA	65.8	9.33 ± 2.28 bAB	46.7	10.50 ± 2.64 abAB	52.5	2.00 ± 0.52 bcC	10.0
Common Bean	7.67 ± 2.42 aA	38.3	12.67 ± 1.65 aA	63.3	11.17 ± 1.64 aA	55.8	11.00 ± 1.88 abA	55.0	9.00 ± 1.92 bA	45.0	0.00 ± 0.00 cB	0.0
Cotton	6.50 ± 2.77 aC	32.5	15.67 ± 0.42 aA	78.3	17.00 ± 2.16 aA	85.0	17.67 ± 1.23 aA	88.3	14.67 ± 2.01 abAB	73.3	7.83 ± 1.40 abBC	39.2
Bell pepper	4.50 ± 0.72 aB	22.5	11.83 ± 1.08 aA	59.2	16.17 ± 1.54 aA	80.8	16.33 ± 1.09 abA	81.7	17.50 ± 1.09 aA	87.5	10.33 ± 2.25 aAB	51.7
Tomato	7.67 ± 2.42 aB	38.3	14.33 ± 1.33 aAB	71.7	15.83 ± 1.94 aA	79.2	15.00 ± 1.75 abAB	75.0	12.00 ± 2.83 abAB	60.0	11.17 ± 2.52 aAB	55.8
$F_{\text{(host; temperature; host×temperature)}}$	11.20; 24.34; 1.79											
$P_{\text{(host; temperature; host×temperature)}}$	<0.001; <0.001; 0.026											
$DF_{\text{residuals}}$	150											

[1] Means ± 1 SE followed by the same lowercase letter in the column and the same uppercase letter in the row do not differ significantly from each other according to Tukey's test (α = 0.05). Statistical analysis was performed on data transformed using the Box–Cox method.

In cotton, the highest number of viable individuals was observed between 23 °C and 32 °C, with the number of individuals varying between 14.67 ± 2.01 and 17.67 ± 1.23. For bell pepper and tomato, the greatest number of viable individuals was found between 23 °C and 35 °C. For these plants, 20 °C resulted in the lowest viability, with only 4.50 ± 0.72 and 7.67 ± 2.42 viable individuals in bell pepper and tomato, respectively (Table 2).

At temperatures of 20 °C, 23 °C, and 26 °C, no difference in the number of viable individuals was observed among host plants. However, at 29 °C, the lowest viability was observed in soybean, while at 32 °C and 35 °C, the lowest viability was recorded in common bean.

The lowest temperature threshold (Tb) differed slightly among host plants, with the lowest observed in soybean (9.81 °C), followed by tomato (9.89 °C), common bean (10.00 °C), cotton (10.37 °C), and bell pepper (11.17 °C; Table 3). The lowest thermal constant (K) was observed in bell pepper (333.33 degree-days), while the highest was observed in soybean (384.62 degree-days). Common bean, cotton, and tomato showed similar values of K (370.37 degree-days). Somewhat surprisingly, the development of whiteflies reared at 35 °C was longer than that of those reared at 32 °C, which deviates from the expected thermal performance curve. This anomaly suggests that 35 °C may exceed the optimal thermal range for development, potentially inducing heat stress. Therefore, to calculate the regression, the values at 35 °C were discarded. All the curves fit the equation well, with R^2 values greater than 0.90 for all host plants (Figure 1).

Table 3. Temperature threshold (Tb) and thermal constant (K) of white fly *Bemisia tabaci* (Gennadius) MED egg–adult stage.

Host	Tb (°C)	Thermal Constant (K)
Soybean	9.81	384.62
Common Bean	10.00	370.37
Cotton	10.37	370.37
Bell Pepper	11.17	333.33
Tomato	9.89	370.37

Based on the Kaplan–Meier curves, the fastest development times in soybean were observed at temperatures of 32 °C and 29 °C, where insects completed their development in approximately 18.5 to 20 days. At the highest tested temperature, 35 °C, the development period increased by about 5 days compared to 32 °C and 29 °C. The slowest development occurred at 20 °C, where insects took an average of 41.17 days to complete their development (χ^2 = 53.767, *df* = 5, *p* < 0.001; Figure 2A).

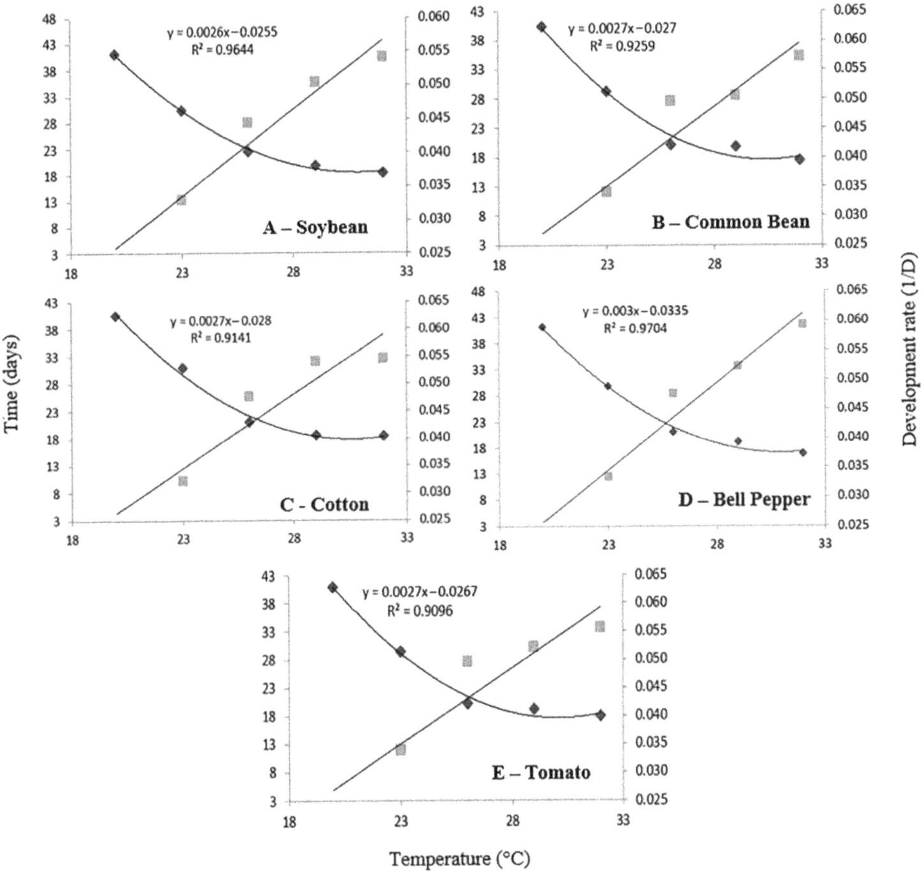

Figure 1. Temperature-dependent development time and rate of *Bemisia tabaci* MED on five host plants.

Similarly, when common bean was used as the host plant, 20 °C resulted in the longest development time, with insects taking about 40.33 days. No significant differences were observed among the temperatures of 26 °C, 29 °C, and 32 °C, with development durations ranging from 18.83 to 20.17 days ($\chi^2 = 0.172$, $df = 4$, $p = 0.967$). No whiteflies survived at 35 °C (Figure 2B).

In cotton, the shortest development times were observed at 26 °C, 29 °C, and 32 °C, with insects completing development in approximately 18.5 to 21 days, with no significant differences among these temperatures ($\chi^2 = 5.298$, $df = 5$, $p = 0.102$). At 20 °C, development was significantly slower, with insects taking more than twice as long (Figure 2C).

For bell pepper and tomato, the fastest development times were observed at the highest tested temperatures (Figure 2D,E). At 32 °C and 35 °C, insects reared on bell pepper completed development in 17 to 19.5 days ($\chi^2 = 46.885$, $df = 5$, $p < 0.001$). On tomato, development at 29 °C, 32 °C, and 35 °C took between 19 and 19.67 days ($\chi^2 = 34.593$, $df = 5$, $p < 0.001$). For both host plants, the lowest temperature (20 °C) significantly extended the development period, with durations of 41.67 days for bell pepper and 41.0 days for tomato.

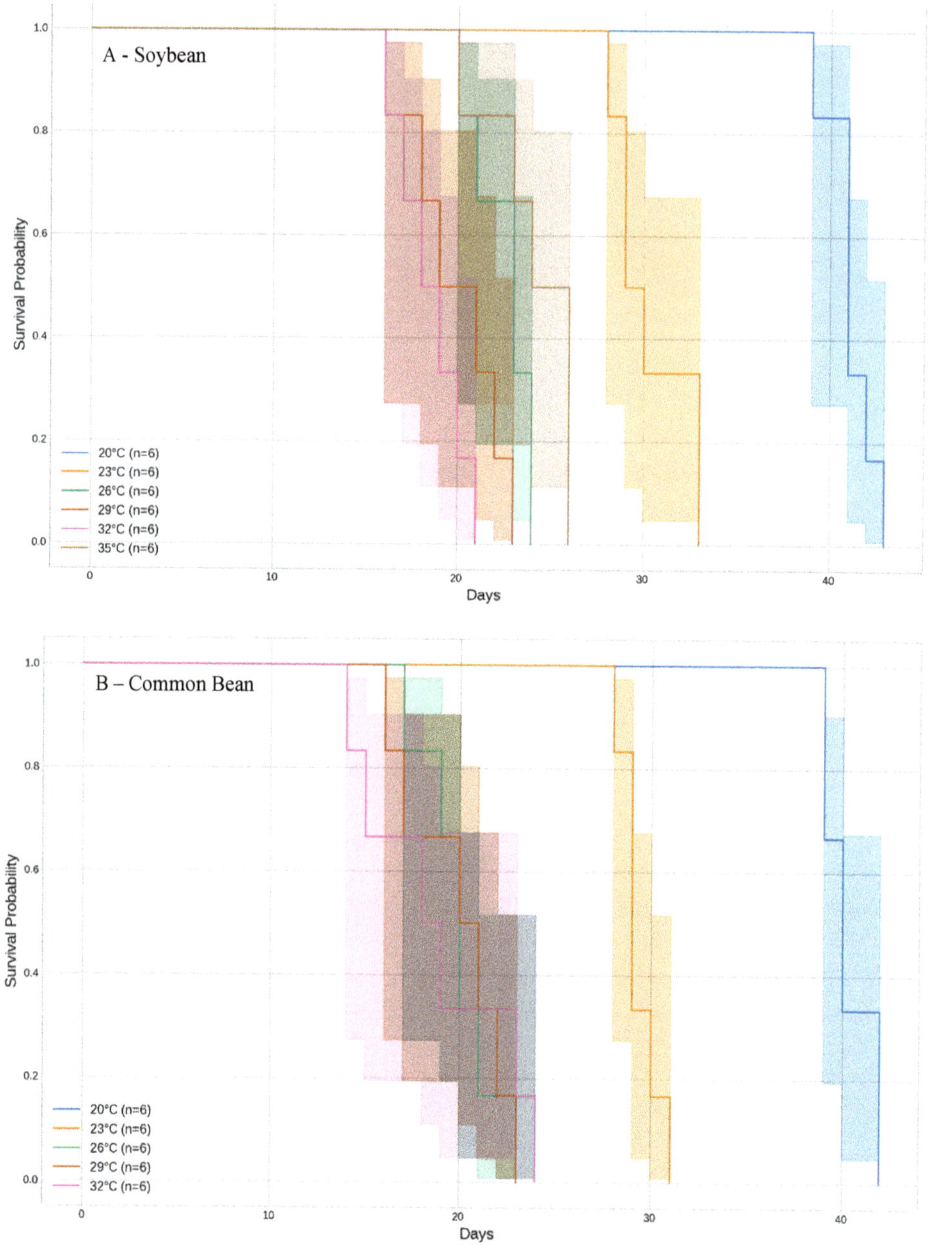

Figure 2. *Cont.*

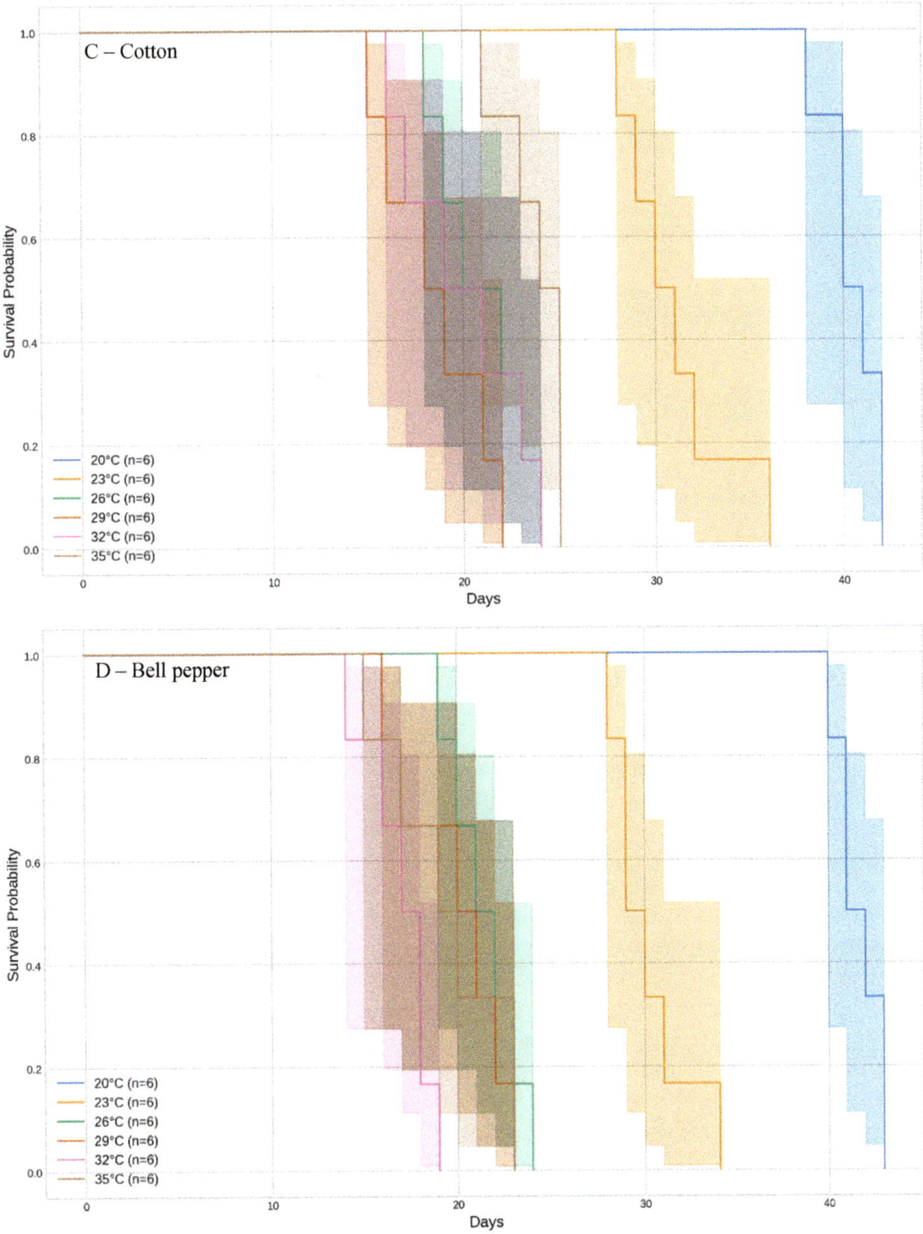

Figure 2. *Cont.*

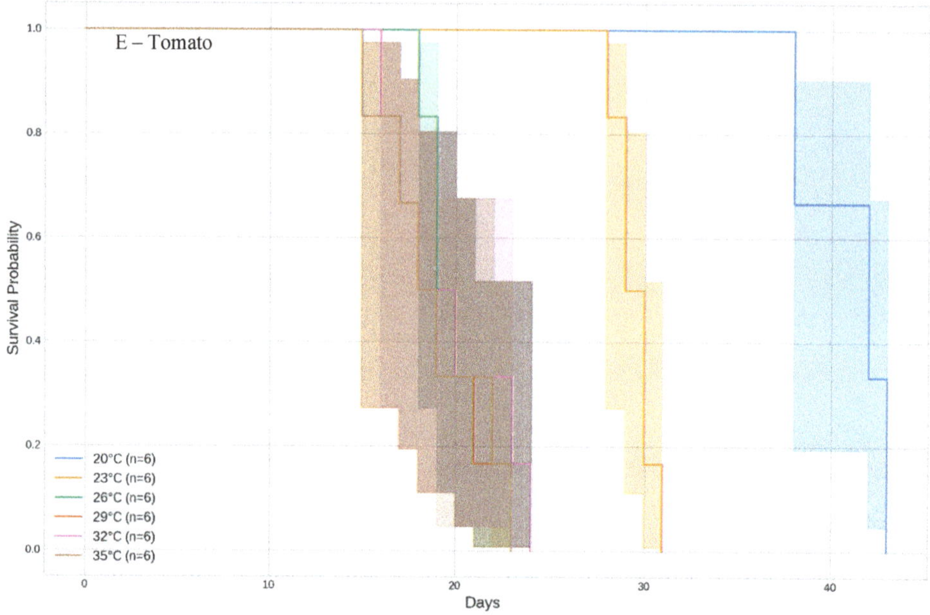

Figure 2. Development time of whitefly *Bemisia tabaci* (Gennadius) MED across different temperature regimes on five host plants, analyzed using Kaplan–Meier curves.

4. Discussion

Although MED is present in Brazil, it has not yet become widespread [15,22]. Studies on the biotic and abiotic factors affecting MED's development and distribution warrant further investigation, particularly given the economic significance of legume crop production in the country.

Temperature and host plant interactions are critical factors influencing *B. tabaci* MED development [15,25,27,45,46]. Our study revealed that high temperatures of 35 °C were nearly unsuitable for MED development on soybean and common bean, with minimal survival to adulthood. In contrast, on tomato, over 50% of nymphs successfully reached adulthood at 35 °C, demonstrating significant host-effects in terms of thermal range. Similar results were reported by Bonato et al. [27], where the optimum temperate for immature development was estimated to be 32.5 °C, and temperatures higher than 35 °C were detrimental to *B. tabaci* MED survivorship, reducing the survival of immatures and longevity. Additionally, these authors reported that the population found on tomato was more tolerant to high temperatures (>33 °C). The optimal development observed here, between 26–32 °C across most host plants, indicates that these temperatures represent the greatest risk for rapid population growth and potential crop damage. This information can assist monitoring and management strategies, particularly in greenhouse settings, where temperature can be manipulated.

Furthermore, the differences in temperature thresholds (Tb) and thermal constants (K) among host plants suggest that development models should be host-specific for more accurate predictions of whitefly phenology. Similar results have been reported in the literature for whitefly MED, with temperature thresholds ranging from 8 °C in tomato [27] to 10.2 °C in poinsettia [46]. Thermal constants have also been reported to vary, from 327 in poinsettia [47], to as high as 400 in tomato [46]. These differences among host plants may be explained by the host's leaf surface morphology—particularly trichome density, type, and length—as well as its metabolites [48,49], in addition to the plant's stomatal morphology and distribution [50], which can alter the microclimate at the leaf surface, thereby affecting the thermal biology of whitefly development [51].

Our experimental data indicate that temperatures between 23 °C and 32 °C are most favorable for MED development on soybean, and between 20 °C and 32 °C for common bean—the main legumes cultivated in Brazil. When considering Brazil's climatological normal, ranging from 27 °C in the northern region to 18.4 °C in the south [52], most regions of Brazil, including the central-west, where 50% of the soybean production is concentrated [10], and the south-central region, where more than 70% of common beans are produced [11], are climatically suitable for MED colonization (Figure 3).

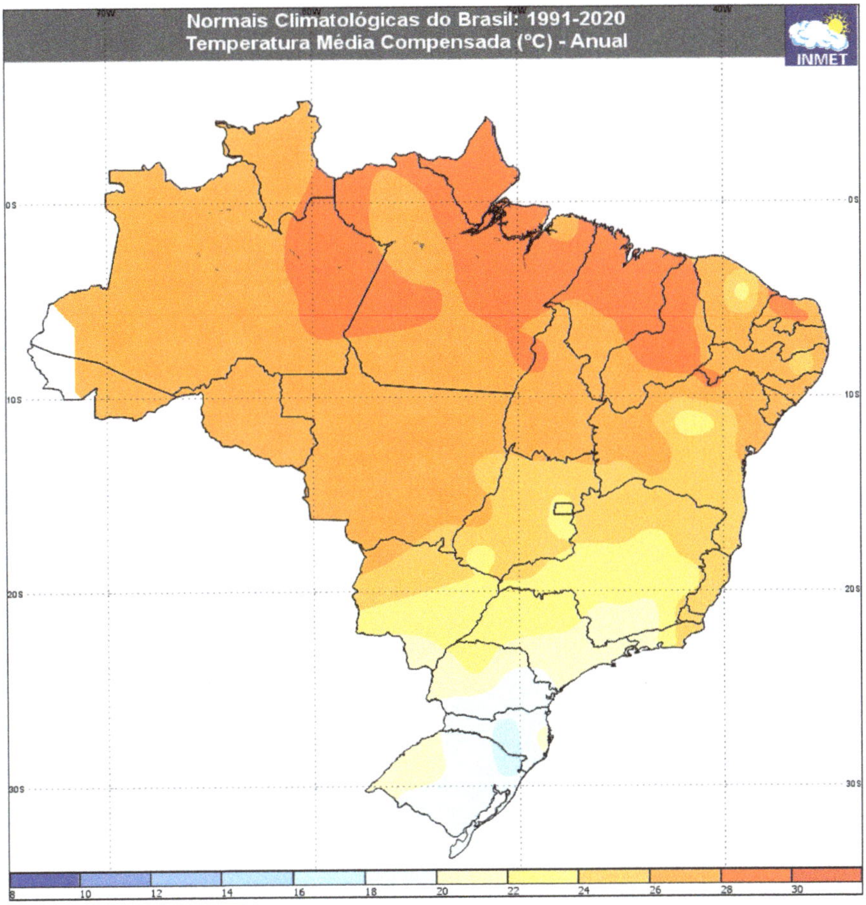

Figure 3. Climatological thermal averages in Brazil: 1991–2020 (https://clima.inmet.gov.br/temp) (Accessed on 20 April 2023).

While the lethal high temperature for MED has been reported at 45–46 °C [25], our data showed minimal survival on legumes at 35 °C. Notably, endosymbionts play a significant role in this thermal adaptation. Infection with *Rickettsia* has been shown to induce the expression of thermotolerance-associated genes [53], and *Cardinium* (present in 20% of our sampled individuals) has been linked to increased thermal tolerance [21]. *Cardinium* also influences protein levels associated with growth and energy metabolism [35], potentially explaining some of our observed variation in differential development across host plants.

Additionally, the presence of *Hamiltonella* (found in 80% of our samples) is often associated with increased whitefly growth rates, especially on nutritionally poor host plants [54], while *Cardinium* decreases the whitefly detoxification metabolism ability and decrease the defense response of the host plant [55]. Considering that the association of these two endosymbionts impacts performance and reproduction on whitefly MED [56] and play important role in the host plant–whitefly interaction, future studies should examine endosymbionts after feeding trials at different temperatures and with different host plants to discover if the insects that survived warmer temperatures had these bacteria.

These findings gain greater significance when considered alongside climate change projections for Brazil, which indicate annual mean temperature increases of up to 2.2 °C by 2050 [57]. Projections also estimate that the impacts of climate change on biodiversity in Brazil will vary spatially, with central and northern regions expected to experience the most severe temperature increases [58]. While these areas are already within the optimal temperature range for whitefly development, the projected changes could push temperatures during extreme heat events beyond the upper thermal limits for whitefly survival—particularly for the MED species on certain hosts. In such cases, extreme heat could occasionally serve as a natural population control mechanism in the hottest regions, especially in legume-producing areas. This is particularly significant when contrasted with the MEAM1 biotype, which tends to exhibit reduced fitness at higher temperatures [25,59]. Therefore, MED's greater tolerance to high temperatures compared to MEAM1 may confer a competitive advantage in terms of distribution and population growth, particularly under field or greenhouse conditions where elevated temperatures are common [25].

It is also crucial to consider the strategies growers adopt when producing legumes in Brazil. Soybean growers frequently practice crop succession with cotton, especially in the central-west region [60], while common bean growers often diversify with vegetables including tomatoes and bell peppers [61,62]. In regions where temperatures increasingly exceed 32 °C due to climate change, we might expect shifts in host plant preference or performance. Crops like tomatoes, which supported better whitefly MED survival at high temperatures in our study, could face increased pest pressure. This potential shift in host plant suitability could have significant implications for agricultural practices, particularly in regions where farmers practice crop succession or rotation involving both legumes and solanaceous crops.

These practices also create complex landscapes that influence whitefly population dynamics. Previous research has shown that, in competitive scenarios without insecticide exposure, MED and MEAM1 remain in equilibrium [14]. However, intensive insecticide use (particularly in cotton cultivation where applications can exceed 20 per season [63]) favors MED. This pattern has been observed in Israel, where MED surpassed MEAM1 during periods of intensive cotton cultivation and insecticide use, while MEAM1 regained dominance when these practices were reduced [23,64].

In the context of integrated pest management, these findings are also important for understanding and exploring the interactions between whitefly MED and its natural enemies [65]. For example, temperature plays a significant role in the development and fecundity of *Encarsia acaudaleyrodis* Hayat, an important biological control agent of whiteflies. Temperatures of 32 °C shorten the development period from egg to adult; however, they also reduce oviposition [66]. The highest intrinsic rate of population increase was observed at around 25 °C, indicating that this moderate temperature is favorable for the biological control activity of *E. acaudaleyrodi* [65,66]. Information about the optimal temperatures for the parasitoid, combined with the favorable temperature range for MED development observed in this study (26–32 °C), can be valuable for laboratory mass-rearing programs, helping to optimize production efficiency.

Despite recent surveys indicating that MED is not yet the predominant species in Brazil [15], our findings highlight the need for continuous monitoring. The complex interactions between temperature, host plants, endosymbionts, and agricultural practices could rapidly shift the competitive balance between whitefly biotypes. Effectively managing whitefly populations requires a comprehensive understanding of these symbiotic relationships and their influence on insect biology and stress tolerance [2]. Given Brazil's dimensions and cropping diversity, continuous sampling efforts are necessary to accurately describe MED's distribution and potential spread.

5. Conclusions

In conclusion, the potential for MED to expand its range in Brazil depends on a complex interaction of factors, including temperature tolerance, host plant suitability, endosymbiont associations, and agricultural practices. Our findings suggest that the conditions favorable to MED exist across much of Brazil, particularly in cropping systems with intensive insecticide use. Future research should focus on mapping the varying compositions of endosymbiont distributions in field populations and modeling how climate change might further alter the competitive dynamics between these economically important pest biotypes on legumes.

Author Contributions: Conceptualization, D.d.L.A., R.C.d.O. and C.M.; methodology, D.d.L.A., R.C.d.O., C.M., D.M.S., F.B.d.S. and R.K.-S.; software, D.d.L.A., D.M.S. and R.H.; validation, D.d.L.A. and R.H.; formal analysis, D.d.L.A. and R.H.; investigation, D.d.L.A., D.M.S. and F.B.d.S.; resources, R.C.d.O. and R.K.-S.; data curation, D.d.L.A. and R.H.; writing—original draft preparation D.d.L.A., R.H., R.C.d.O. and W.W.H.; writing—review and editing, D.d.L.A., R.H., R.C.d.O. and W.W.H.; visualization, D.d.L.A., R.H., R.C.d.O. and W.W.H.; supervision, R.C.d.O., R.K.-S., C.M. and W.W.H.; project administration, D.d.L.A. and R.C.d.O.; funding acquisition, R.C.d.O. All authors have read and agreed to the published version of the manuscript.

Funding: This study was funded in part by the Coordenação de Aperfeiçoamento de Pessoal de Nível Superior—Brasil (CAPES)—Finance Code 001; by the Fundação de Amparo à Pesquisa do Estado de São Paulo (FAPESP; process numbers 2017/21588-7, 2018/02317-5, 2019/10736-0, and 2018/19782-2); and by the Conselho Nacional de Desenvolvimento Científico e Tecnológico (CNPq; 304126/2019-5). Regiane C. de Oliveira holds a CNPq fellowship.

Data Availability Statement: The original contributions presented in the study are included in the article; further inquiries can be directed to the corresponding author.

Acknowledgments: The authors would like to acknowledge the Department of Entomology and Plant Pathology at Oklahoma State University for all the support given to this research and the financial support provided by the following agencies: Coordenação de Aperfeiçoamento de Pessoal de Nível Superior—Brasil (CAPES) and Conselho Nacional de Desenvolvimento Científico e Tecnológico—CNPq. During the preparation of this manuscript, the author(s) used Microsoft 365 Copilot to assist with grammar and language editing. The author(s) reviewed and edited the content as needed and take full responsibility for the final version of the manuscript.

Conflicts of Interest: The author Cristiane Müller was employed by Corteva Agriscience. The remaining authors declare that the research was conducted in the absence of any commercial or financial relationships that could be construed as a potential conflict of interest.

References

1. Navas-Castillo, J.; Fiallo-Olivé, E.; Sánchez-Campos, S. Emerging Virus Diseases Transmitted by Whiteflies. *Annu. Rev. Phytopathol.* **2011**, *49*, 219–248. [CrossRef] [PubMed]
2. Kavallieratos, N.G.; Wakil, W.; Eleftheriadou, N.; Ghazanfar, M.U.; El-Shafie, H.; Simmons, A.M.; Dimase, M.; Smith, H.A.; Chandler, D. Integrated Management System of the Whitefly *Bemisia tabaci*: A Review. *Entomologia* **2024**, *44*, 1117–1133. [CrossRef]

3. Boykin, L.M.; Bell, C.D.; Evans, G.; Small, I.; De Barro, P.J. Is Agriculture Driving the Diversification of the *Bemisia tabaci* Species Complex (Hemiptera: Sternorrhyncha: Aleyrodidae)?: Dating, Diversification and Biogeographic Evidence Revealed. *BMC Evol. Biol.* **2013**, *13*, 228. [CrossRef] [PubMed]
4. De Barro, P.J.; Liu, S.-S.; Boykin, L.M.; Dinsdale, A.B. *Bemisia tabaci*: A Statement of Species Status. *Annu. Rev. Entomol.* **2011**, *56*, 1–19. [CrossRef]
5. Jiu, M.; Hu, J.; Wang, L.-J.; Dong, J.-F.; Song, Y.-Q.; Sun, H.-Z. Cryptic Species Identification and Composition of *Bemisia tabaci* (Hemiptera: Aleyrodidae) Complex in Henan Province, China. *J. Insect Sci.* **2017**, *17*, 78. [CrossRef]
6. Lee, W.; Park, J.; Lee, G.-S.; Lee, S.; Akimoto, S. Taxonomic Status of the *Bemisia tabaci* Complex (Hemiptera: Aleyrodidae) and Reassessment of the Number of Its Constituent Species. *PLoS ONE* **2013**, *8*, e63817. [CrossRef]
7. MacLeod, N.; Canty, R.J.; Polaszek, A. Morphology-Based Identification of *Bemisia tabaci* Cryptic Species Puparia via Embedded Group-Contrast Convolution Neural Network Analysis. *Syst. Biol.* **2022**, *71*, 1095–1109. [CrossRef]
8. Mugerwa, H.; Seal, S.; Wang, H.-L.; Patel, M.V.; Kabaalu, R.; Omongo, C.A.; Alicai, T.; Tairo, F.; Ndunguru, J.; Sseruwagi, P.; et al. African Ancestry of New World, *Bemisia tabaci*-Whitefly Species. *Sci. Rep.* **2018**, *8*, 2734. [CrossRef]
9. Wan, F.-H.; Yang, N.-W. Invasion and Management of Agricultural Alien Insects in China. *Annu. Rev. Entomol.* **2016**, *61*, 77–98. [CrossRef]
10. Brazil Soybean Area, Yield and Production. Available online: https://ipad.fas.usda.gov/countrysummary/Default.aspx?id=BR&crop=Soybean (accessed on 24 June 2025).
11. Companhia Nacional de Abastecimento (CONAB). *Boletim da Safra de Grãos—9º Levantamento—Safra 2024/25*; Grãos; Conab: Brasília, Brazil, 2025; p. 135.
12. Filho, A.B.; Inoue-Nagata, A.K.; Bassanezi, R.B.; Belasque, J.; Amorim, L.; Macedo, M.A.; Barbosa, J.C.; Willocquet, L.; Savary, S. The Importance of Primary Inoculum and Area-Wide Disease Management to Crop Health and Food Security. *Food Sec.* **2016**, *8*, 221–238. [CrossRef]
13. Ferreira Rodrigues, R.H.; Silva, L.B.; Silva, M.C.F.; Lopes, J.W.B.; Araujo Lima, E.; Sobreira Barbosa, R.; Oliveira Almeida, L.F. Population Fluctuation and Distribution of *Bemisia tabaci* MEAM1 (Hemiptera: Aleyrodidae) in Soybean Crops. *Front. Agron.* **2022**, *4*, 958498. [CrossRef]
14. Watanabe, L.F.M.; Bello, V.H.; De Marchi, B.R.; Silva, F.B.d.; Fusco, L.M.; Sartori, M.M.P.; Pavan, M.A.; Krause-Sakate, R. Performance and Competitive Displacement of *Bemisia tabaci* MEAM1 and MED Cryptic Species on Different Host Plants. *Crop Prot.* **2019**, *124*, 104860. [CrossRef]
15. Fernandes, D.S.; Okuma, D.; Pantoja-Gomez, L.M.; Cuenca, A.; Corrêa, A.S. *Bemisia tabaci* MEAM1 Still Remains the Dominant Species in Open Field Crops in Brazil. *Braz. J. Biol.* **2022**, *84*, e256949. [CrossRef]
16. Ferreira, A.L.; Ghanim, M.; Xu, Y.; Pinheiro, P.V. Interactions between Common Bean Viruses and Their Whitefly Vector. *Viruses* **2024**, *16*, 1567. [CrossRef]
17. Bevilaqua, J.G.; Padilha, G.; Pozebon, H.; Marques, R.P.; Cargnelutti Filho, A.; Ramon, P.C.; Boeni, L.; Castilhos, L.B.; Da Luz, G.R.; Brum, A.L.S.D.S.; et al. A Sustainable Approach to Control Whitefly on Soybean: Integrating Entomopathogenic Fungi with Insecticides. *Crop Prot.* **2023**, *164*, 106145. [CrossRef]
18. Horowitz, A.R.; Ghanim, M.; Roditakis, E.; Nauen, R.; Ishaaya, I. Insecticide Resistance and Its Management in *Bemisia tabaci* Species. *J. Pest Sci.* **2020**, *93*, 893–910. [CrossRef]
19. Dai, T.M.; Wang, Y.S.; Liu, W.X.; Lü, Z.C.; Wan, F.H. Thermal Discrimination and Transgenerational Temperature Response in *Bemisia tabaci* Mediterranean (Hemiptera: Aleyrodidae): Putative Involvement of the Thermo-Sensitive Receptor BtTRPA. *Environ. Entomol.* **2018**, *47*, 204–209. [CrossRef]
20. Pan, H.; Preisser, E.L.; Chu, D.; Wang, S.; Wu, Q.; Carrière, Y.; Zhou, X.; Zhang, Y. Insecticides Promote Viral Outbreaks by Altering Herbivore Competition. *Ecol. Appl.* **2015**, *25*, 1585–1595. [CrossRef]
21. Yang, K.; Yuan, M.-Y.; Liu, Y.; Guo, C.-L.; Liu, T.-X.; Zhang, Y.-J.; Chu, D. First Evidence for Thermal Tolerance Benefits of the Bacterial Symbiont *Cardinium* in an Invasive Whitefly, *Bemisia tabaci*. *Pest Manag. Sci.* **2021**, *77*, 5021–5031. [CrossRef]
22. Bello, V.H.; da Silva, F.B.; Watanabe, L.F.M.; Vicentin, E.; Muller, C.; de Freitas Bueno, R.C.O.; Santos, J.C.; De Marchi, B.R.; Nogueira, A.M.; Yuki, V.A.; et al. Detection of *Bemisia tabaci* Mediterranean Cryptic Species on Soybean in São Paulo and Paraná States (Brazil) and Interaction of Cowpea Mild Mottle Virus with Whiteflies. *Plant Pathol.* **2021**, *70*, 1508–1520. [CrossRef]
23. Tang, X.-T.; Cai, L.; Shen, Y.; Xu, L.-L.; Du, Y.-Z. Competitive Displacement between *Bemisia tabaci* MEAM1 and MED and Evidence for Multiple Invasions of MED. *Insects* **2019**, *11*, 35. [CrossRef] [PubMed]
24. Bai, J.; Liu, X.-N.; Lu, M.-X.; Du, Y.-Z. Transcriptional Profiling of MED Exposed to Thermal Stress and Verification of HSP70 Expression. *Entomol. Res.* **2021**, *51*, 251–262. [CrossRef]
25. Xiao, N.; Pan, L.-L.; Zhang, C.-R.; Shan, H.-W.; Liu, S.-S. Differential Tolerance Capacity to Unfavourable Low and High Temperatures between Two Invasive Whiteflies. *Sci. Rep.* **2016**, *6*, 24306. [CrossRef]
26. Chu, D.; Tao, Y.; Zhang, Y.; Wan, F.; Brown, J.K. Effects of Host, Temperature and Relative Humidity on Competitive Displacement of Two Invasive *Bemisia tabaci* Biotypes [Q and B]. *Insect Sci.* **2012**, *19*, 595–603. [CrossRef]

27. Bonato, O.; Lurette, A.; Vidal, C.; Fargues, J. Modelling Temperature-Dependent Bionomics of *Bemisia tabaci* (Q-Biotype). *Physiol. Entomol.* **2007**, *32*, 50–55. [CrossRef]
28. Muñiz, M.; Nombela, G. Differential Variation in Development of the B- and Q-Biotypes of *Bemisia tabaci* (Homoptera: Aleyrodidae) on Sweet Pepper at Constant Temperatures. *Environ. Entomol.* **2001**, *30*, 720–727. [CrossRef]
29. Nuno, M.M.S.A.; Cividanes, F.J. Exigências Térmicas de *Bemisia tabaci* (Genn.) Biótipo B (Hemiptera: Aleyrodidae). *Neotrop. Entomol.* **2002**, *31*, 359–363. [CrossRef]
30. Xue, Y.; Lin, C.; Wang, Y.; Liu, W.; Wan, F.; Zhang, Y.; Ji, L. Predicting Climate Change Effects on the Potential Distribution of Two Invasive Cryptic Species of the *Bemisia tabaci* Species Complex in China. *Insects* **2022**, *13*, 1081. [CrossRef]
31. Guo, C.; Zhu, Y.; Zhang, Y.; Keller, M.A.; Liu, T.-X.; Chu, D. Invasion Biology and Management of Sweetpotato Whitefly (Hemiptera: Aleyrodidae) in China. *J. Integr. Pest Manag.* **2021**, *12*, 2. [CrossRef]
32. Alvarez, D.d.L.; Hayashida, R.; Cavallaro, M.C.; Santos, D.M.; Santos, L.M.; Müller, C.; Watanabe, L.F.M.; Bello, V.H.; Krause-Sakate, R.; Hoback, W.W.; et al. Susceptibility of *Bemisia tabaci* Gennadius (Hemiptera: Aleyrodidae) Mediterranean Populations Found in São Paulo, Brazil to 11 Insecticides and Characterization of Their Endosymbionts. *Insects* **2024**, *15*, 670. [CrossRef]
33. Ghanim, M.; Kontsedalov, S. Susceptibility to Insecticides in the Q Biotype of *Bemisia tabaci* Is Correlated with Bacterial Symbiont Densities. *Pest Manag. Sci.* **2009**, *65*, 939–942. [CrossRef] [PubMed]
34. Moraes, L.A.d.; Muller, C.; Bueno, R.C.O.d.F.; Santos, A.; Bello, V.H.; De Marchi, B.R.; Watanabe, L.F.M.; Marubayashi, J.M.; Santos, B.R.; Yuki, V.A.; et al. Distribution and Phylogenetics of Whiteflies and Their Endosymbiont Relationships after the Mediterranean Species Invasion in Brazil. *Sci. Rep.* **2018**, *8*, 14589. [CrossRef] [PubMed]
35. Milenovic, M.; Ghanim, M.; Hoffmann, L.; Rapisarda, C. Whitefly Endosymbionts: IPM Opportunity or Tilting at Windmills? *J. Pest Sci.* **2022**, *95*, 543–566. [CrossRef]
36. Shah, S.H.J.; Malik, A.H.; Zhang, B.; Bao, Y.; Qazi, J. Metagenomic Analysis of Relative Abundance and Diversity of Bacterial Microbiota in *Bemisia tabaci* Infesting Cotton Crop in Pakistan. *Infect. Genet. Evol.* **2020**, *84*, 104381. [CrossRef] [PubMed]
37. Bravo-Pérez, D.; Hernández-Zepeda, C.; Chaidez-Quiroz, C.; Pérez-Brito, D.d.l.C.; González-Gómez, J.-P.; Minero-García, Y.; Rosiles-González, G.; Carrillo-Jovel, V.H.; Moreno-Valenzuela, O.A. Composition of the Whiteflies Microbiome in Populations with and without Insecticide Applications in Yucatan Mexico. *Biologia* **2024**, *79*, 2569–2579. [CrossRef]
38. Walsh, P.S.; Metzger, D.A.; Higuchi, R. Chelex 100 as a Medium for Simple Extraction of DNA for PCR-Based Typing from Forensic Material. *Biotechniques* **1991**, *10*, 506–513. [CrossRef]
39. Simon, C.; Frati, F.; Beckenbach, A.; Crespi, B.; Liu, H.; Flook, P. Evolution, Weighting, and Phylogenetic Utility of Mitochondrial Gene Sequences and a Compilation of Conserved Polymerase Chain Reaction Primers. *Ann. Entomol. Soc. Am.* **1994**, *87*, 651–701. [CrossRef]
40. Bosco, D.; Loria, A.; Sartor, C.; Cenis, J.L. PCR-RFLP Identification of *Bemisia tabaci* Biotypes in the Mediterranean Basin. *Phytoparasitica* **2006**, *34*, 243–251. [CrossRef]
41. Marubayashi, J.M.; Kliot, A.; Yuki, V.A.; Rezende, J.A.M.; Krause-Sakate, R.; Pavan, M.A.; Ghanim, M. Diversity and Localization of Bacterial Endosymbionts from Whitefly Species Collected in Brazil. *PLoS ONE* **2014**, *9*, e108363. [CrossRef]
42. Haddad, M.L.; Parra, J.R.P. *Métodos para Estimar os Limites Térmicos e a Faixa Ótima de Desenvolvimento das Diferentes Fases do Ciclo Evolutivo dos Insetos*; Agricultura e Desenvolvimento; Escola Superior de Agricultura "Luiz de Queiroz": Piracicaba, Brazil, 1984; 14p.
43. Kaplan, E.L.; Meier, P. Nonparametric Estimation from Incomplete Observations. *J. Am. Stat. Assoc.* **1958**, *53*, 457–481. [CrossRef]
44. Mantel, N. Propriety of the Mantel—Haenszel Variance for the Log Rank Test. *Biometrika* **1985**, *72*, 471–472. [CrossRef]
45. Tsagkarakou, A.; Tsigenopoulos, C.S.; Gorman, K.; Lagnel, J.; Bedford, I.D. Biotype Status and Genetic Polymorphism of the Whitefly Bemisia Tabaci (Hemiptera: Aleyrodidae) in Greece: Mitochondrial DNA and Microsatellites. *Bull. Entomol. Res.* **2007**, *97*, 29–40. [CrossRef] [PubMed]
46. Bradshaw, C.D.; Hemming, D.; Baker, R.; Everatt, M.; Eyre, D.; Korycinska, A. A Novel Approach for Exploring Climatic Factors Limiting Current Pest Distributions: A Case Study of *Bemisia tabaci* in North-West Europe and Assessment of Potential Future Establishment in the United Kingdom under Climate Change. *PLoS ONE* **2019**, *14*, e0221057. [CrossRef] [PubMed]
47. Enkegaard, A. The Poinsettia Strain of the Cotton Whitefly, *Bemisia tabaci* (Homoptera: Aleyrodidae), Biological and Demographic Parameters on Poinsettia (*Euphorbia Pulcherrima*) in Relation to Temperature. *Bull. Entomol. Res.* **1993**, *83*, 535–546. [CrossRef]
48. Chen, G.; Klinkhamer, P.G.L.; Escobar-Bravo, R.; Leiss, K.A. Type VI Glandular Trichome Density and Their Derived Volatiles Are Differently Induced by Jasmonic Acid in Developing and Fully Developed Tomato Leaves: Implications for Thrips Resistance. *Plant Sci.* **2018**, *276*, 87–98. [CrossRef]
49. Glas, J.J.; Schimmel, B.C.J.; Alba, J.M.; Escobar-Bravo, R.; Schuurink, R.C.; Kant, M.R. Plant Glandular Trichomes as Targets for Breeding or Engineering of Resistance to Herbivores. *Int. J. Mol. Sci.* **2012**, *13*, 17077–17103. [CrossRef]
50. Harrison, E.L.; Arce Cubas, L.; Gray, J.E.; Hepworth, C. The Influence of Stomatal Morphology and Distribution on Photosynthetic Gas Exchange. *Plant J.* **2020**, *101*, 768–779. [CrossRef]

51. Lin, P.-A.; Chen, Y.; Ponce, G.; Acevedo, F.E.; Lynch, J.P.; Anderson, C.T.; Ali, J.G.; Felton, G.W. Stomata-Mediated Interactions between Plants, Herbivores, and the Environment. *Trends Plant Sci.* **2022**, *27*, 287–300. [CrossRef]
52. Brazilian Climate Data. Available online: https://clima.inmet.gov.br (accessed on 20 April 2023).
53. Brumin, M.; Kontsedalov, S.; Ghanim, M. Rickettsia Influences Thermotolerance in the Whitefly *Bemisia tabaci* B Biotype: *Rickettsia* Influence on Thermotolerance. *Insect Sci.* **2011**, *18*, 57–66. [CrossRef]
54. Su, Q.; Xie, W.; Wang, S.; Wu, Q.; Liu, B.; Fang, Y.; Xu, B.; Zhang, Y. The Endosymbiont *Hamiltonella* Increases the Growth Rate of Its Host *Bemisia tabaci* during Periods of Nutritional Stress. *PLoS ONE* **2014**, *9*, e89002. [CrossRef]
55. Liu, Y.; Yang, K.; Wang, J.; Chu, D. *Cardinium* Infection Alters Cotton Defense and Detoxification Metabolism of Its Whitefly Host. *Insect Sci.* **2023**, *30*, 473–485. [CrossRef] [PubMed]
56. Shan, H.-W.; Liu, S.-S. The Costs and Benefits of Two Secondary Symbionts in a Whitefly Host Shape Their Differential Prevalence in the Field. *Front. Microbiol.* **2021**, *12*, 739521. [CrossRef] [PubMed]
57. Spano, D.; Armiento, M.; Aslam, M.F.; Bacciu, V.; Bigano, A.; Bosello, F.; Breil, M.; Buonocore, M.; Butenschön, M.; Cadau, M.; et al. G20 Climate Risk Atlas. Impacts, Policy and Economics in the G20 2021. Available online: https://www.cmcc.it/g20. (accessed on 22 May 2025).
58. Malecha, A.; Manes, S.; Vale, M.M. Climate Change and Biodiversity in Brazil: What We Know, What We Don't, and Paris Agreement's Risk Reduction Potential. *Perspect. Ecol. Conserv.* **2025**, *23*, 77–84. [CrossRef]
59. Elbaz, M.; Weiser, M.; Morin, S. Asymmetry in Thermal Tolerance Trade-offs between the B and Q Sibling Species of *Bemisia tabaci* (Hemiptera: Aleyrodidae). *J. Evol. Biol.* **2011**, *24*, 1099–1109. [CrossRef]
60. Lima, F.F.D.; Alves, L.R.A. Portfolio Theory Approach to Plan Areas for Growing Cotton, Soybean, and Corn in Mato Grosso, Brazil. *Rev. Econ. Sociol. Rural* **2023**, *61*, e258224. [CrossRef]
61. Stratton, A.E.; Wittman, H.; Blesh, J. Diversification Supports Farm Income and Improved Working Conditions during Agroecological Transitions in Southern Brazil. *Agron. Sustain. Dev.* **2021**, *41*, 35. [CrossRef]
62. Rivas, M.; Vidal, R.; Neitzke, R.S.; Priori, D.; Almeida, N.; Antunes, I.F.; Galván, G.A.; Barbieri, R.L. Diversity of Vegetable Landraces in the Pampa Biome of Brazil and Uruguay: Utilization and Conservation Strategies. *Front. Plant Sci.* **2023**, *14*, 1232589. [CrossRef]
63. Quintão, F.C.S.; Dias Da Silva Furtado, J.; Mendes Diniz Tripode, B.; Miranda, J.E. Inseticidas para controle do bicudo do algodoeiro—Eficiência, período residual e perdas por escorrimento. In *Pesquisa e Inovação nas Ciências que Alimentam o Mundo*; Agrárias; Editora Artemis: Curitiba, Brazil, 2020; Volume IV, pp. 55–65, ISBN 978-65-87396-25-5.
64. Horowitz, A.R.; Ishaaya, I. Dynamics of Biotypes B and Q of the Whitefly *Bemisia tabaci* and Its Impact on Insecticide Resistance. *Pest Manag. Sci.* **2014**, *70*, 1568–1572. [CrossRef]
65. Abubakar, M.; Koul, B.; Chandrashekar, K.; Raut, A.; Yadav, D. Whitefly (*Bemisia tabaci*) Management (WFM) Strategies for Sustainable Agriculture: A Review. *Agriculture* **2022**, *12*, 1317. [CrossRef]
66. Zandi-Sohani, N.; Shishehbor, P. Temperature Effects on the Development and Fecundity of *Encarsia Acaudaleyrodis* (Hymenoptera: Aphelinidae), a Parasitoid of *Bemisia tabaci* (Homoptera: Aleyrodidae) on Cucumber. *BioControl* **2011**, *56*, 257–263. [CrossRef]

Disclaimer/Publisher's Note: The statements, opinions and data contained in all publications are solely those of the individual author(s) and contributor(s) and not of MDPI and/or the editor(s). MDPI and/or the editor(s) disclaim responsibility for any injury to people or property resulting from any ideas, methods, instructions or products referred to in the content.

agronomy

MDPI

Article

Towards an Integrated *Orobanche* Management: Understanding Farmers' Decision-Making Processes Using a Discrete Choice Experiment

Ahmed Yangui [1,*], Taheni Mlayeh [1], Zouhaier Abbes [2] and Mohamed Kharrat [2]

[1] Agricultural Economic Laboratory (LER–LR16INRAT07), National Institute for Agricultural Research of Tunisia INRAT, University of Carthage, El Menzah IV, Tunis 1004, Tunisia; mlayehtahani@gmail.com

[2] Field Crop Laboratory (LGC–LR16INRAT02), National Institute for Agricultural Research of Tunisia INRAT, University of Carthage, El Menzah IV, Tunis 1004, Tunisia; zouhaier.abbes@isste.ucar.tn (Z.A.); kharrat.mohamed@inrat.ucar.tn (M.K.)

* Correspondence: ahmed.yangui@inrat.ucar.tn

Abstract: Controlling the *Orobanche* weed parasite is a major challenge for farmers, and the individual application of various management practices has not yet proven to be successful in addressing this issue. To develop an effective strategy for managing this parasitic weed, an Integrated *Orobanche* Management (IOM) approach has become a priority. Using a Discrete Choice Experiment (DCE) methodology, we analyze the trade-off in farmers' preferences between different attributes of IOM scenarios and estimate their willingness to pay (WTP). A sample of 124 Tunisian faba bean farmers participated in the study. The findings indicate that Tunisian farmers are open to adopt an IOM that includes *Orobanche*-tolerant faba bean varieties, and that the cost of technical package does not seem to be an obstacle. Nevertheless, farmers feel to be rewarded for delaying the sowing date from November to December. Furthermore, the study highlights that farmers show no clear preferences for the use of herbicide, specifically glyphosate, as well as for the practice of intercropping with fenugreek. While increasing faba bean yields remains a priority, farmers are willing to pay more for IOM scenarios that reduce the *Orobanche* plant shoot count. In conclusion, there is significant heterogeneity in farmers' preferences, their financial situation, and the severity of *Orobanche* infestation significantly influencing their decision. Policy recommendations are derived from our results.

Keywords: *Orobanche*; Integrated *Orobanche* Management; discrete choice experiment; random parameter logit; farmers' preferences; willingness to pay

1. Introduction

Broomrapes species (*Orobanche crenata*, *O. cumana*, *O. foetida*, *Phelipanche ramosa*, etc.) are parasitic weeds that affect a wide range of important crops, including faba bean, chickpea, pea, lentil, sunflower, tomato, tobacco, eggplant, potato, carrot, and other crops. These aggressive and destructive weeds have a significant impact on agriculture, particularly in Middle East and North Africa (MENA) regions, including Tunisia [1–3]. The parasite depends entirely on its host for nutrition, leading to significant losses in both yield and quality of the affected crops. Despite the availability of various control technologies, including preventive measures [1], genetic approaches [3–5], biological [6–8], physical [9,10], chemical [11,12], and agronomic methods [13–17], the rapid spread of *Orobanche* remains a major threat. This is due to the diversity of host crops and *Orobanche* species, which continue to challenge farmers worldwide. Accordingly, the parasite affects agriculture on multiple fronts—economically, socially, and environmentally [2,18].

Abang et al. [2] and Abbes et al. [4] highlighted that while individual control measures may lead to improvement in crop yields, a reduction in infestation rates, and some decrease in *Orobanche* seed banks in the soil, none of the currently available control methods have proven to be completely effective methods in combatting *Orobanche*. This may be attributed to the limited knowledge and adoption of diverse control methods by farmers, many of whom continue to rely on ineffective management strategies or have ceased cultivating infected crops altogether. The reasons for this include the fact that not all control methods are universally acceptable, applicable, or effective, and many of the technologies developed have not been effectively disseminated [2]. Therefore, it is crucial to consider the socio-demographic characteristics of farmers and the socio-economic features of individual farming systems in order to implement effective management practices for combating the *Orobanche* parasite.

To develop an effective approach for controlling *Orobanche* and minimizing the associated damages and crop losses worldwide, particularly in MENA regions, including Tunisia, it is crucial to meet three key objectives in all farming systems: (i) preventing the spread of *Orobanche* seeds into non-infested areas; (ii) reducing the *Orobanche* seed bank in the soil; (iii) preventing *Orobanche* reproduction [2,19–24].

To address this challenge, an integrated weed management (IWM) program is essential for effectively combating *Orobanche*, which could be termed Integrated *Orobanche* Management (IOM). IWM is an approach that combines multiple methods simultaneously, such as chemical, biological, and mechanical techniques, specific crop management practices, and preventive measures, to effectively manage and reduce the impact of weeds in agriculture systems [25,26]. However, the adoption of these strategies has been slow among farmers. The reasons for farmers' reluctance to embrace IWM, particularly in the case of *Orobanche*, are not well understood and remain underexplored [1]. Few systematic surveys have been conducted to help decision-makers, researchers, and extension services understand farmers' perception of the *Orobanche* problem from the perspective of farming systems and from their socio-economic context. This understanding is crucial for defining, planning, and implementing an effective IOM strategy [27–30].

Rogers [31] commented that promoting preventive practices is often ineffective because these methods do not provide immediate benefits in the short-term. Furthermore, Llewellyn et al. [32] mentioned that marketing strategies for agrichemical products and herbicide product performance guarantees have reduced the perceived need for alternative weed control methods, like IWM.

From a socio-economic perspective, Sirinivas et al. [1] found that the low-level adoption of IOM among FCV tobacco farmers is partly due to a lack of understanding of the biological cycle of *Orobanche* and ignorance of existing effective management practices to control the parasite. They also pointed out that weak adoption of IWM is due to the low level of scientific orientation and failure to address the specific needs and socio-economic issues of farmers. Consequently, to increase farmers' knowledge and improve their adoption of recommended *Orobanche* management practices, it is essential to consider factors such as farmers' age, education level, landholding size, access to training, extension services, and their overall scientific orientation. In the same vein, Chèze et al. [33] pinpoint that the farmer' willingness to change practices is significantly influenced by their income, particularly if they earn some money from farms outside. According to Danne et al. [34], there is favorable relationship between farm size and change practice behaviors. Indeed, they explained that the need for skilled labor and reliance on herbicides for timely weed control both rise with farm size.

In addition, Wilson et al. [35] and Llewellyn et al. [36] demonstrated that the short-term complexity of achieving direct effective results through the adoption of IWM, along with the associated cost in terms of time, learning, and money, have proven to be significant barriers to its adoption. Previous studies have confirmed that farmers' knowledge, awareness, and on-farm management decisions are heavily influenced by their prior values, beliefs, and experiences, which are often based on personal needs rather than best agricultural practices [37–40]. Therefore, failing to understand and consider these factors when implementing agricultural outreach strategies is likely to hinder the effective promotion and adoption of IWM. Without addressing these underlying variables, the progress of IWM may be limited.

Consequently, to contribute to the development of an integrated and sustainable management program for controlling *Orobanche* and facilitate its adoption by farmers, this study aims to assess Tunisian farmers' preferences and willingness to pay (WTP) for attributes of integrated *Orobanche* management (IOM) practices through a discrete choice experiment (DCE). Specifically, the primary objectives of this study are (i) to estimate farmers' willingness to pay for the adoption of different IOM scenarios, including agronomic and chemical methods, as well as the use of resistant varieties, and to identify the key determinant attributes; (ii) to understand the decision-making process of farmers regarding on-farm management and how it influences their adoption of IOM; (iii) to identify the socio-demographic and socio-economic factors that affect farmers' decisions to adopt IOM; (iv) ultimately, to provide policymakers with insights into farmer behavior that can be used to design more effective outreach strategies, encouraging wider adoption of the IOM approach.

This work contributes to the limited literature addressing the socio-economic factors influencing farmers' behavior in adopting the IWM approach. It is the first study focusing on the decision-making process of farmers to increase the adoption of IOM, both globally and particularly in Tunisia. Furthermore, it is the first using a DCE to analyze the trade-off in farmers' preferences regarding the attributes designing the IOM.

The DCE methodology is based on the economic theory of consumer choice, which has been extended to cover the non-market valuation and the medicine framework. Recently, this approach has been increasingly used to investigate farmers' preferences and the determinant factors affecting their decision-making process [33,34,41–46]. In a DCE, respondents are presented with a series of choice situations representing different scenarios or alternatives. They are then asked to select the option they prefer in each case [47]. Therefore, the DCE is a suitable tool for presenting farmers with practically relevant decisions, where they must balance and choose between different components of an integrated management approach to control *Orobanche*.

2. The Tunisian Context Related to *Orobanche* Control

Over a century ago, Boeuf [48] first reported *Orobanche* infestation in Tunisia. The primary host crop for *Orobanche*, particularly *O. crenata* and *O. foetida*, is faba bean (*Vicia faba*), with yield losses in heavy infested area as much as 50–100% [3,49–51]. The areas infested by *Orobanche* and the severity of the infestation have been steadily increasing [5,52].

Consequently, a breeding program focused primarily on developing faba bean varieties tolerant/resistant to *O. foetida* was initiated several years ago at National Institute of Agricultural Research of Tunisia (INRAT). Since then, many faba bean varieties partially resistant to *Orobanche* have been registered in the Tunisian Official Catalogue of Plant Varieties such as "Najeh", "Chorouk", "Chams" and "Zaher" [5,50,53]. However, recent climate change has exacerbated *Orobanche* damages to other important crops, such as sunflower, chick-pea, and lentil, which has discouraged farmers from cultivating susceptible crops, particularly faba bean [5].

In response, the research program initiated by Field Crop Laboratory at INRAT has intensified its research activities in various areas of *Orobanche* control, including chemical treatments [3,54], agronomic management practices (e.g., the effect of sowing date and intercropping with trap crops like fenugreek) [19,55,56], and biological methods [8,57–59]. While several potential methods have been studied and disseminated to farmers, none of these approaches have proven completely effective in controlling *Orobanche*. The most successful strategy to combat this parasite is through an integrated management approach, primarily based on selecting genetic material with tolerance/resistance to *Orobanche* [2,4,5].

Unfortunately, Tunisian farmers adopting an IOM remain a minority and the behavioral factors that are crucial for understanding farmers' decision-making are still insufficiently researched [60]. Additionally, it is puzzling that, to date, no studies have evaluated the benefit–cost of different IOM scenarios from farmers' perspectives or examined which potential methods should be incorporated into these scenarios to increase adoption rates among farmers. For these reasons, our study aims to first examine faba bean farmers' behaviors towards the adoption of different IOM scenarios.

3. Methodology

3.1. Integrated Orobanche Management Attributes and DCE Design

DCE is the ideal methodology for assessing the trade-offs between the different scenarios. In this study, DCE was used to present various scenarios of IOM to farmers, based in six key attributes, variety, chemical treatment, sowing date, intercropping (with two levels each attribute), technical package cost (with three levels), and output (with four levels) (Table 1). The selection of these attributes was informed by the literature, discussions with experts involved in the *Orobanche* control program at the Field Crop Laboratory at INRAT, and the results of experimental studies aimed at identifying effective IOM strategies for controlling the parasite. The experimental studies were conducted over two faba bean-growing seasons and their main objective was to assess the agronomic impact of different scenarios for IOM, based on a combination of factors such as variety, chemical treatment, intercropping, and sowing date. Additionally, it is worth noting that the selected attributes are the most practical from farmers' perspective, especially regarding preventive practices (which farmers tend to reject) and biological methods, which are unpopular and difficult to carry out.

Given the number of attributes and their levels, a full factorial design would generate 192 ($3 \times 2^4 \times 4 = 192$) possible IOM scenarios, which would be too complex for farmers to assess. To reduce the number of combinations farmers needed to evaluate and avoid cognitive overload, we adopted the approach outlined by Street and Burgess [61] and generated an orthogonal fractional factorial design with 16 scenarios. These 16 IOM scenarios were then presented as the first option in each choice set for the respondents.

Each choice set included five alternatives: three IOM scenarios, one status quo, and one opt-out alternative. The status quo option represents the current situation of faba bean farmers in areas heavily infested by *Orobanche*. In this scenario, most of the farmers do not invest significant efforts in controlling *Orobanche*. They continue to use local susceptible faba bean varieties, typically sowing them at the beginning of November, or perhaps earlier, without applying any chemical treatments to control the parasite. Additionally, they do not use trap crops to manage *Orobanche*. The average cost of faba bean cropping season per hectare range from 750 to 850 Tunisian dinars (TND), and typically, their yields do not exceed 1.5 quintals per hectare (qx/ha). The other two IOM scenarios were generated using the generators (111110) and (011113) [61], resulting in a 100% efficient main-effect design. Figure 1 illustrates an example of a choice set. To make it easier for participants, we divided the 16 choice sets into two blocks. Each participant was presented with eight choice decisions, drawn randomly from one of the two blocks.

Table 1. The attributes and attribute levels of IOM scenarios.

Attributes		Levels		Status Quo
Variety		Non-tolerant		Local variety
		Tolerant		
Chemical Treatment		TRT with Glyphosate		Without any TRT to control *Orobanche*
		TRT With Imazamox + Bentazone		
Sowing Date		November		November
		December		
Intercropping		Without fenugreek		Without intercropping
		with fenugreek		
Package Cost (TND/ha)		[900; 1000]		[750; 850] TND/ha
		[1000; 1100]		
		[1100; 1200]		
Output		Yield increase	Decrease in *Orobanche* plant	[0–1.5] qx/ha
	Output 1	↑ 100–400%	↓ [50–75%]	
	Output 2	↑ 100–400%	↓ [75–95%]	
	Output 3	↑ 500–1000%	↓ [50–75%]	
	Output 4	↑ 500–1000%	↓ [75–95%]	

"↑" represents the increase of yields by 100–400% or by 500–1000%; "↓" represents the decrease of *Orobanche* plant by 50–75% or by 75–95%.

3.2. Theoretical Framework and Model Specification

The DCE modeling framework is based on both random utility theory (RUT) [62] and Lancaster consumer theory [63]. According to RUT, the value of a good is the sum of the values of its characteristics. In the context of DCE, the utility provided by alternative j ($j = 1\ldots J$) from choice set s ($s = 1\ldots S$) to individual i ($i = 1\ldots N$) is given by the following:

$$U_{ijs} = V_{ijs} + \varepsilon_{ijs} \tag{1}$$

where V_{ijs} is a deterministic component and ε_{ijs} is the stochastic component. In a traditional model, the (indirect) utility can be described as a function of alternative attributes as follows:

$$V_{ijs} = \beta_{ijks} * X_{kjs} \tag{2}$$

where X_{kjs} is the vector of attributes related to alternative j; β_{ikjs} is the vector of marginal utilities of the individual i related to the k attributes in alternatives j from the choice set s.

Assuming that the individuals are fully rational in their choices, the participants should be choosing the alternatives that provide them with the greatest utility. Therefore, the probability of respondents i choosing the alternative j out the total set of alternatives is:

$$P_{ij} = Prob\left[U_{ij} > U_{ik}\right] = Prob\left[V_{ij} + \varepsilon_{ij} > V_{ik} + \varepsilon_{ik}\right] \forall\, j \neq k\, \in s \tag{3}$$

	Identification Number: ………..		**Choice set N° 1**		**BLOC N° 1**
	STATUS QUO	Scenario N° 1	Scenario N° 2	Scenario N° 3	Opt-out
Variety	Local	Non Tolerant	Tolerant	Non Tolerant	None of these alternatives
Chemical Treatment	Without any TRT to control *Orobanche*	TRT with Glyphosate	TRT with Corum	TRT with Corum	
Sowing Date	November	Mid-December	Mid-November	Mid-November	
Intercropping	Without intercropping	With fenugreek	Without fenugreek	Without fenugreek	
Package cost (TND/ha)	[750; 850]	[1000; 1100]	[1100; 1200]	[1100; 1200]	
Output	Yield: [0–1.5] qx/ha + 12–15 Orob. plant/m^2	↑ 500–1000% Yield + ↓ - [50–75%] P.Orob	↑ 500–1000% Yield + ↓ - [50–75%] P.Orob	↑ 100–400% Yield + ↓ - [75–95%] P.Orob	
	☐	☐	☐	☐	☐
	Please select the most preferred option for you (put a cross in the corresponding box)				

Figure 1. An example of a choice set of IOM scenarios from the DCE design. Source: authors' own elaboration.

By assuming that the stochastic component is distributed following type I extreme value, we obtain the familiar multinomial logit model where the probability of respondents i choosing option j from the specific choice set s is:

$$P_{ij} = \frac{e^{\mu V_{ij}}}{\sum_{k=1}^{J} e^{\mu V_{ik}}} \forall j \in s \tag{4}$$

3.3. A Random Parameter Logit Model (RPL)

To estimate the marginal utility associated with each attribute, a random parameter logit (RPL) model was employed. Also known as mixed logit [64], the RPL model has the advantage of accounting for preference heterogeneity among farmers, unlike other multinomial logit models such as the conditional logit model. Specifically, the RPL model assumes that the parameter estimates associated with each attribute are not constant but instead vary across individuals, following a specific distribution, such as normal, triangular, lognormal, etc. [64].

In this model, the individual-specific parameter estimates, β_{ij}, are expressed as:

$$\beta_{ij} = \beta_j + \sigma_j \vartheta_{ij} \tag{5}$$

In this formulation β_j is the sample mean for the alternative j, ϑ_{ij} is individual specific heterogeneity, with mean zero and standard deviation equal to one [65].

The probability that individual i chooses the alternative j in a particular choice set s is given by:

$$Prob_i\{j \text{ is chosen}\} = \int L_{ij}(\beta_{ij}) f(\beta_i/\theta) d\beta_i, \text{ with } j \in s \tag{6}$$

where $f(\beta_i/\theta)$ is the density function of the coefficients β_i and θ , referring to the moments of the parameter distributions, and $L_{ij}(\beta_{ij})$ is given by the follow equation:

$$L_{ij}(\beta_{ij}) = \frac{e^{\beta_{ij} X_{ij}}}{\sum_{k=1}^{J} e^{\beta_{ik} X_{ik}}} k \in s \tag{7}$$

3.4. Willingness to Pay (WTP)

The WTP for product or service attribute represents the monetary value that an individual is willing to pay for a unit increase in a given attribute. It can be calculated as the negative ratio of the partial derivative of the utility function with respect to the attribute of interest, divided by the derivative of the utility function with respect to the monetary variable:

$$WTP_{Attribute} = -\frac{\frac{\partial U_{ijs}}{\partial Attribute}}{\frac{\partial U_{ijs}}{\partial Monetary\ attribute}} = -\frac{\beta_{Attribute}}{\beta_{Monetary\ attribute}} \tag{8}$$

3.5. Empirical Model

The deterministic component V_{ijs} was commonly specified as linear in parameters including variables that represent the attributes of the IOM scenario concept and the characteristics of respondents. In the empirical specification, the deterministic component is given by:

$$\begin{gathered} V_{ijs} = \beta_{NOP} NOP + \beta_{STQ} STQ + \beta_{VTOL} VTOL_{ijs} + \beta_{TGLY} TGLY_{ijs} + \beta_{DEC} DEC_{ijs} + \\ \beta_{WFENG} WFENG_{ijs} + \beta_{output2} OUTPUT2_{ijs} + \beta_{output3} OUTPUT3_{ijs} + \beta_{output4} OUTPUT4_{ijs} + \\ \beta_{tpcost} TPCOST_{ijs} \end{gathered} \tag{9}$$

In Equation (9), the attribute levels were coded as dummy variables, except the cost attribute is specified as continue variable (Table 2).

Table 2. Definitions and categories of the different variables included in the empirical model.

Variable	Category	Definition
NOP	Dummy Variable	Takes the value 1 when the respondents select the opt-out option and do not prefer the other presented alternatives; 0 otherwise.
STQ	Dummy Variable	Takes the value 1 when the respondents select the status quo alternatives and do not have willingness to change their actual situation; 0 otherwise.
VTOL	Dummy Variable	Takes the value 1 when the tolerant variety attribute level is included in the presented IOM scenarios; 0 otherwise.
TGLY	Dummy Variable	Takes the value 1 when the chemical treatment with glyphosate attribute level is included in the presented IOM scenarios; 0 otherwise.
DEC	Dummy Variable	Takes the value 1 when the December sowing date attribute level is included in the presented IOM scenarios; 0 otherwise.
WFENG	Dummy Variable	Takes the value 1 when the intercropping practice using fenugreek attribute level is included in the presented IOM scenarios; 0 otherwise.
OUTPUT2	Dummy Variable	Takes the value 1 when the output of the corresponding IOM scenario consists of increasing yields by 100 to 400% and decreases in *Orobanche* plants m^{-2} to around [75–95%]; 0 otherwise.
OUTPUT3	Dummy Variable	Takes the value 1 when the output of the corresponding IOM scenario consists of increasing yields by 500 to 1000% and decreases in *Orobanche* plants m^{-2} to around [50–75%]; 0 otherwise.
OUTPUT4	Dummy Variable	Takes the value 1 when the output of the corresponding IOM scenario consists of increasing yields by 500 to 1000% and decreases in *Orobanche* plants m^{-2} to around [75–95%]; 0 otherwise.
TPCOST	Continue Variable	The cost of the corresponding IOM scenario could be one of the three alternatives: [900, 1000], [1000, 1100], and [1100, 1200] TND/ha.

In addition, an extended model was estimated to test the effect of two variables on farmers' decision-making: (i) an economic variable "debt level", this variable assesses the percentage of the farmers' debts relative to their incomes for the current crop season. The debt level is measured on scale of five points (<10%; 10–25%; 26–50%; 50–70%; >70%); (ii) and an agronomic variable "*Orobanche* infestation rate": this variable evaluates the infestation rate of *Orobanche* on the total area of infested crops. The infestation rate is measured on scale of five points (<5%; 6–25%; 26–50%; 51–75%; 76–100%). The goal is to understand whether a farmer's financial situation (debt level) or the severity of *Orobanche* infestation plays a significant role in their willingness to adopt new practices, and if so, how these factors shape their preferences for different IOM scenarios.

3.6. Presentation of the Questionnaire and Data Collection

The questionnaire was divided into five sections. Sections 1 and 2 focused on collecting socio-demographic and agro-economic data, respectively, from the sample. Information on crops infested by *Orobanche* and the level of infestation level was gathered in Section 3. Section 4, farmers' risk preferences were assessed, with related questions included in this part of the questionnaire. Section 5 presented the choice set cards for the DCE. This section detailed 16 choice sets, which were organized into two blocks. The order of the choice sets was randomized to minimize hypothetical bias associated with learning and fatigue effect.

The survey, conducted using face-to-face interviews, took place from March to June 2024, with the majority of interviews held directly on the farms. At the beginning of each interview, the main objective of study was clearly explained. The DCE approach and IOM were then described in detail and simplified for the farmers. Next, the attributes and their levels were explained. To minimize hypothetical bias, we conducted a practices exercise with the farmers using an example of a choice card, which was different from the actual cards used in the experiment. The participants were asked to select their preferred option. Following this, the farmers were informed that the DCE section of the interview would begin. After completing the choice sets, the farmers were asked socio-demographic, agro-economic, and attitudinal questions from the remaining sections of the questionnaire.

4. Results and Discussion

4.1. Data Descriptive Statistics of the Sample

The data used in this study were obtained from a survey carried out on a sample of Tunisian' farmers. A filter question was used to select the sample, specifying that the Agricultural Utilized Area (SAU) managed by farmers should be infested with *Orobanche*. As a result, 124 farmers (A small sample size is a common limitation in DCE studies targeting farmers, as this population is often more difficult to reach [33]. Several published DCE studies focusing on farmers have had small sample size. For example, Schulz et al. [66] surveyed 128 German farmers, Greiner [67] recruited 104 Australian farmers, Beharry-Borg et al. [68] collected data from 97 English farmers, Hudson and Lusk [69] included 49 American farmers, Jaeck and Lifran [70] studied 104 French farmers, and Chèze et al. [33] surveyed 90 French farmers) were recruited from nine regions to participate in this study. Over 57% of our sample is from the northwestern part of Tunisia, particularly from Beja, Jendouba, and Bizerte, which represent 27%, 19%, and 11% of the sample, respectively. This region is the main area for seed legume production, principally faba bean [71].

Tables 3 and 4 present the descriptive statistics of the sample, including socio-demographic and agro-economic data, respectively. The majority of our sample are men, with only 2.42% of the sample being women farmers (Table 3). The average age of the participants is 53.46 years, with standard deviation of 11.71, reflecting the aging population of farmers. In fact, nearly 37% of the sample consists of farmers over 60, while just 13% are younger farmers. Half of the sample is either illiterate or has only a primary education, while just 11.41% have completed university education. Most participants have no formal agriculture training, and agriculture is their primary, if not sole, source of income. However, from Riemens et al. [72] and Sharma et al. [73] perspectives, this profile of farmers is not the good one to enhance the adoption of IWM. The authors display that full time young farmers, or farmers with limited experience, are more likely to adopt IPM strategy. This indicates that the other profile of farmers, experienced aged farmers based on their own experiences, can slow down the adoption of this strategy.

Table 4 shows that the average farm size of the participants is approximately 34 hectares, with a higher standard deviation value is around of 75, reflecting significant variability in farms size and, consequently, in farming system among participants. All farmers' categories—from very small to large farmers—are represented in our sample. Meanwhile, it is important to note that the group most representative of the real Tunisian context is relatively tiny farmers represents 28% of the sample. However, it is important to note that between the factors that complicate the adoption of IWM was the restricted capacity to trial it on a small scale [74]. An important factor influencing farmers' adaptation decisions and their vulnerability is their land tenure system. Table 4 also displays that the two main land tenure system adopting by our sample are the ownership and the combined system (i.e., ownership with rented land, ownership with sharecropping, or combination such as sharecropping with rented land), accounting for 45.53% and 47.15% of the sample, respectively.

Table 3. Socio-demographic profile of the farmer respondents.

Variables and Levels	Frequency	Percentages
Gender		
Male	121	97.58
Female	3	2.42
Age		
Average (SD)	53.46 (11.71)	
<40 years old	16	12.90
40 to 60 years old	63	50.81
≥60 years old	45	36.29
Education		
Illiterate	16	13.11
Primary level	45	36.89
Secondary level + professional training	47	38.52
University level	14	11.48
Agriculture Training		
No	93	75.61
Yes	30	24.39
Other Professional Activity		
No	102	82.26
Yes	22	17.74

Additionally, one of the main preventive practices for limiting the spread of *Orobanche* on farms is controlling animal grazing movement. Therefore, it is important to understand the extent to which farmers engage in animal husbandry activities. Indeed, 63% of the sample practice mixed farming, combining cropping system with livestock farming. The results shown in Table 4 reflect the heterogeneity of the farming systems adopted by Tunisian farmers. However, this variability could be a limiting factor for adopting IWM, as it may have lower compatibility with existing farming systems [75]. Furthermore, it is noteworthy that over 60% of participants perceive their financial health positively and report little or no debt relative to their income levels. This suggest that most of the sample view their financial situation as satisfactory. Consequently, financial constraints are unlikely to prevent farmers from adopting *Orobanche* control practices or IOM approach.

Table 5 compares the key socio-economic characteristics of our sample with those of the broader population of Tunisian farmers. The table shows that our sample is representative in terms of age, useful agriculture area, and education level.

Additional descriptive statistics related to faba bean cropping and *Orobanche* infestation rate are displayed in Table 6. The data revealed that 20% of our sample has stopped growing faba bean, due mainly to *Orobanche* infestation. Moreover, approximately 70% of the participants reported that the infestation level in their area is moderate to heavy. However, it is surprising to observe that more than 76% of the participants have little to no information about *Orobanche* management practices or how to control this plant parasite.

Table 4. Agro-economic profile of farmer respondents.

Variables and Levels	Frequency	Percentages
Useful Agriculture Area (UAA)		
Average (SD)	33.94 (75.31)	
<5 ha	35	28.23
5 to 10 ha	19	15.32
10 to 20 ha	27	21.77
20 to 50 ha	29	23.39
>50 ha	14	11.29
Land Tenure		
Ownership	56	45.53
Rented	5	4.07
Sharecropping	4	3.25
Combined	58	47.15
Animal Husbandry		
Yes	79	63.71
No	45	36.29
Financial Health		
Highly unsatisfactory	11	8.87
Unsatisfactory	27	21.77
Not bad	63	50.81
Satisfactory	18	14.52
Higly satisfactory	5	4.03
Debt Level		
[0–10%]	63	50,81
[10–25%]	17	13.71
[25–50%]	24	19.35
[50–70%]	16	12.90
>70%	4	3.23

Table 5. Comparison of sample socio-economics profile with Tunisian farmers statistics.

Variables and Levels	Sample Statistics (%)	Tunisian Farmers' Statistics (%)
Age		
Average (SD)	53.46	50–60
<40 years old	12.90	13
40 to 60 years old	50.81	44
≥60 years old	36.29	43
UAA: 0 to 20 ha	65.32	83
Education		
Proportion that finished or not the secondary level	88.52	Around to 80%
Proportion having a higher education level	11.48	Around to 10%

Table 6. Current faba bean farming, *Orobanche* infestation, and farmer respondents' knowledge level of *Orobanche* management practices.

Variables and Levels	Frequency	Percentages
Faba bean cropping		
Yes	86	69.35
No	13	10.48
I stopped	25	20.16
Orobanche infestation level		
<5%	21	16.94
6 to 25%	18	14.52
26 to 50%	24	19.35
51 to 75%	37	29.84
76 to 100%	24	19.35
Knowledge level		
Not at all informed	64	52.46
More and less informed	29	23.77
informed	19	15.57
Fairly informed	6	4.92
Well informed	4	3.28

4.2. Farmers' Preferences Towards IOM Attributes

In the DCE data analysis, the RPL model is used to estimate farmers' marginal utilities associated with attribute and attribute levels. Moreover, the model allows for correlation between alternatives by estimating the full covariance matrix structure. Following Burton [76], when using an RPL model with dummy variables, such as in our case, it is important to account for the correlation between random parameters to avoid an increase in Type-I errors.

Table 7 presents the parameter estimates for three estimated models. The first model (model 1) estimates a standard RPL, the second model (model 2) accounts for the correlation between alternatives but excludes interaction factors, and the third model (model 3) estimates a RPL that includes both correlation between alternatives and interaction factors.

As commonly assumed in the literature [65], the marginal utility associated with the cost attribute of the technical package is considered constant, while the other attributes are treated as random parameters following a normal distribution (Table 7).

The goodness-of-fit indicators show that the model 3 provides the best to the data. This model has the lowest values for both Akaike Information Criterion (AIC) and log-likelihood criteria (Ll), in absolute values. Therefore, for the remainder of the paper, we will focus on the results from this model.

Model 3 provides estimates of the mean of marginal utilities associated with the attributes' levels included in the DCE, as well as the standard deviations for the random parameters considered (VTOL, TGLY, DEC, WFENG, OUTPUT2, OUTPUT3, and OUTPUT4). To further explore sources of heterogeneity and better understand farmers' choices, we analyzed potentially influencing factors by examining the interactions between "debt level (Debt)" and "infestation rate (TINFEST)" with the attribute levels (VTOL, OUTPUT2, OUTPUT3, OUTPUT4 and TPCOST).

Table 7. Estimated parameters results of the three RPL models, standard RPL, RPL with correlation, and RPL with correlation and interactions factors.

	Model 1	**Model 2**	**Model 3**
Variables		Coefficients (SE)	
Opt-Out (NOP)	−3.121 *** (0.955)	−4.029 *** (0.975)	−4.535 *** (1.198)
Status Quo (STQ)	−4.280 *** (0.643)	−5.202 *** (0.677)	−5.139 *** (0.699)
Variety Tolerant (VTOL)	0.522 ** (0.221)	0.319 * (0.189)	1.113 *** (0.304)
Chemical Treatment—Glyphosate (TGLY)	−0.003 (0.135)	0.053 (0.126)	0.118 (0.151)
Sowing date (mid-December) (DEC)	−0.876 *** (0.174)	−1.141 *** (0.203)	−1.089 *** (0.183)
Intercropping–Fenugreek (WFENG)	−0.062 (0.132)	−0.493 ** (0.167)	−0.097 (0.164)
Output 2	1.465 *** (0.194)	1.837 *** (0.303)	2.790 *** (0.464)
Output 3	0.287 (0.255)	0.938 ** (0.323)	1.569 *** (0.484)
Output 4	2.514 *** (0.235)	2.784 *** (0.300)	3.406 *** (0.475)
Technical Package Cost (TPCOST)	−0.001 ** (0.0008)	−0.001 ** (0.0008)	−0.002 ** (0.001)
Variables		Standard deviations (SE)	
VTOL	1.886 *** (0.181)	2.474 *** (0.233)	2.449 *** (0.229)
TGLY	0.751 (0.529)	0.432 ** (0.187)	1.145 *** (0.215)
DEC	1.412 *** (0.186)	1.732 *** (0.213)	1.953 *** (0.241)
WFENG	0.961 *** (0.171)	1.661 *** (0.225)	2.113 *** (0.253)
Output 2	0.190 (0.950)	2.424 *** (0.399)	1.956 *** (0.360)
Output 3	1.675 *** (0.390)	2.229 *** (0.349)	2.491 *** (0.331)
Output 4	1.504 *** (0.381)	2.532 *** (0.326)	2.921 *** (0.420)
Interaction Factors		Coefficients (SE)	
Debt Level (Debt) × VTOL	--	--	−0.905 ** (0.392)
Debt × Output2	--	--	−1.004 ** (0.489)
Debt × Output3	--	--	−0.100 (0.534)
Debt × Output4	--	--	−0.175 (0.502)
Debt × TPCOST	--	--	0.001 ** (0.0005)
Infestation Level (TINFEST) × VTOL	--	--	0.741 ** (0.361)
TINFEST × Output2	--	--	−0.089 (0.471)
TINFEST × Output3	--	--	−0.996 * (0.531)
TINFEST × Output4	--	--	−0.298 (0.479)
TINFEST × TPCOST	--	--	−0.001 ** (0.0005)
	Goodness of fit		
Number of Observations		4960	
Degree of Freedom (Df)	17	38	48
Akaike Information Criterion (AIC)	1893.163	1805.713	1769.802
Log-likelihood Ll (null)	−1070.215	−1070.215	−1056.648
Log-likelihood Ll (model)	−929.5813	−864.713	−836.9009

***, ** and * indicate that the corresponding parameter is statistically significant at the 1%, 5% or 10% level, respectively.

The results show that the coefficients for both STQ (Status Quo) and NOP (Opt-Out) choices are negative and highly significant. This suggests that the farmers in the sample tend to reject the status-quo or for the no-option alternative when presented with IOM scenarios. In general, participants seem willing to accept the various IOM scenarios and their associated characteristics.

The statistically significant positive coefficients for the variables relative to faba bean tolerant varieties to *Orobanche* (VTOL), as well as they associated with the outputs of the different IOM scenarios (output 2, output 3, and output 4), indicate that farmers are more likely to accept IOM scenarios that include, primarily, tolerant varieties for *Orobanche*. These results confirm findings of previous studies, which emphasize that host plant resistance is a key element in the fight against *Orobanche*, although it is not an effective standalone control strategy [2,4,5]. Lamichane et al. [77] confirm that a possible way to enhance effectiveness IWM is by incorporating resistant crop varieties, along with cultural practices, physical and mechanical tactics, chemical control, and others methods.

Additionally, the results reveal that farmers do not present preferences for using Glyphosate as a chemical treatment to control *Orobanche*. The coefficient associated with the attribute level (TGLY) is not statistically significant. Danne et al. [34] highlight that farmers have no clear preferences for glyphosate use. Intensive research has identified herbicides with good potential for controlling *Orobanche* [11], including Glyphosate. However, several obstacles limit the successful use of herbicide, such as its limited persistence, high costs, lack of it approval as herbicide for *Orobanche* control, like our case in Tunisia [2]. Despite promising results from research conducted by Kharrat and Halila [3], which demonstrated the effectiveness of very low doses of Glyphosate in controlling *Orobanche*, this selective herbicide has not yet been registered for use against the parasite in Tunisia. Doole and James [75] highlighted that the drivers of herbicides broad scale use by farmers are their high observable efficacy, low cost, low complexity, and significant flexibility. However, Tunisian farmers seem to be wary of using this selective herbicide, as applying incorrect doses could lead to serious crop yield losses. This underscores the critical role of extension services in providing farmers with the necessary, accurate information and effective practices for managing *Orobanche* [2]. Abang et al. [2] add that after 15 years promoting the use of Glyphosate to control O. *Crenata* in faba bean, in Morocco, only 15% of the interviewed extension agents were able to give a correct description of its application technology.

Similarly, the coefficient associated with the attribute level for intercropping with fenugreek (WFENG) is not statistically significant. This result indicates that the majority of the participants do not appear to be convinced of using fenugreek as a trap crop to control *Orobanche*. Despite the effectiveness of growing trap crops, particularly fenugreek, in reducing *Orobanche* shoot counts [2,13,19], Tunisian farmers seem to be unaware of this practice. This lack of awareness may be due to the challenge of managing both crops simultaneously, faba bean and fenugreek, during the sowing and harvest phases.

Moreover, the results reveal negative marginal utilities for sowing date change attribute in faba bean cropping. The estimated parameters for this attribute level (DEC) is negative and highly statistically significant. In line with studies on chickpea cropping, Rubiales et al. [14], Van Hezewijk [78], and Kebreab and Murdoch [79] have demonstrated that *Orobanche* infection is favored by early sowing dates (October-November), mild winters, and rainy autumns and springs. Furthermore, Mesa-Garcia and Garcia-Torres [80] highlighted that the number of *Orobanche* plants successfully attaching to the host, as well as the duration of the underground stage of *Orobanche*, decreases as broad bean planting is delayed. However, it appears that the recruited farmers do not consider this practice to manage *Orobanche* and may not be fully informed about the optimal conditions for delaying the sowing dates.

Upon examining the estimated parameters associated with the output attributes presented in Table 7, it is clear that participants value the outputs levels of different IOM scenarios. All the marginal utilities for OUTPUT2, OUTPUT3, and OUTPUT4 are positive and highly significant. Additionally, it is noteworthy that the marginal utilities for OUTPUT2 and OUTPUT4 are more important than the marginal utility associated with OUTPUT3. These results suggest that, given the serious damage suffered by faba bean farmers, increasing yields is a very important goal. However, reducing the *Orobanche* shoot count appears to be even greater significance.

In the same vein, the results indicate that the cost of technical package is not a serious barrier for adopting IOM scenarios for *Orobanche* control. Table 7 shows that the coefficient associated with the attribute "technical package cost (TPCOST)" is negative and significant, but very low, at approximately −0.002. This suggests that, although participants show a slight negative utility towards the cost of technical package, it is not a major deterrent, likely due to the serious problems they face with *Orobanche*.

Furthermore, Section 2 of Table 7 includes the standards deviation of all random parameters considered in the model: VTOL, TGLY, DEC, WFENUG, OUTPUT2, OUTPUT3, and OUTPUT4. All the standard deviation coefficients are important and highly significant. This indicates that preferences for the IOM scenarios attributes and their respective levels vary across the participants. To better understand the source of this heterogeneity, two potentially influencing factors were introduced in the model interacting with various attribute levels that define the IOM scenarios.

4.3. Influence of Farmers' Debt Level

The primary concern of this study regarding the adoption the IOM scenarios by farmers is to redesign cropping systems in a way that minimizes the impact of the *Orobanche* parasite, ensuring that its presence does not negatively affect yields and farmers' profitability. Nonetheless, IWM, and particularly IOM, has been reported to be expensive due to its combination of multiple methods, including chemical, biological, mechanical, and specific crop management techniques [25,72,81]. Swanton et al. [82] demonstrated that IWM systems can be perceived as unreliable, increasing management risk. Furthermore, there is no clear, direct economic benefit from IWM, and sustained support for its adoption has been limited.

Given this, it can hypothesize that farmers' financial conditions might act as a constraint to adopting such an approach. The literature suggests that the annual income level of farmers is a key factor governing the decision to adopt IWM [83,84]. Specifically, higher income levels increase the likelihood of farmers adopting IWM. However, Houngbo et al. [45] present a contradictory finding, showing that farmers who are more likely to apply integrated pest management (IPM) tend to have relatively low incomes. Increased debt level raise farming costs, reduce profitability, and negatively affect annual incomes. Therefore, consistent with the finding of Houngbo et al. [45] it can be argued that farmers with higher debt levels (and relatively low incomes) are more likely to adopt IOM strategies.

The results presented in Table 7 confirm the findings of Houngbo et al. [45]. Specifically, the interaction terms "DEBTxVTOL" and DEBTxOUTPUT2" are negative and statistically significant. This indicates that farmers with very low levels of debt are less likely to choose IOM scenarios involving *Orobanche*-tolerant varieties. Several factors may explain this result, including the higher cost of tolerant varieties compared to non-tolerant and local ones, limited access to these varieties in the market, and the weak dissemination efforts aimed at promoting them [2].

For these farmers, increasing faba bean yields is a top priority. They are less inclined to choose alternatives that offer only modest improvements in yields, even if such alternatives reduce *Orobanche* shoot count. Furthermore, this group of farmers appears to be more tolerant to the technical package cost than those with higher levels of debt. The coefficient of the interaction term "DEBTxTPCOST" is positive and statistically significant (Table 7), suggesting that these farmers do not face significant financial constraints in adopting IOM strategy, unlike the majority of farmers, for whom cost is a major obstacle.

Therefore, a targeted dissemination strategy aimed at educating this group of farmers about the principles and practices of IOM, particularly the role of Orobacnhe-tolerant varieties, could enhance their acceptance and adoption of IOM practices.

4.4. Influence of Infestation Rate

Orobanche infestation continues to rise, a significant threat to the livelihoods of millions of farmers. In Severe cases, infestation can lead to 100% crop failure. Therefore, the objective of this section is to understand how *Orobanche* infestation rates influence farmers' preferences for IOM attributes and affect their decision-making process. Horowitz and Lichtenberg [85] emphasized that the level of infestation is one of the key sources of risk in making informed decisions about pest/parasite control practices. Similarly, Lopez-Granados and Garcia-Torres [86] highlighted *Orobanche* infestation as a critical factor in determining weed management strategies.

The results presented in Table 7 underscore the significant influence of *Orobanche* infestation rates on farmers' perception. Specifically, the coefficient for the interaction term "TINFEST × VTOL" is positive and statistically significant. This finding suggests that farmers currently experiencing high *Orobanche* infestation rates have positive preferences for the inclusion of tolerant varieties in IOM scenarios. This outcome aligns with the conclusions of Horowitz and Lichtenberg [85] and Lopez-Granados and Garcia-Torres [86], confirming that higher infestation rates increase farmers' willingness to adopt strategies that incorporate *Orobanche*-tolerant varieties.

Nonetheless, the estimated parameter for the interaction term "TINFEST × TPCOST" is negative and statistically significant. This indicates that farmers with higher *Orobanche* infestation rate are more sensitive to the technical package cost of IOM scenarios. In contrast, the interaction terms between infestation rate and the output attribute levels "TINFEST × OUTPUT2", TINFEST × OUTPUT3", and "TINFEST × OUTPUT4" are not statistically significant. This suggests that farmers experiencing significant *Orobanche* infestation do not have strong preferences for any particular output level of IOM scenarios. Instead, their primary concern is to adopt effective practices that can immediately reduce *Orobanche* shoot plants.

Consequently, this section allow us to conclude that it is essential to provide special attention to the farmer segment that is suffering high *Orobanche* infestation rates in order to successfully disseminate the many faba bean varieties that are tolerant or resistant to *Orobanche*. It is important to emphasize how growing these varieties can reduce *Orobanche* infestation when used as a determinant factor in an IOM strategy.

4.5. Willingness to Pay

Table 8 presents the farmers' willingness to pay (WTP) estimates derived from the considered estimated model (model 3), along with their confidence intervals (IC). The results indicate that Tunisian faba bean growers are willing to pay 418 Tunisian Dinars (TND) per hectare (ha) for *Orobanche*-tolerant faba bean varieties over sensitive varieties. Furthermore, growers are willing to pay a premium ranging from 590 TND/ha to 1280 TND/ha to adopt

IOM scenarios with outputs, OUTPUT2, OUTPUT3 and OUTPUT4 compared to the base IOM scenario with OUTPUT1.

Table 8. Farmers' willingness to pay (WTP) for IOM attributes.

	VTOL	DEC	Output 2	Output 3	Output 4
WTP	418.009	−409.149	1047.665	589.457	1278.833
IC	[10.13; 825.88]	[−748.85; −69.43]	[128.62; 1966.70]	[2.67; 1176.24]	[197.35; 2360.30]

IC: Confidence Interval.

It is important to note that the IOM output attribute is a composite measure combining both yield improvements and *Orobanche* plant shoot reduction (see Table 1). Accordingly, the results demonstrate that farmers are willing to pay a premium or around 590 TND/ha for an IOM scenario that significantly increases faba bean yields from 500% to 1000% while maintaining the same reduction level in *Orobanche* plant shoot count (up to 75%) as the base level (OUTPUT1). However, the premium that farmers are willing to pay for the IOM scenario with OUTPUT2, which offers a 95% reduction in *Orobanche* shoot plant count while maintaining the same level of yield improvement, is approximately 1050 TND/ha. This premium is almost double the amount they are willing to pay for OUTPUT3. These findings suggest that farmers are highly concerned with *Orobanche* infestation rates.

Moreover, the highest premium farmers are willing to pay is for the IOM scenario with OUTPUT4, at approximately 1279 TND/ha. This premium is more than double the amount for OUTPUT3 and slightly higher than the premium for OUTPUT2. Accordingly, farmers are concerned about crop yield; however, their primary concern is the infestation rate of *Orobanche*. In conclusion, the farmers in our sample are willing to pay a substantial premium for IOM scenarios that focus first on reducing *Orobanche* plant shoot count, and then, improving crop yields.

Finally, it is important to note that our sample perceives being compensated for changing the sowing date from November to December. Table 8 displays that farmers' WTP for DEC attribute is negative and significant, with an amount of −409 TND/ha. This confirms our results, indicating that recruited farmers are unaware about the good effects of this agronomic practice and the appropriate conditions needed for success.

5. Conclusions

Several factors play a crucial role in farmers' decision-making when selecting an *Orobanche* management approach. The present study examines Tunisian farmers' preferences and their WTP for attributes of integrated approach to control this parasitic that continues to threaten their incomes. The findings show that our sample of faba bean growers consists mostly of older men with low educational level, many of whom are illiterate. While they are familiar with *Orobanche* and its potential damage, they have limited knowledge about parasite life cycle and the practices that could be used to control it. Approximately, 20% of the farmers in our sample have ceased growing faba beans due to the damage caused by *Orobanche*, and half of the sample suffers from moderate to severe infestations. These results suggest that the number of farmers abandoning faba bean cultivation is likely to increase unless an effective control strategy is implemented. For this strategy to succeed, it must involve all relevant stakeholders, particularly the farmers and extension services.

Tunisian farmers appear to be interested in the integrated scenarios for controlling *Orobanche* presented in this study, however, there is significant preferences heterogeneity across the sample regarding the different attributes and attribute levels defining the IOM. The majority of farmers expressed positive preferences for the inclusion of *Orobanche*-

tolerant faba bean varieties in the IOM scenarios. In contrast, they were more reluctant to combine control methods with a delayed sowing date. Farmers did not show significant preferences for the use of chemical treatments or intercropping practices with trap crops like fenugreek to control *Orobanche*. This reluctance could primarily be attributed to the lack of technical knowledge and the gap in their understanding of these two practices. Most farmers have difficulty to manage properly the chemical products recommended for *Orobanche* control, as they lack knowledge about the correct doses, timing, and number of applications required. Similarly, they are unfamiliar with the proper sowing and harvesting techniques for intercropping practices with fenugreek. As a result, farmers appear to be hesitant to adopt these control methods. Furthermore, the study found that farmers were more willing to pay an important premium ranging from 1050 to 1280 TND/ha, for IOM scenarios that focused primarily on reducing *Orobanche* plant shoot counts, rather than improving faba bean yields.

The findings suggest that, to facilitate the adoption of IOM approach by farmers, a target promotion strategy is necessary, with a focus on two key attributes: the availability of *Orobanche* tolerant faba bean varieties, and the effectiveness of IOM scenarios in controlling *Orobanche* and reducing crop losses in infested areas. Furthermore, following Gonzalez-Andujar [25], there is a still need to strengthen the efforts of all stakeholders involved, such as research institutions, extension services, and interprofessional groups, to disseminate research findings. It is crucial to simplify information and clarify the technical aspects of control methods like intercropping with trap crops, delaying sowing date, and applying chemical treatments, to ensure farmers can effectively implement them. Additionally, decision-makers should prioritize accelerating the herbicide registration process for *Orobanche* control, which could help overcome the barriers that have hindered the successful use of some herbicides like Glyphosate and other containing efficient molecules from Imidazolinone families.

To conclude, it is noteworthy to emphasize that a successful communication strategy targeting farmers should be developed using a participatory approach. This strategy must address several key pillars: (i) understanding the vulnerability of farmers, which can manifest in various forms like financial insecurity and relatively low incomes, low education levels, high *Orobanche* infestation, diverse land tenure systems, and local farming practices; (ii) identifying distinct farmers groups, understanding their concerns, selecting appropriate communication tools, and tailoring messages to each group's specific needs; (iii) crafting clear and relevant messages that raise farmers' awareness about IOM scenarios, while considering their local conditions, concerns and priorities; (iv) diversifying knowledge transfer methods and communication tools (roundtable, conferences, information days, scientific coffee, simplified brochures, recurring short messages, applications, social media, farmer-to-farmer network, etc.); and (v) ensuring that adequate resources (financial, technical, and logistical) are allocated to effectively implement and sustain the strategy.

Future research should focus on defining the characteristics of the appropriate communication strategy taking into account the priorities, context, socio-economics characteristics, and farmers' psychological behaviors. This shift towards a more comprehensive approach to control *Orobanche* will play a crucial role in ensuring the long-term viability of farming systems and sustain farmers' resilience. However, additional future research focusing on assessing the long-term cost-effectiveness and financial sustainability of IOM for farmers is crucial. Ultimately, various factors are determinant in farmer's process decisions; however, little attention has been paid to the effect of farmers' risk perception or risk aversion.

Author Contributions: Conceptualization, A.Y. and T.M.; methodology, A.Y.; software, A.Y.; validation, A.Y., Z.A. and M.K.; formal analysis, A.Y.; investigation, A.Y., T.M., Z.A. and M.K.; resources, M.K.; data curation, A.Y. and T.M.; writing—original draft preparation, A.Y.; writing—review and editing, A.Y., Z.A. and M.K.; visualization, M.K.; supervision, M.K.; project administration, M.K. All authors have read and agreed to the published version of the manuscript.

Funding: This research was funded by Ministry of Higher Education and Scientific Research within the framework of the project Zero Parasitic Prima Section 2.

Data Availability Statement: The original contributions presented in the study are included in the article, further inquiries can be directed to the corresponding author.

Acknowledgments: The authors would like to thank the Ministry of Higher Education and Scientific Research and the Ministry of Agriculture, Hydraulic Resources and Maritime Fisheries and the project Zero Parasitic Prima Section 2. Authors gratefully acknowledge the General Directorate of Plant Health and Control of Agricultural Inputs (DGSVCIA), specifically Mona Mahfdhi and Salwa Ben Fraj. Furthermore, the authors present their sincere gratitude to engineers and technical staff from the regional extension services offices, specifically, Rim Ben Salem (CRDA Zaghouane), Olfa Rebhi (CRDA Bizerte), Emna Ben Massaoud (CRDA Nabeul), Nourredine Zouabi (CTV Amdoune, Beja), Fadhel Sallemi (CRDA Jendouba), Yamenta Gannouni (CTV BouSalem, Jendouba), Mostapha Khmiri (CTV Fernana, Jendouba), Adel Zouebi (CTV Sbikha), and Dalel Toumi Azouz (IRESA) for their technical support.

Conflicts of Interest: The authors declare no conflicts of interest.

References

1. Srinivas, A.; Damodar Reddy, D.; Hema, B.; Kasthuri Krishna, S. Overcoming *Orobanche* challenges: A study of knowledge and adoption of management practices among FCV Tabaco farmers in Andhra Pradesh, India. *J. Sci. Res. Rep.* **2024**, *7*, 357–365. [CrossRef]
2. Abang, M.M.; Bayaa, B.; Abu-Irmaileh, B.; Yahyaoui, A. A participatory farming system approach for sustainable broomrape (*Orobanche* spp.) management in the Near East and North Africa. *Crop Prot.* **2007**, *26*, 1723–1732. [CrossRef]
3. Kharrat, M.; Halila, M.H. Orobanche species on faba bean (*Vicia faba* L.) in Tunisia: Problem and management. In *Biology and Management of Orobanche, Proceedings of the Third International Workshop on Orobanche and Related Striga Research, Amsterdam, The Netherlands, 8–12 November 1993*; Pieterse, A.H., Verkleij, J.A.C., ter Borg, S.J., Eds.; Royal Tropical Institute: Amsterdam, The Netherlands, 1994; pp. 639–643.
4. Abbes, Z.; Kharrat, M.; Delavault, P.; Simier, P.; Chaïbi, W. Field evaluation of the resistance of some faba bean (*Vicia faba* L.) genotypes to the parasitic weed *Orobanche foetida* Poiret. *Crop Prot.* **2007**, *26*, 1777–1784. [CrossRef]
5. Amri, M.; Trabelsi, I.; Abbes, Z.; Kharrat, M. Release of a new faba bean variety "Chourouk" resistant to the parasitic plants *Orobanche foetida* Poir. *and Orobanche crenata Forsk. in Tunisia. Int. J. Agric. Biol.* **2019**, *21*, 499–505.
6. Kroschel, J.; Klein, O. Biological control of *Orobanche* spp. in the Near East and North Africa by inundative releases of the herbivore Phytomyza orobanchia. In *Integrated Management of Orobanche Food Legumes in the Near East and North Africa, Proceedings of the Expert Consultation on IPM for Orobanche in Food Legume Systems in the Near East and North Africa, Rabat, Morocco, 7–9 April 2003*; Dahan, R., El Mourid, M., Eds.; ICARDA, INRA, FAO: Rome, Italy, 2004; pp. 55–66.
7. Klein, O.; Kroschel, J. Biological control of *Orobanche* spp. with Phytomyza orobanchia, a review. *Biocontrol* **2002**, *47*, 245–277. [CrossRef]
8. Mabrouk, Y.; Zougui, L.; Sifi, B.; Belhadj, O. The potential of Rhizobium strains for biological control of *Orobanche crenata. Biol. Brast.* **2007**, *62*, 139–143. [CrossRef]
9. Abu-Irmaileh, B.E. Soil solarization. In *Weed Management for Developing Countries—Addendum 1*; FAO Plant Production and Protection Paper No. 120; Labrada, R., Ed.; FAO: Rome, Italy, 2003; pp. 211–222.
10. Sauerborn, J.; Saxena, M.C. Effect of soil solarization on *Orobanche* spp. infestation and other pests in faba bean and lentil. In *Proceedings of the 4th International Symposium on Parasitic Flowering Plants*; Weber, H.C., Forstreuter, W., Eds.; FRG: Marburg, Germany, 1987; pp. 733–744.
11. Garcia-Torres, L.; Lopez-Granados, F.; Jurado-Exposito, M.; Diaz-Sanchez, J. Chemical control of Orobanche in legumes: Achievements and constraints. In Proceedings of the 4th International Workshop on Orobanche, Albena, Bulgaria, 23–26 September 1998; Wegmann, K., Musselman, L.J., Eds.; Kluwer Academic Publishers: Albena, Bulgaria, 1998; pp. 391–399.
12. Jurado-Exposito, M.; Garcia-Torres, L.; Castejon-Munoz, M. Broad bean and lentil seed treatments with imidazolinones for the control of broomrape (*Orobanche crenata*). *J. Agric. Sci.* **1997**, *129*, 307–314. [CrossRef]
13. Abebe, G.; Sahile, G.; Al-Tawaha, A.M. Evaluation of potential trap crops on *Orobanche* soil seed bank and tomato yield in the Central Rift Valley of Ethiopia. *World J. Agric. Sci.* **2005**, *1*, 148–151.

14. Rubiales, D.; Alcantara, C.; Perez-de-Luque, A.; Gil, J.; Sillero, J.C. Infection of chickpea (*Cicer arietinum*) by crenate broomrape (*Orobanche crenata*) as influenced by sowing date and weather conditions. *Agronomie* **2003**, *23*, 359–362. [CrossRef]
15. Thomas, H.; Sauerborn, J.; Singh, L. Impact and management of Orobanche in cropping systems in Nepal. In Proceedings of the 4th International Workshop on Orobanche, Albena, Bulgaria, 23–26 September 1998; Wegman, K., Musselman, L.J., Joel, D.M., Eds.; Kluwer Academic Publishers: Albena, Bulgaria, 1998; pp. 413–418.
16. Nassib, S.M.; Hussein, A.H.A.; Saber, H.A.; El Deeb, M.A. Effect of N, P, and K nutrients with a reduced rate of glyphosate on control and yield of faba bean in middle Egypt. *Egypt. J. Appl. Sci.* **1992**, *7*, 720–730.
17. Wilhelm, S.; Strokan, R.C.; Sagan, J.E.; Carpenter, T. Large scale soil fumigation against broomrape. *Phytopathology* **1959**, *49*, 530–532.
18. Habimana, S.; Nduwumuremyi, J.D.; Chinama, R. Management of *Orobanche* in field crops—A review. *J. Soil Sci. Plant Nutr.* **2014**, *14*, 43–62.
19. Abbes, Z.; Trabelsi, I.; Kharrat, M.; Amri, M. Intercropping with fenugreek (*Trigonella foenum-graecum*) enhanced seed yield and reduced *Orobanche foetida* infestation in faba bean (*Vicia faba*). *Biol. Agric. Hort.* **2019**, *35*, 238–247. [CrossRef]
20. Trabelsi, I.; Abbes, Z.; Amri, M.; Kharrat, M. Study of some resistance mechanisms to *Orobanche* spp. Infestation in faba bean (*Vicia faba* L.) breeding lines in Tunisia. *Plant Prod. Sci.* **2016**, *19*, 562–573. [CrossRef]
21. Ahmed, N.R.; Abdallah, N.G.; Abd-Elraoufn, R.M. Intercropping fenugreek (*Trigonella foenum graecum* L.) on the faba bean (*Vicia faba*) to reduce the incidence of (*Orobanche crenata*). *World Rural Obs.* **2015**, *7*, 88–99.
22. Fernández-Aparicio, M.; Emeran, A.A.; Rubiales, D. Control of *Orobanche crenata* in legumes intercropped with fenugreek (*Trigonella foenum-graecum*). *Crop Prot.* **2008**, *27*, 653–659. [CrossRef]
23. Rubiales, D.; Perez-de-Luque, A.; Fernandez-Aparico, M.; Sillero, J.; Roman, B.; Kharrat, M.; Khalil, S.; Joel, D.; Riches, C. Screening techniques and sources of resistance against parasitic weeds in grain legumes. *Euphytica* **2006**, *147*, 187–199. [CrossRef]
24. Sillero, J.C.; Cubero, J.I.; Fernández-Aparicio, M.; Rubiales, D. Search for resistance to crenate broomrape (*Orobanche crenata*) in Lathyrus. *Lathyrus Lathyrism Newslett.* **2005**, *4*, 7–9.
25. Gonzalez-Andujar, J.L. Integrated weed management: A shift towards more sustainable and holistic practices. *Agronomy* **2023**, *13*, 2646. [CrossRef]
26. Kudsk, P. *Advances in Integrated Weed Management*, 1st ed.; Burley Dodds Science Publishing: Cambridge, UK, 2022.
27. Michalis, E.; Yangui, A.; Ragkos, A.; Kharrat, M.; Chachalis, D. Opinions and perceptions on sustainable weed management: A comparison between Greek and Tunisian Farmers. *Proceeding* **2024**, *94*, 48.
28. Ellis-Jones, J.; Schulz, S.; Douthwaite, B.; Hussaini, M.A.; Oyewole, B.D.; Olanrewaju, A.S.; White, R. An assessment of integrated Striga hermonthica control and early adoption by farmers in northern Nigeria. *Exp. Agric.* **2004**, *40*, 353–368. [CrossRef]
29. Emechebe, A.M.; Ellis-Jones, J.; Schulz, S.; Chikoye, D.; Douthwaite, B.; Kureh, I.; Tarawali, G. Farmer's perception of the Striga problem and its control in northern Nigeria. *Exp. Agric.* **2004**, *40*, 215–232. [CrossRef]
30. Lutzeyer, H.J.; Kroschel, J.; Sauerborn, J. Orobanche crenata in legume cropping: Farmers' perception, difficulties and prospects of control—A case study in Morocco. In *Biology and Management of Orobanche, Proceedings of the Third International Workshop on Orobanche and Related Striga Research, Amsterdam, The Netherlands, 8–12 November 1993*; Pieterse, A.H., Verkleij, J.A.C., Borg, S.J., Eds.; Royal Tropical Institute: Amsterdam, The Netherlands, 1994; pp. 432–441.
31. Rogers, E.M. *Diffusion of Innovations*, 5th ed.; Free Press: New York, NY, USA, 2003.
32. Llewellyn, R.S.; Lindner, R.K.; Pannell, D.J.; Powles, S.B. Resistance and the herbicide resource: Perceptions of Western Australian grain growers. *Crop Prot.* **2002**, *21*, 1067–1075. [CrossRef]
33. Chèze, B.; David, M.; Martinet, V. Understanding farmers' reluctance to reduce pesticide use: A choice experiment. *Ecol. Econ.* **2020**, *167*, 106349. [CrossRef]
34. Danne, M.; Musshoff, O.; Schulte, M. Analysing the importance of glyphosate as part of agricultural strategies: A discrete choice experiment. *Land Use Policy* **2019**, *86*, 189–207. [CrossRef]
35. Wilson, R.S.; Hooker, N.; Tucker, M.; LeJeune, J.; Doohan, D. Targeting the farmers decision making process: A pathway to increased adoption of integrated weed management. *Crop Prot.* **2009**, *28*, 756–764. [CrossRef]
36. Llewellyn, R.S.; Pannell, D.J.; Lindner, R.K.; Powles, S.B. Targeting key perceptions when planning and evaluating extension. *Aust. J. Exp. Agric.* **2005**, *45*, 1627–1633. [CrossRef]
37. Eckert, E.; Bell, A. Invisible force: Farmers' mental models and how they influence learning and actions. *J. Ext.* **2005**, *43*, 3. Available online: https://tigerprints.clemson.edu/joe/vol43/iss3/3 (accessed on 6 January 2005).
38. Eckert, E.; Bell, A. Continuity and change: Themes of mental model development among small-scale farmers. *J. Ext.* **2006**, *44*, 4.
39. Corselius, K.L.; Simmons, S.R.; Flora, C.B. Farmer perspectives on cropping systems diversification in northwestern Minnesota. *Agric. Hum. Values* **2003**, *20*, 371–383. [CrossRef]
40. Czapar, G.F.; Curry, M.P.; Wax, L.M. Grower acceptance of economic thresholds for weed management in Illinois. *Weed Technol.* **1997**, *11*, 828–831. [CrossRef]
41. Setiawan, B.; Maulana Noor, Y.A.; Faizal, F.; Rayeza, N.F.; Rohman, M.S. The level of rice farmers' adoption of sustainable agricultural practices standard in Indonesia: A Discrete Choice Experiments. *J. Sustain. Sci. Manag.* **2024**, *19*, 110–124. [CrossRef]

42. Fu, H.; Peng, Y.; Zheng, L.; Liu, Q.; Zhou, L.; Zhang, Y.; Kong, R.; Turvey, C.G. Heterogeneous choice in WTP and WTA for renting land use rights in rural china: Choice experiments from the field. *Land Use Policy* **2022**, *119*, 106–123. [CrossRef]
43. Miriti, P.; Regassa, M.D.; Ojiewo, C.O.; Melesse, M.B. Farmers' preferences and willingness to pay for traits of sorghum varieties: Informing product development and breeding programs in Tanzania. *J. Crop Improv.* **2022**, *37*, 253–272. [CrossRef]
44. Hervé-Sossou, C.H.; Midingoyi, S.G.; Codjo, V. Cashew growers' preferences for market information system design in Benin: A choice experiment approach. *Paki. J. Agri. Sci.* **2022**, *59*, 19–28.
45. Houngbo, S.; Zannou, A.; Zossou, E.; Calmette, S.G.Z.; Aoudji, A.; Sinzogan, A.A.; Sikirou, R.; Ahanchédé, A. Farmers' Preferences and Willingness to Pay for Attributes of Integrated Pest Management Methods Against Spodoptera frugiperda (*Lepidoptera: Noctuidae*) in Benin. *J. Integr. Pest Manag.* **2022**, *12*, 45. [CrossRef]
46. Zossou, S.R.C.; Adegbola, P.Y.; Oussou, B.T.; Dagbenonbakin, G.; Mongbo, R. Modelling smallholder farmers' preferences for soil fertility management technologies in Benin: A stated preference approach. *PLoS ONE* **2021**, *16*, e0253412. [CrossRef]
47. Louviere, J.J.; Street, D. Stated-preference methods. In *Handbook of Transport Modelling*; Hensher, D.A., Button, K.J., Eds.; Pergamon Press: Amsterdam, The Netherlands, 2000; pp. 131–143.
48. Boeuf, F. Les *Orobanches* en Tunisie. *J. Agric. Prat.* **1905**, *9*, 11–14.
49. Trabelsi, I.; Abbes, Z.; Amri, M.; Kharrat, M. Performance of faba bean genotypes with *Orobanche foetida* and *Orobanche crenata* infestation in Tunisia. *Chil. J. Agric. Res.* **2015**, *75*, 27–34. [CrossRef]
50. Kharrat, M.; Abbes, Z.; Amri, M. A new faba nean small seeded variety Najeh tolerant to *Orobanche* registered in the Tunisian cataloguye. *Tun. J. Plant Prot.* **2010**, *5*, 125–130.
51. Abbes, Z.; Kharrat, M.; Chaïbi, W. Seed Germination and tubercle development of *Orobanche foetida* and *Orobanche crenata* in presence of different plant species. *Tunis. J. Plant Prot.* **2008**, *3*, 101–109.
52. Kharrat, M.; Souissi, T. Research on *Orobanche foetida* and *O. crenata* in Tunisia. In *Integrated Management of Orobanche in Food Legumes in the Near East and North Africa, Proceedings of the Expert Consultation on IPM for Orobanche in Food Legume Systems in the Near East and North Africa, Rabat, Morocco, 7–9 April 2003*; Dahan, R., El Mourid, M., Eds.; ICARDA, INRA, FAO: Rome, Italy, 2004; pp. 106–110.
53. Abbes, Z.; Kharrat, M.; Shaaban, K.; Bayaa, B. Comportement de différentes accessions améliorées de féverole (*Vicia faba* L.) vis-à-vis d'*Orobanche crenata* Forsk. et *Orobanche foetida* Poir. *Cah. Agric.* **2010**, *19*, 194–199. [CrossRef]
54. Abbes, Z.; Mkadmi, M.; Trabelsi, I.; Amri, M.; Kharrat, M. *Orobanche foetida* control in faba vean by foliar application of Benzothiadiazole (BTH) and Salicylic Acid. *Bulg. J. Agri. Sci.* **2014**, *20*, 1439–1443.
55. Abbes, Z.; Sellami, F.; Amri, M.; Kharrat, M. Effet of Sowing date on *Orobanche foetida* infection and seed yield od resistant and susceptible faba bean cultivars; Acta Phyt. *Ento. Hung.* **2010**, *45*, 267–275.
56. Thebti, S.; Bouallegue, A.; Rzigui, T.; En-Nahli, Y.; Horchani, F.; Hosni, T.; Kharrat, M.; Amri, M.; Abbes, Z. Potential Physiological Tolerance mechanism in faba bean to *Orobanche* sp. Parasitism. *Front. Plant Sci.* **2014**, *15*, 1497303.
57. Bouraoui, M.; Abbes, Z.; L'taief, B.; Alshahami, M.O.; Hachana, A.; Bouaziz, S. Exploring the biochemical dynamics in faba bean (*Vivia faba* L. minor) in reponse to *Orobanche foetida* Poir. parasitism under inoculation with different rhizobia strains. *PLoS ONE* **2024**, *19*, e0304673. [CrossRef] [PubMed]
58. Triki, E.; Trabelsi, I.; Amri, M.; Nefzi, F.; Kharrat, M.; Abbes, Z. Effect of Bensothiadiazole and Salicylic acid resistance inducers on *Orobanche foetida* infestation in *Vicia faba*. *Tun. J. Plant Prot.* **2018**, *13*, 113–125.
59. Bouraoui, M.; Abbes, Z.; Rouissi, M.; Abdi, N.; Hemissi, I.; Kouki, S.; Sifi, B. Effect of rhizobia inoculation, N and P supply on *Orobanche foetida* parasitising faba bean (*Vicia faba* minor) under field conditions. *Bioc. Sci. Tech.* **2016**, *26*, 776–791. [CrossRef]
60. Finger, R.; Sok, J.; Ahovi, E.; Akter, S.; Bremmerj, J.; Dachbrodt-Saaydeh, S.; de Lauwere, C.; Kreft, C.; Kudsk, K.; Lambarraa-Lehnhardt, F.; et al. Towards sustainable crop protection in agriculture: A framework for research and policy. *Agri. Syst.* **2024**, *219*, 104037. [CrossRef]
61. Street, D.; Burgess, L. *The Construction of Optimal Stated Discrete Choice Experiments: Theory and Methods*; A John Wiley & Sons: Hoboken, NJ, USA, 2007.
62. McFadden, D. Conditional logit analysis of qualitative choice behavior. In *Frontiers in Econometrics*; Zambreka, P., Ed.; Academic Press: New York, NY, USA, 1973; pp. 105–142.
63. Lancaster, K.J. A new approach to consumer theory. *J. Political Econ.* **1966**, *74*, 132–157. [CrossRef]
64. Train, K. *Discrete Choice Methods with Simulation*; Cambridge University Press: Cambridge, UK, 2003.
65. Hensher, D.A.; Greene, W.H. Mixed logit models: State of practice. *Transportation* **2003**, *30*, 133–176. [CrossRef]
66. Schulz, N.; Breustedt, G.; Latacz-Lohmann, U. Assessing farmers' willingness to accept "greening": Insights from a discrete choice experiment in Germany. *J. Agric. Econ.* **2014**, *65*, 26–48. [CrossRef]
67. Greiner, R. Factors influencing farmers' participation in contractual biodiversity conservation: A choice experiment with northern Australian pastoralists. *Aust. J. Agric. Resour. Econ.* **2016**, *60*, 1–21. [CrossRef]
68. Beharry-Borg, N.; Smart, J.; Termansen, M.; Hubacek, K. Evaluating farmers' likely participation in a payment programme for water quality protection in the UK uplands. *Reg. Environ. Chang.* **2013**, *13*, 633–647. [CrossRef]

69. Hudson, D.; Lusk, J. Risk and transaction cost in contracting: Results from a choice based experiment. *J. Agric. Food Ind. Organ.* **2004**, *2*, 1046. [CrossRef]
70. Jaeck, M.; Lifran, R. Farmers' preferences for production practices: A choice experiment study in the Rhone river delta. *J. Agric. Econ.* **2014**, *65*, 112–130. [CrossRef]
71. Institut National des Grandes Cultures (INGC). *Etude D'élaboration d'une Stratégie de Développement et de Restructuration du Secteur des Oléo-Protéagineux*; Institut National des Grandes Cultures (INGC): Bou Salem, Tunisia, 2017; pp. 71–94.
72. Riemens, M.; Sonderskov, M.; Moonen, A.C.; Storkey, J. An integrated weed management framework: A pan-European perspective. *Euro. J. Agron.* **2022**, *133*, 126443. [CrossRef]
73. Sharma, A.; Bailey, A.; Fraser, I. Technology adoption and pest control strategies among uk cereals farmers: Evidence from parametric and nonparametric count data models. *J. Agric. Econ.* **2011**, *62*, 73–92. [CrossRef]
74. Pannell, D.J.; Marshall, G.R.; Barr, N.; Curtis, A.; Vanclay, F.; Wilkinson, R. Understanding and promoting adoption of conservation practices by rural landholders. *Aust. J. Exp. Agric.* **2006**, *46*, 1407–1424. [CrossRef]
75. Doole, G.J.; James, T.K. Profitable management of a finite herbicide resource. *Crop Prot.* **2023**, *172*, 106314. [CrossRef]
76. Burton, M. Model invariance when estimating random parameters with categorical variables. In Proceedings of the 2019 Conference (63rd), Melbourne, Australia, 12–15 February 2019.
77. Lamichane, J.R.; Dachbrodt-Saaydeh, S.; Kudsk, P.; Messean, A. Towards a reduced reliance on conventional pesticides in European agricultural. *Plant Dis.* **2016**, *100*, 10–24. [CrossRef]
78. Van Hezewijk, M.J.; Van Beem, A.P.; Verkleij, J.A.C. Germination of *Orobanche crenata* seeds, as influenced by conditioning temperature and period. *Can. J. Bot.* **1993**, *71*, 786–792. [CrossRef]
79. Kebreab, E.; Murdoch, A.J. Simulation of integrated control strategies for *Orobanche* spp. Based on a life cycle model. *Expl. Agric.* **2001**, *37*, 37–51. [CrossRef]
80. Mesa-Garcia, J.; Garcia-Torres, L. Status of *Orobanche crenata* in faba bean in the Mediterranean re-gion and its control. *Opti. Medit.* **1991**, *10*, 75–78.
81. Imoloame, E.O.; Ayanda, I.F.; Yusuf, O.J. Integrated weed management practices and sustainable food production among farmers in Kwara state, Nigeria. *Open Agric.* **2021**, *6*, 124–134. [CrossRef]
82. Swanton, C.J.; Mahoney, K.J.; Chandler, K.; Gulden, R.H. Integrated weed management: Knowledge-based weed management systems. *Weed Sci.* **2008**, *56*, 168–172. [CrossRef]
83. Bakker, L.; Sok, J.; Van der Werf, W.; Bianchi, F.J.J.A. Kicking the habit: What makes and breaks farmers' intentions to reduce pesticide use? *Ecol. Econo.* **2021**, *180*, 106868. [CrossRef]
84. Ghorbani, M.; Kulshreshtha, S. An environmental and economic perspective on integrated weed management in Iran. *Weed Tech.* **2013**, *27*, 352–361. [CrossRef]
85. Horowitz, J.; Lichtenberg, E. Risk-reducing and risk-increasing effects of pesticides. *J. Agri. Econ.* **1994**, *45*, 82–89. [CrossRef]
86. Lopez–Granados, F.; Garcia-Torres, L. Short- and long-term economic implications of controlling *crenata* broomrape (*Orobanche crenata* Forsk.) in braod bean (*Vicia faba* L.) under various management strategies. *Crop Prot.* **1998**, *17*, 139–143. [CrossRef]

Disclaimer/Publisher's Note: The statements, opinions and data contained in all publications are solely those of the individual author(s) and contributor(s) and not of MDPI and/or the editor(s). MDPI and/or the editor(s) disclaim responsibility for any injury to people or property resulting from any ideas, methods, instructions or products referred to in the content.

agronomy

MDPI

Article

The Efficacy of Pre-Emergence Herbicides Against Dominant Soybean Weeds in Northeast Thailand

Ultra Rizqi Restu Pamungkas , Sompong Chankaew , Nakorn Jongrungklang , Tidarat Monkham and Santimaitree Gonkhamdee *

Department of Agronomy, Faculty of Agriculture, Khon Kaen University, Khon Kaen 40002, Thailand; ultrarizqirestupamungkas.u@kkumail.com (U.R.R.P.); somchan@kku.ac.th (S.C.); nuntjo@kku.ac.th (N.J.); tidamo@kku.ac.th (T.M.)

* Correspondence: gsanti@kku.ac.th; Tel.: +66-8914-81616

Abstract: Soybean production in Thailand faces significant challenges from malignant weed competition, potentially reducing yields by up to 37% and incurring annual economic losses of approximately USD 3.8 billion. Pre-emergence herbicides are critical for integrated weed management, but their efficacy varies depending on local conditions and soybean varieties. This study evaluates the performance of three pre-emergence herbicides, pendimethalin (1875 g a.i. ha^{-1}), s-metolachlor (900 g a.i. ha^{-1}), and flumioxazin (125 g a.i. ha^{-1}), on weed control efficiency (WCE), soybean growth, phytotoxicity, and yield in Northeast Thailand using a randomised complete block design with two varieties (CM60 and Morkhor60) across rainy (2023) and dry (2024/2025) seasons. Herbicide performance varied seasonally: s-metolachlor showed optimal rainy season results (61.54% weed control efficiency at 63 days after herbicide application (DAA), with a yield of 1036 kg ha^{-1}), while flumioxazin excelled in dry conditions (64.32% WCE, <4% phytotoxicity, and 1243 kg ha^{-1} yield). Pendimethalin performed poorly under wet conditions but improved in drier weather. Among five dominant weed species, *Cyperus rotundus* proved the most resilient. CM60 demonstrated superior herbicide tolerance and yield stability, particularly under rainy conditions. These results emphasise that season-specific herbicide selection and variety matching are crucial for herbicide resistance management and effective weed control in Thailand's rainfed soybean systems.

Keywords: weed management; weed species; summed dominance ratio; phytotoxicity

1. Introduction

Global demand for soybeans continues to grow, particularly for plant-based protein products [1,2]. Thailand's soybean production is concentrated in the northern and north-eastern regions, where improved varieties like CM60 have been widely adopted. CM60 is recognised for its rust resistance and high yield potential of 1875 kg ha^{-1} with 44% protein content [3]. Recently, Khon Kaen University has developed a new soybean variety called Morkhor60, which shows promising adaptation to Thailand's growing conditions, with yields of 1250–2063 kg ha^{-1} across seasons [4]. However, domestic soybean production in Thailand is marginal, with annual output stagnating at around 50,000–60,000 metric tonnes due to unattractive returns compared to other field crops such as corn and cassava. As a result, Thailand continues to rely heavily on imports to meet its soybean requirements, with recent annual imports exceeding 3.4 million metric tonnes. This dependence on foreign supply is further compounded by government policy, as the cultivation of genetically modified soybeans remains strictly prohibited. These factors highlight the importance of improving domestic soybean productivity and sustainability, particularly in key agricultural regions such as Northeastern Thailand [5]. A major constraint to soybean productivity is weed competition, with potential losses ranging from 20 to 90% globally without proper control measures, which is often economically more significant than losses due to insects, pathogens, or other biotic constraints altogether [6]. Weeds such as *Imperata cylindrica* and *Cyperus* spp. compete aggressively for nutrients, water, and light while also harbouring pests and reducing post-harvest quality [7]. The critical period for weed control (CPWC) in soybeans spans 18–31 days after emergence (DAE), with weed-free conditions required for up to 61 DAE to prevent yield declines [8,9]. Traditional manual weeding, although common in Thailand, is labour-intensive (40–60% of production costs) and increasingly impractical due to labour shortages [10,11].

Pre-emergence herbicides, such as pendimethalin, s-metolachlor, and flumioxazin, offer sustainable solutions by providing residual soil activity to suppress early weed emergence [12]. These herbicides are particularly critical in light of the widespread development of glyphosate resistance and the limited efficacy of post-emergence options [13,14]. For instance, pendimethalin (1339 g a.i. ha^{-1}) effectively controls grasses and broad-leaved weeds without phytotoxicity in soybeans under suitable conditions [15], while flumioxazin (125 g a.i. ha^{-1}) can achieve up to 96.8% weed control efficiency [16]. However, herbicide efficacy and crop safety are highly variable and contingent on interactions among the herbicide, crop variety, and environmental conditions [17]. Notably, the planting season, whether rainy or dry, profoundly influences edaphoclimatic factors, such as soil moisture and rainfall patterns, which critically impact herbicide phytotoxicity and efficacy [18,19]. Under suboptimal conditions, herbicides can cause significant injury to soybean plants, leading to reduced emergence (19–73%) and yield losses [18,20].

Varietal differences further compound this variability. Early maturation soybean cultivars exhibit heightened susceptibility to herbicide injury compared to medium- or long-cycle cultivars, likely due to less time for stress recovery [21–23]. Concerns persist that very early maturation cultivars may be inherently more susceptible [24,25], emphasising the critical need for variety-specific evaluations under local conditions. Despite their advantages, pre-emergence herbicides remain significantly understudied for key local Thai varieties, such as CM60 and Morkhor60, particularly in relation to the distinct environmental pressures of Thailand's rainy and dry seasons. Therefore, this study aims to evaluate the efficacy and selectivity of several pre-emergence herbicides against weeds in soybean fields, with a specific focus on the varieties CM60 and Morkhor60. The evaluation considers weed control efficiency, soybean growth, phytotoxicity, yield, and yield components, explicitly accounting for the critical influence of planting season on herbicide performance.

2. Materials and Methods

2.1. Experimental Site Description and Experimental Design

Field experiments were conducted during the rainy season of 2023 (planting date: 29 July 2023) and the dry season of 2024/2025 (planting date: 26 October 2024) at the Agronomy Field, Faculty of Agriculture, Khon Kaen University (16.47028° N, 102.80863° E) to study the effect of pre-emergence herbicides and different soybean varieties on soybean productivity, associated weed species, and some physiological characteristics under experimental site conditions. Meteorological data, including rainfall (mm), relative humidity (%), and maximum and minimum temperatures (°C), were monitored throughout both growing seasons and sourced from the Agricultural Weather Station, Faculty of Agriculture, Khon Kaen University (Figure 1). Composite soil samples (0–30 cm depth) were collected to analyse the physical and chemical properties (Table 1), revealing a sandy loam texture typical of the experimental site.

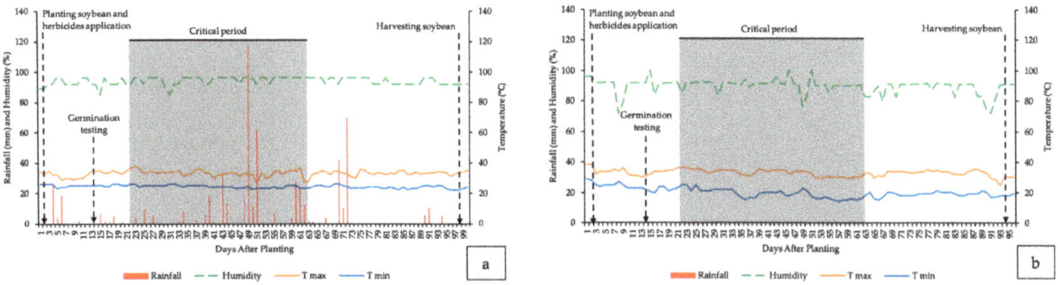

Figure 1. Weather information and experiment duration during (**a**) the rainy season of 2023 and (**b**) the dry season of 2024/2025. The average temperature during the rainy season in 2023 was 28.6 °C, and the average temperature was 26.4 °C during the dry season of 2024/2025.

Table 1. Physical and chemical characteristics of the 0–30 cm soil layer in the experimental area during the 2023 and 2024/2025 soybean crop seasons.

Soil Properties	Rainy Season of 2023	Dry Season of 2024/2025
Physical Characteristics		
Sand (%)	73.36	73.90
Silt (%)	20.87	17.40
Clay (%)	5.77	8.70
Soil Texture	Sandy Loam	Sandy Loam
Chemical Characteristics		
pH (1:1)	6.19	5.40
EC (1:5) (dS/m)	0.04	0.04
OM (%)	0.44	0.61
Total N (%)	0.04	0.03
Available P (mg/kg)	15.00	70.00
K (mg/kg)	103.96	71.56
Ca (mg/kg)	590.34	225.18
Mg (mg/kg)	62.39	40.05
CEC (c mol/kg)	3.00	3.36

The experimental site was prepared through primary tillage using a disc harrow to a depth of 30 cm, followed by land levelling to ensure uniform surface conditions. A basal application of 9.38 kg N ha^{-1}, 4.09 kg P ha^{-1}, and 7.79 kg K ha^{-1} was administered during field preparation, followed by a top dressing 45 days after planting (DAP) supplying 23.48 kg N ha^{-1}, 10.22 kg P ha^{-1}, and 19.45 kg K ha^{-1} based on recommendations from the Department of Agriculture, Thailand, and the site's history of continuous soybean cultivation. Supplemental nitrogen was applied despite the soybean's nitrogen-fixing capacity to ensure sufficient nitrogen availability during early growth, as biological fixation may be limited by soil acidity and environmental factors. Seeds were not inoculated with N-fixing bacteria because native rhizobia populations were assumed adequate due to prior soybean cropping, as well as to avoid confounding effects on yield and phytotoxicity.

Irrigation was managed using a sprinkler system, with scheduling based on soil moisture monitoring. Soybean seeds were sown at five seeds per hill and thinned to three plants per hill at 14 DAP. The experiment employed a split-plot design within a randomised complete block design (RCBD) with four replications. The main plots consisted of five weed management treatments: weed-free control (hand weeding), weedy control (weedy), and three pre-emergence herbicides (pendimethalin, s-metolachlor, and flumioxazin) applied one DAP (Table 2). Subplots comprised two soybean cultivars (Morkhor60 and CM60), with individual plots measuring 5 × 5 m and plant spacing of 0.50 m between rows and 0.25 m within rows.

Table 2. Herbicides and rates used for the treatment of herbicide selectivity.

Treatment	Class	Active Ingredient	Trade Name Manufacturer	Doses (g a.i. ha^{-1}) */
Hand Weeding	–	–	–	–
Weedy	–	–	–	–
Pendimethalin	K1	45.5% *w/v* CS	Prowl CS/BASF (Bangkok, Thailand)	1875
S-metolachlor	K3	96% *w/v* EC	Dualgold/Syngenta (Samut Prakan, Thailand)	900
Flumioxazin	E	50% WP	Zumizoya/TJC (Samut Prakan, Thailand)	125

*/ grams of active ingredient per hectare; hand weeding = plots were kept free of weeds throughout the experiment via manual hoeing; weedy = plots were maintained without weed control. Class = classification of herbicide mechanisms of action according to the Weed Science Society of America (WSSA) and the Herbicide Resistance Action Committee (HRAC) (K1: inhibitors of microtubule assembly; K3: inhibitors of synthesis of very long-chain fatty acids; E: inhibitors of protoporphyrinogen oxidase). CS: capsule suspension; EC: emulsifiable concentrate; WP: wettable powder; and *w/v*: weight per volume.

Pre-emergence herbicides were uniformly applied using a calibrated 15 L knapsack sprayer with a flooding fan nozzle (500 L ha^{-1} spray volume) one day after planting (DAP) during the rainy season of 2023 (30 July 2023) and the dry season of 2024/2025 (27 October 2024). During spraying operations, the air temperature, relative humidity, and soil temperature at a depth of 5 cm were continuously monitored using a mini digital temperature humidity metre (UNI-T UT333) and a glass thermometer. Applications occurred during the early morning under low wind conditions (<5 km h^{-1}), with solutions vigorously agitated before use. Real-time measurements during the 2023 rainy season applications showed averages of 34 ± 4 °C (air), 48 ± 9% RH, and 24 ± 1 °C (soil), while the 2024/2025 dry season exhibited averages of 35 ± 2 °C (air), 62 ± 6% RH, and 31 ± 0.6 °C (soil).

The experiment was conducted on soils with naturally acidic pH levels (5.40–6.19), reflecting typical conditions in Northeastern Thailand's rainfed farmlands, where sandy, weathered soils and tropical climate result in pH values of 4.5–6.0 and liming is uncommon [26,27]. While optimal soybean growth occurs at pH levels of 6.5–7.0 [28,29], some genotypes—including those used in this study—are adapted to acidic, sandy soils and can perform well without pH adjustment [4]. Thus, maintaining these pH levels ensures the experiment's relevance to local farming practices and provides realistic insights into herbicide efficacy and crop response.

2.2. Data Collection

2.2.1. Studies on Weeds

Weed parameters, including weed density (plants m^{-2}), weed biomass (g m^{-2}), the summed dominance ratio (SDR), and weed control efficiency (WCE), were evaluated using a rectangular sampling frame measuring 0.5 × 1 m (0.5 m^2). Following the methodology of Aekrathok et al. [30], two random samplings were conducted per plot for each evaluation interval at 21, 35, 49, and 63 days after herbicide application (DAA), with sampling locations systematically varied to avoid re-sampling previously assessed areas. All weed specimens within each quadrat were identified at the species level, counted to obtain weed density, and then oven-dried at 80 °C for 72 h to determine the dry biomass. The SDR values were calculated based on established formulas by Janiya and Moody [31], as cited in Hasan et al. [32]. The five most dominant weed species, as determined via the SDR analysis, were further characterised (Table 3).

$$\text{SDR} = \frac{\text{Relative density (RD)+relative dry weight (RDW)}}{2} \quad (1)$$

where

$$\text{RD} = \frac{\text{Density of a given species}}{\text{Total density}} \times 100 \quad (2)$$

$$\text{RDW} = \frac{\text{Dry weight of a given species}}{\text{Total dry weight}} \times 100 \quad (3)$$

Table 3. Scientific, common, and family names of dominant weeds in soybean crops during the 2023 and 2024/2025 growing seasons.

Categories	Scientific Name	Common Name	Family
Broad-leaved weeds	*Trianthema portulacastrum* L. *Oldenlandia corymbosa* L.	Horse purslane Diamond flower	Aizoaceae Rubiaceae
Grassy weeds	*Dactyloctenium aegyptium* (L.) P. B. *Digitaria ciliaris* (Retz.) Koeler.	Crowfoot grass Summer grass	Gramineae Gramineae
Sedge weeds	*Cyperus rotundus* L.	Purple nutsedge	Cyperaceae

Weed control efficiency measures the effectiveness of any weed control treatment in comparison to no weed control treatment. Mathematically, it can be expressed as follows [32,33]:

$$\text{WCE (\%)} = \frac{\text{Weed biomass in weedy plots} - \text{Weed biomass in treated plots}}{\text{Weed biomass in weedy plots}} \times 100 \quad (4)$$

2.2.2. Agronomic Trait

The soybean germination rate (%) was calculated as the proportion of emerged seeds relative to the total sown, evaluated at 3, 5, 7, and 14 DAA [34]. Five plants were randomly selected to measure plant height and the number of soybean nodes at the critical period of soybean plants at 21, 35, 49, and 63 DAA. Herbicide phytotoxicity symptoms on soybeans were observed periodically using the European Weed Research Council (EWRC) rating scale, with damaged plant rates of 0% (no effect on plants) to 100% (total loss of plants and yield), as described by [35].

2.2.3. Yield and Yield Component

Harvesting occurred at physiological maturity, which was determined as 90–95 DAP for CM60 and 95–100 DAP for Morkhor60, depending on the respective environmental conditions during the season of planting. At this stage, five plants per plot were randomly sampled to quantify the pod number (branches and main stem), seed number per pod, and 100-seed weight (g) (determined from a random subsample of harvested seeds). The plant population was determined by counting the number of plants within the harvest area and extrapolating to plants per hectare (plants ha^{-1}). Grain yield was measured from a harvest area of 2 $\times$ 2 m and converted to kg ha^{-1}. Grain yield loss (%) was calculated as the percentage reduction in yield compared to the hand weeding treatment using the following formula:

$$\text{Grain yield loss (\%)} = \frac{\text{Yield in hand weeding} - \text{Yield in weed control}}{\text{Yield in hand weeding}} \times 100 \quad (5)$$

2.3. Statistical Analysis

The data were analysed using analysis of variance (ANOVA) to investigate variations in treatment effects, such as weed control efficiency, soybean toxicity, yield, and yield components. All traits were analysed for homogeneity of variance using Levene's test method, and the results are shown in Table S1. The relationships between the analysed variables (traits) were tested using Pearson correlation coefficients (Figure S1). The data underwent statistical analysis using the Statistix® version 10.0 (1985–2013) tool (Analytical Software, Tallahassee, FL, USA). Treatment mean differences were determined using the LSD at a 5% probability level.

3. Results

3.1. Effects of Pre-Emergence Herbicides in Weed Data

3.1.1. Weed Control Efficiency

The efficacy of pre-emergence herbicides against weeds in soybean crops exhibited pronounced seasonal variability (Table 4 and Figure 2). Manual hand weeding consistently achieved 100% weed control efficiency (WCE) at all assessment intervals (21–63 DAA), serving as the experimental control benchmark. All herbicide treatments demonstrated highly significant differences ($p < 0.01$) compared to the weedy check in both seasons. In the rainy season of 2023, s-metolachlor (87.21 ± 3.96%, ab) and flumioxazin (76.44 ± 12.55%, b) showed statistically similar efficacy at 21 DAA ($p > 0.05$), with both outperforming pendimethalin (54.54 ± 8.26%, c; $p < 0.05$). The superior early performance of s-metolachlor despite heavy rainfall likely accelerated degradation (Figure 2a), although by 63 DAA, its WCE declined to 61.54 ± 6.78% (b) while remaining significantly higher than that of flumioxazin (29.96 ± 11.59%, c; $p < 0.05$). Pendimethalin's efficacy plummeted to 10.03 ± 13.39% (d) by 63 DAA, which is consistent with its known susceptibility to leaching under heavy rainfall conditions (Figure 2a).

During the dry season of 2024/2025, s-metolachlor maintained strong initial control (85.32 ± 2.14%, b at 21 DAA), although its efficacy was statistically comparable to pendimethalin (65.50 ± 4.80%, c; $p > 0.05$) and flumioxazin (66.30 ± 6.98%, c) at this stage. Remarkably, pendimethalin's persistence improved under drier conditions, achieving 81.49 ± 5.90% (b) WCE by 63 DAA, which was significantly higher than those of s-metolachlor (64.88 ± 3.92%, c) and flumioxazin (64.32 ± 9.93%, c; $p < 0.05$) (Figure 2b). This reversal highlights how soil moisture modulates herbicide longevity, with pendimethalin's lipophilic nature favoring adsorption in dry soils.

The soybean variety CM60 consistently enhanced herbicide performance in later growth stages, achieving significantly higher WCE than Morkhor60 at 49 DAA (55.94 ± 8.16% vs. 33.49 ± 9.72%; $p < 0.01$) and 63 DAA (48.46 ± 8.39% vs. 32.15 ± 10.76%; $p < 0.05$) during the rainy season (Figure 2c). This trend persisted during the dry season (63 DAA: 68.51 ± 8.22% vs. 55.77 ± 8.35%; $p < 0.01$) (Figure 2d), suggesting that CM60's competitive traits (e.g., canopy closure) may suppress late-season weed resurgence. No significant interactions ($p > 0.05$) between herbicides and varieties were detected, indicating consistent varietal effects across treatments.

Table 4. Effects of pre-emergence herbicides and soybean varieties on weed control efficiency (%) in the rainy and dry seasons.

Variable	Weed Control Efficiency (%)							
	Rainy Season of 2023				Dry Season of 2024/2025			
	21 DAA 1/	35 DAA	49 DAA	63 DAA	21 DAA	35 DAA	49 DAA	63 DAA
Weed Control (WC)								
Hand weeding	100.00 ± 0.00 a 2/	100.00 ± 0.00 a	100.00 ± 0.00 a	100.00 ± 0.00 a	100.00 ± 0.00 a	100.00 ± 0.00 a	100.00 ± 0.00 a	100.00 ± 0.00 a
Weedy	0.00 ± 0.00 */ d	0.00 ± 0.00 d	0.00 ± 0.00 c	0.00 ± 0.00 d	0.00 ± 0.00 d	0.00 ± 0.00 d	0.00 ± 0.00 c	0.00 ± 0.00 d
Pendimethalin	54.54 ± 8.26 c	41.48 ± 12.37 c	20.22 ± 12.86 c	10.03 ± 13.39 d	65.50 ± 4.80 c	65.94 ± 5.24 c	77.41 ± 7.66 b	81.49 ± 5.90 b
S-metolachlor	87.21 ± 3.96 ab	79.75 ± 2.35 ab	48.30 ± 11.04 b	61.54 ± 6.78 b	85.32 ± 2.14 b	82.59 ± 4.07 b	72.79 ± 3.64 b	64.88 ± 3.92 c
Flumioxazin	76.44 ± 12.55 b	65.17 ± 8.18 b	55.06 ± 8.37 b	29.96 ± 11.59 c	66.30 ± 6.98 c	80.16 ± 5.33 b	78.50 ± 6.83 b	64.32 ± 9.93 c
Variety (Var)								
Morkhor60	60.27 ± 9.24	54.63 ± 8.41	33.49 ± 9.72 b	32.15 ± 10.76 b	63.65 ± 8.43	67.40 ± 8.40	63.47 ± 8.45	55.77 ± 8.35 b
CM60	67.01 ± 8.89	59.93 ± 9.30	55.94 ± 8.16 a	48.46 ± 8.39 a	63.20 ± 7.97	64.08 ± 8.12	68.01 ± 8.28	68.51 ± 8.22 a
F-test								
Weed Control (WC)	** 4/	**	**	**	**	**	**	**
Variety (Var)	ns	ns	**	* 3/	ns	ns	ns	**
WC × Var	ns	ns	**	ns 5/	ns	ns	ns	ns
CV WC	30.94	32.71	51.91	43.10	18.33	16.96	25.20	19.04
CV WC × Var	25.14	31.52	25.60	59.76	16.18	17.55	18.84	22.47

*/ mean ± SE, 1/ Days after application, 2/ the same letters are not significantly different by LSD at $p < 0.05$, 3/ significant at $p < 0.05$, 4/ significant at $p < 0.01$, and 5/ not significant.

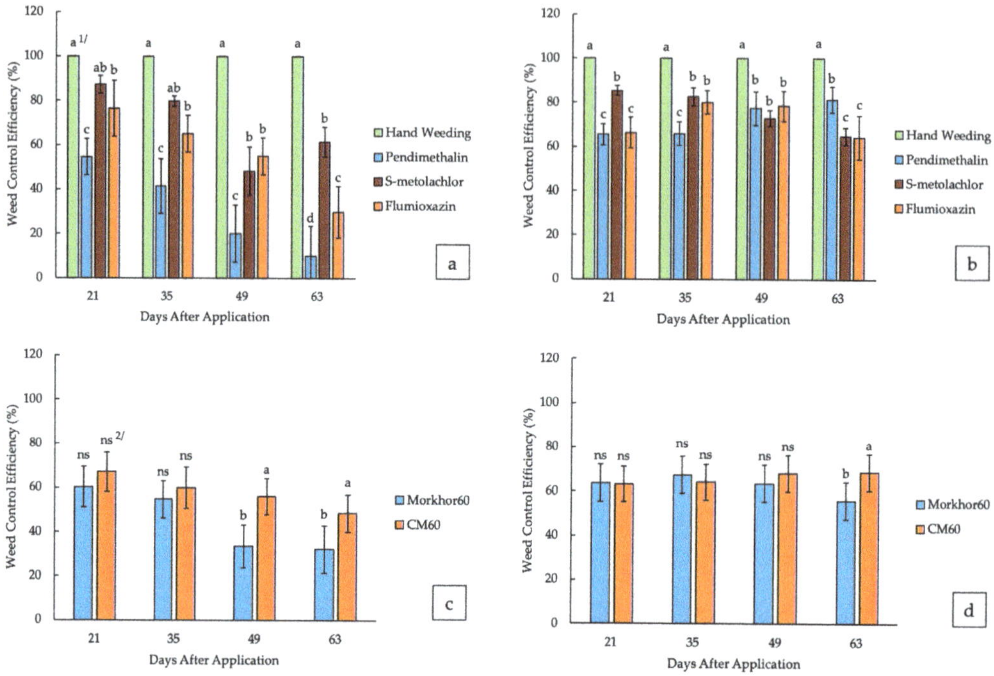

Figure 2. Effect of weed control treatments and soybean varieties on the weed control efficiency of each dominant weed species at 21, 35, 49, and 63 days after application. (**a**) Weed control treatments during the rainy season of 2023; (**b**) weed control treatments during the dry season of 2024/2025; (**c**) soybean varieties during the rainy season of 2023; and (**d**) soybean varieties during the dry season of 2024/2025. [1/] Means followed by the same letter are not significantly different according to LSD at $p < 0.05$, [2/] not significant.

3.1.2. Summed Dominance Ratio (%)

Analysis of the summed dominance ratio (SDR) in untreated (weedy) plots showed clear seasonal shifts in weed community composition in Northeast Thailand, with weeds categorised as broad-leaved, grasses, and sedges (Table 5). SDR quantifies each species' contribution to the total weed flora, highlighting the most problematic weeds under natural conditions. Across both seasons, the five most dominant species were *Digitaria ciliaris* (Retz.) Koeler, *Trianthema portulacastrum* L., *Cyperus rotundus* L., *Dactyloctenium aegyptium* (L.) P. B., and *Oldenlandia corymbosa* L. The specific names and families of each dominant weed species are shown in Table 3.

Table 5. Summed dominance ratio (%) in weedy plots at 21, 35, 49, and 63 DAA during the rainy season of 2023 and the dry season of 2024/2025.

Weed Species	Summed Dominance Ratio of Weed Species (%)							
	Rainy Season of 2023				Dry Season of 2024/2025			
	21 DAA [1/]	35 DAA	49 DAA	63 DAA	21 DAA	35 DAA	49 DAA	63 DAA
Broad-Leaved								
Oldenlandia corymbosa L.	- [2/]	2.19	11.72	6.51	0.03	4.16	9.53	12.03
Trianthema portulacastrum L.	49.65	34.88	11.46	4.18	2.88	1.7	1.38	1.55
Lindernia ciliate (Colsm.) Pennell	-	-	0.37	0.98	0.11	-	0.93	0.19

Table 5. *Cont.*

Weed Species	Summed Dominance Ratio of Weed Species (%)							
	Rainy Season of 2023				Dry Season of 2024/2025			
	21 DAA [1/]	35 DAA	49 DAA	63 DAA	21 DAA	35 DAA	49 DAA	63 DAA
Cleome rutidosperma	0.23	0.15	0.00	0.00	0.08	0.09	0.25	0.03
Praxelis clematidea R.M.King & H.Rob.	4.80	0.75	2.47	1.66	-	0.04	0.24	0.00
Indigofera hirsuta L.	-	-	-	-	-	0.00	0.14	0.04
Alternanthera sessilis	0.55	0.35	0.27	1.02	0.05	0.23	0.09	0.13
Ipomoea gracillis R.Br.	-	-	-		0.04	0.00	0.04	0.03
Amaranthus viridis L.	1.14	0.46	0.29	0.37	-	0.00	0.00	0.00
Xanthium strumarium L.	-	-	-	-	0.00	0.00	0.00	-
Wrighia arborea (Dennst.) Mabb.	-	-	-	-	0.03	-	0.00	-
Borreria alata (Aubl.) DC.	-	-	-	-	-	-	0.00	-
Ipomoea pestigridis L.	-	-	-	-	0.03	0.04	-	-
Sida cordifolia L.	-	-	-	-		0.00	-	-
Phyllanthus amarus Schumach. & Thonn.	-	-	0.00	-	-	-	-	-
Grasses								
Digitaria ciliaris (Retz.) Koeler	29.09	38.28	38.19	37.04	88.51	88.28	81.58	81.24
Dactyloctenium aegyptium (L.) P. B.	0.43	6.02	23.61	32.39	0.00	0.08	0.86	1.65
Eleusine indica (L.) Gaertn.	-	0.67	-	0.00	0.55	0.19	0.21	0.10
Cynodon dactylon (L.) Pers.	-	-	-	-	0.00	1.68	0.00	0.12
Eragrostis pectinacea (Michx.) Nees	-	-	-	-	-	-	0.08	-
Sedges								
Cyperus rotundus L.	14.11	16.26	6.44	12.9	7.70	3.53	3.60	2.48
Cyperus compressus L.	-	-	5.17	2.98	-	-	-	-
Cyperus esculentus L.	-	-	-	-	-	-	0.46	0.22
Cyperus difformis L.	-	-	-	-	-	-	0.60	0.21
Total	100.00	100.00	100.00	100.00	100.00	100.00	100.00	100.00

[1/] Days after application, [2/] No weed species were detected during the observation period.

During the 2023 rainy season, *T. portulacastrum* was initially dominant (49.65% at 21 DAA) but declined rapidly to 4.18% by 63 DAA. Meanwhile, *D. ciliaris* increased from 29.09% to 37.04% over the same period, becoming the main competitor later in the season. *C. rotundus* maintained a notable presence (14.11–16.26% early, above 6% later), while *D. aegyptium* rose sharply in late season (from 0.43% to 32.39% at 63 DAA). *O. corymbosa* also increased at specific times, peaking at 11.72% at 49 DAA. In contrast, during the 2024/2025 dry season, *D. ciliaris* overwhelmingly dominated throughout (over 88% at 21 and 35 DAA, above 81% at 63 DAA), reflecting its adaptability to dry conditions. The other four species contributed much less: *O. corymbosa* was the only broad-leaved species to increase notably late in the season (from 0.025% to 12.03%), while *C. rotundus*, *D. aegyptium*, and *T. portulacastrum* remained at low but persistent levels.

These findings are highly relevant for the development of effective pre-emergence herbicide programmes. The rapid shift from broad-leaved to grass weed dominance in the rainy season and the persistent grass dominance in the dry season suggest that herbicide selection must prioritise the residual control of *D. ciliaris* and *D. aegyptium* while also considering the early emergence of broad-leaved weeds like *T. portulacastrum*. The continued presence of *C. rotundus* further emphasises the need for herbicides with activity against sedges or for integrated management approaches.

3.1.3. Weed Density and Weed Biomass

An analysis of weed density and weed biomass in the field trials revealed clear differences among weed control treatments and across seasons, providing important insights into the efficacy of pre-emergence herbicides against dominant weed species in soybean cultivation in Northeast Thailand. In both the rainy and dry seasons, hand weeding consistently resulted in zero weed density and biomass for all assessed species at every observation interval, confirming its effectiveness as a benchmark for complete weed suppression. In contrast, untreated (weedy) plots exhibited the highest weed densities and biomasses, reflecting the natural competitive pressure from the local weed flora (Tables 6 and 7).

In the rainy season of 2023, the application of pre-emergence herbicides significantly reduced both weed density and biomass compared to the untreated control, although the degree of suppression varied depending on the herbicide and species. S-metolachlor was particularly effective in reducing the densities of both broad-leaved and grass weeds, with *T. portulacastrum* density dropping to 12.12 plants m^{-2} and *D. ciliaris* to 2.75 plants m^{-2} by 63 DAA (Table 6). The corresponding biomasses were 35.52 g m^{-2} and 9.66 g m^{-2}, respectively, indicating substantial, but not complete, suppression. Flumioxazin also reduced the densities and biomasses, particularly of broad-leaved weeds, but was less consistent against grasses and sedges at the later stages. Pendimethalin showed limited efficacy during the rainy season, particularly against sedges, as reflected by high *C. rotundus* densities (94.24 plants m^{-2} at 63 DAA) and biomasses (148.1 g m^{-2} at 63 DAA). These patterns closely parallel the WCE data (Table 4 and Figure 2), where s-metolachlor and flumioxazin provided higher and more sustained weed control efficiency compared to pendimethalin, especially in the early and mid-season assessments.

In the dry season of 2024/2025, overall weed pressure for broad-leaved species was lower but much higher for grasses, particularly *D. ciliaris*, which reached densities of 363.88 plants m^{-2} at 21 DAA and maintained high biomass throughout (peaking at 238.59 g m^{-2} at 49 DAA in the untreated plots) (Table 7). Pre-emergence herbicides again reduced weed density and biomass, with s-metolachlor and flumioxazin showing the greatest efficacy. For example, s-metolachlor reduced *D. ciliaris* density to 69.88 plants m^{-2} and biomass to 59.93 g m^{-2} at 63 DAA, while flumioxazin achieved similar reductions. Pendimethalin, however, was less effective during the dry season, as evidenced by the higher late-season densities and biomasses of grasses and sedges. These results align with the WCE data (Table 4 and Figure 2), where s-metolachlor and flumioxazin maintained high weed control efficiency, particularly in the early to mid-season, while pendimethalin's efficacy was more variable.

The soybean variety CM60 further modulated outcomes, exhibiting a lower weed biomass for *C. rotundus* (42.63 g m^{-2} at 63 DAA) compared to Morkhor60 (70.79 g m^{-2}) in the rainy season (Table 7), likely due to enhanced canopy closure or herbicide retention, as reflected in its higher WCE (68.51% at 63 DAA in the dry season; Table 4 and Figure 2).

Table 6. Effect of weed control treatments and soybean varieties on weed density at 21, 35, 49, and 63 DAA.

Variable	Weed Density (plants m^{-2})																			
	Broad-Leaved								Grasses								Sedges			
	Trianthema portulacastrum L.				*Oldenlandia corymbosa* L.				*Digitaria ciliaris* (Retz.) Koeler.				*Dactyloctenium aegyptium* (L.) P.B.				*Cyperus rotundus* L.			
	21 DAA [1/]	35 DAA	49 DAA	63 DAA	21 DAA	35 DAA	49 DAA	63 DAA	21 DAA	35 DAA	49 DAA	63 DAA	21 DAA	35 DAA	49 DAA	63 DAA	21 DAA	35 DAA	49 DAA	63 DAA
	Rainy Season of 2023																			
Weed Control																				
Hand weeding	0.00 b [2/]	0.00 c	0.00 b	0.00 b	0.00	0.00	0.00 b	0.00 b	0.00 b	0.00 b	0.00 b	0.00 b	0.00	0.00 b	0.00 b	0.00 b	0.00 c	0.00 c	0.00 b	0.00 b
Weedy	141.13 a	87.50 a	20.37 a	3.75 b	0.00	7.50	33.50 a	7.50 a	88.62 a	78.00 a	47.12 a	29.12 a	1.25	10.00 a	18.50 a	16.12 a	36.87 b	43.75 bc	14.12 b	11.87 b
Pendimethalin	0.87 b	0.62 c	0.00 b	0.75 b	0.00	0.00	0.00 b	0.75 b	0.12 b	1.37 b	4.25 b	2.75 b	0.00	0.25 b	0.12 b	0.12 b	73.50 a	109.12 a	83.87 a	94.24 a
S-metolachlor	26.50 b	35.12 b	26.37 a	12.12 a	0.00	0.00	0.25 b	0.12 b	4.00 b	0.50 b	1.87 b	2.75 b	0.00	0.00 b	0.00 b	2.00 b	5.87 c	9.25 c	20.37 b	18.00 b
Flumioxazin	2.00 b	1.12 c	6.00 b	3.00 b	0.00	0.00	0.00 b	0.00 b	8.50 b	2.37 b	9.87 b	7.37 b	0.00	0.75 b	1.37 b	2.87 b	22.00 bc	61.75 ab	71.75 a	73.87 a
Variety																				
Morkhor60	24.4	21.55	9.60	4.75	0.00	3.00	6.65	1.85	23.50	13.95	12.80	10.95	0.25	3.85 a	3.40	3.30	26.25	46.40	39.60	46.15
CM60	43.75	28.20	11.45	3.10	0.00	0.00	6.85	1.50	17.00	18.95	12.45	7.60	0.25	0.50 b	4.60	5.15	29.05	43.15	37.25	33.05
	Dry Season of 2024/2025																			
Weed Control																				
Hand weeding	0.00 b	0.00 b	0.00 b	0.00 b	0.00	0.00 b	0.00 b	0.00 b	0.00 b	0.00 b	0.00 c	0.00 c	0.00	0.00	0.00	0.00	0.00 d	0.00 c	0.00 b	0.00 c
Weedy	19.88 a	8.63 b	5.88 b	5.00 b	0.13	21.75 a	57.50 a	55.25 a	363.88 a	301.75 a	200.12 a	151.50 a	0.00	0.25	2.00	4.25	14.13 bc	11.38 bc	13.75 b	8.50 bc
Pendimethalin	0.00 b	0.63 b	0.00 b	0.50 b	0.00	0.00 b	0.13 b	0.00 b	5.50 b	5.25 b	12.63 c	18.00 c	0.13	0.00	0.38	0.00	25.88 a	53.88 a	51.75 a	32.63 a
S-metolachlor	7.88 b	22.63 a	16.38 a	12.25 a	0.00	5.00 b	14.88 b	12.75 b	14.50 b	48.75 b	78.00 b	69.88 b	0.00	0.00	1.50	3.25	6.00 cd	3.75 bc	9.75 b	3.13 c
Flumioxazin	0.38 b	0.25 b	0.63 b	1.38 b	0.00	0.00 b	0.00 b	0.00 b	15.75 b	33.00 b	65.00 b	56.00 b	1.63	0.00	3.50	2.63	20.50 ab	19.75 b	25.38 ab	16.88 b
Variety																				
Morkhor60	5.25	8.50	5.10	4.85 a	0.05	4.05	19.70	22.15 a	79.15	80.70	74.60	45.35 b	0.20	0.00	0.15 b	0.00 b	13.65	16.60	15.40	10.30
CM60	6.00	4.35	4.05	2.80 a	0.00	6.65	9.30	5.05 b	80.70	74.80	67.70	72.80 a	0.50	0.10	2.80 a	4.05 a	12.95	18.90	24.85	14.15

The data were measured per square metre (1 m^2). [1/] Days after application; [2/] the same letters are not significantly different based on LSD at $p < 0.05$.

Table 7. Effect of weed control treatments and soybean varieties on weed biomass at 21, 35, 49, and 63 DAA.

Variable	Weed Density (plants m^{-2})																			
	Broad-Leaved								Grasses								Sedges			
	Trianthema portulacastrum L.				*Oldenlandia corymbosa* L.				*Digitaria ciliaris* (Retz.) Koeler.				*Dactyloctenium aegyptium* (L.) P.B.				*Cyperus rotundus* L.			
	21 DAA [1/]	35 DAA	49 DAA	63 DAA	21 DAA	35 DAA	49 DAA	63 DAA	21 DAA	35 DAA	49 DAA	63 DAA	21 DAA	35 DAA	49 DAA	63 DAA	21 DAA	35 DAA	49 DAA	63 DAA
	Rainy Season of 2023																			
Weed Control																				
Hand weeding	0.00 b [2/]	0.00 c	0.00 b	0.00 b	0.00	0.00	0.00 b	0.00	0.00 b	0.00 b	0.00 c	0.00 c	0.00	0.00.b	0.00 b	0.00 b	0.00 b	0.00 c	0.00 c	0.00 c
Weedy	17.34 a	35.25 a	13.16 b	4.46 b	0.00	0.25	2.20 a	2.45	7.27 a	49.16 a	68.21 a	78.02 a	0.10	8.46 a	56.05 a	92.83 b	3.85 b	16.01 c	4.77 c	21.83 c
Pendimethalin	0.03 b	0.97 c	0.00 b	4.38 b	0.00	0.00	0.00 b	0.15	6.25 b	1.47 b	25.68 b	9.00 bc	0.00	0.36 b	0.17 b	0.37 b	13.47 a	53.68 a	58.36 a	148.10 a
S-metolachlor	2.47 b	15.67 b	48.16 a	35.52 a	0.00	0.00	0.67 b	0.01	0.20 b	0.22 b	2.76 bc	9.66 bc	0.00	0.00 b	0.00 b	0.37 b	0.76 b	2.50 c	5.71 c	14.76 c
Flumioxazin	0.18 b	1.20 c	5.31 b	7.08 b	0.00	0.00	0.00 b	0.00	0.66 b	1.73 b	9.77 bc	23.55 b	0.00	0.71 b	7.53 b	7.91 b	4.27 b	34.61 b	40.28 b	98.86 b
Variety																				
Morkhor60	3.24	10.13	14.65	11.29	0.00	0.10	0.84	0.17	1.90	9.48	22.01	27.80 a	0.03	3.06 a	9.05	17.71	4.28	23.19	25.37	70.79
CM60	4.77	11.10	12.00	9.29	0.00	0.00	0.31	0.87	1.35	11.55	20.56	20.29 b	0.00	0.75 b	16.45	22.87	4.66	19.53	18.28	42.63
	Dry Season of 2024/2025																			
Weed Control																				
Hand weeding	0.00 b	0.00	0.00 b	0.00 b	0.00	0.00 b	0.00 b	0.00 b	0.00 b	0.00 b	0.00 d	0.00 c	0.00 b	0.00	0.00	0.00 b	0.00 c	0.00 c	0.00 b	0.00 c
Weedy	0.79 a	0.81	0.79 b	0.57 b	0.01	1.32 a	2.79 a	2.34 a	48.09 a	164.90 a	238.59 a	194.22 a	0.00 b	0.10	2.01	4.15 a	6.21 bc	4.94 bc	4.72 b	2.64 bc
Pendimethalin	0.00 b	2.09	0.00 b	0.03 b	0.00	0.00 b	0.01 b	0.00 b	3.43 b	11.27 b	9.05 cd	15.77 c	0.05 ab	0.00	4.22	0.00 b	15.41 a	40.31 a	28.63 a	11.32 a
S-metolachlor	0.33 ab	3.88	4.91 a	3.74 a	0.00	0.26 b	0.75 b	0.51 b	4.04 b	20.11 b	57.94 b	59.93 b	0.00 b	0.00	0.29	1.83 ab	3.40 c	1.70 bc	3.81 b	0.83 c
Flumioxazin	0.03 b	0.11	0.05 b	0.15 b	0.00	0.00 b	0.00 b	0.00 b	4.01 b	20.17 b	34.41 bc	53.88 b	0.95 a	0.00	1.13	1.47 b	12.88 ab	11.85 b	9.94 b	5.07 b
Variety																				
Morkhor60	0.17	1.53	1.02	1.36	0.01	0.29	0.60	0.91 a	12.49	49.30	46.64 b	56.79 b	0.13	0.00	1.69	0.00 b	7.75	12.25	5.75	3.90
CM60	0.29	1.22	1.28	0.44	0.00	0.34	0.82	0.23 b	11.34	37.27	89.36 a	72.73 a	0.28	0.04	1.37	2.98 a	7.41	11.27	13.09	4.05

The data were measured per square metre (1 m^2). [1/] Days after application, and [2/] the same letters are not significantly different by LSD at $p < 0.05$.

3.2. Effects of Pre-Emergence Herbicides on Agronomic Traits

3.2.1. Germination Rate on Soybean

An analysis of the soybean germination rate at 14 DAA revealed significant differences among weed control treatments and varieties, with notable seasonal variation. During the rainy season of 2023, hand weeding resulted in the highest germination rate (93.43%), followed by the untreated (weedy) control (84.41%), pendimethalin (80.00%), and s-metolachlor (73.04%) (Table 8). Flumioxazin, however, caused a marked reduction in the germination rate (33.65%), indicating considerable phytotoxicity to soybean seedlings under these conditions, consistent with the high level of toxicity associated with flumioxazin, as illustrated in Figure 3a. The F-test for weed control treatments was highly significant ($p < 0.01$), confirming that the choice of pre-emergence herbicide had a strong impact on soybean emergence in the rainy season. Varietal differences were also significant ($p < 0.05$), with CM60 showing a higher germination rate (83.00%) than Morkhor60 (63.21%), suggesting a greater tolerance of CM60 to herbicide application or environmental stress during early establishment. The interaction between weed control and variety was not significant, indicating that the effect of herbicide treatment on germination was generally consistent across varieties.

Table 8. Effects of weed control treatments and soybean varieties on germination (%) at 14 DAA.

Variable	Germination Rate (%) at 14 DAA 1/	
	Rainy Season	Dry Season
Weed Control (WC)		
Hand weeding	93.43 a 2/	95.13
Weedy	84.41 ab	96.25
Pendimethalin	80.00 ab	95.88
S-metolachlor	73.04 b	95.50
Flumioxazin	33.65 c	92.13
Variety (Var)		
Morkhor60	63.21 b	92.10 b
CM60	83.00 a	97.85 a
F-test		
Weed Control (WC)	** 4/	ns 5/
Variety (Var)	* 3/	**
WC × Var	ns	ns
CV WC	24.17	3.49
CV WC × Var	23.33	3.26

1/ Days after application. 2/ The same letters are not significantly different based on LSD at $p < 0.05$, 3/ significant at $p < 0.05$, 4/ significant at $p < 0.01$, and 5/ not significant.

During the dry season of 2024/2025, germination rates were uniformly high across all treatments, ranging from 92.13% (flumioxazin) to 96.25% (weedy), and the F-test for weed control was not significant (Table 8). This suggests that under drier conditions, the phytotoxic effects of pre-emergence herbicides on soybean germination are less pronounced or that environmental factors mitigate potential injury. However, varietal differences were highly significant ($p < 0.01$) during the dry season, with CM60 (97.85%) again outperforming Morkhor60 (92.10%), underscoring the importance of varietal selection for optimal crop establishment.

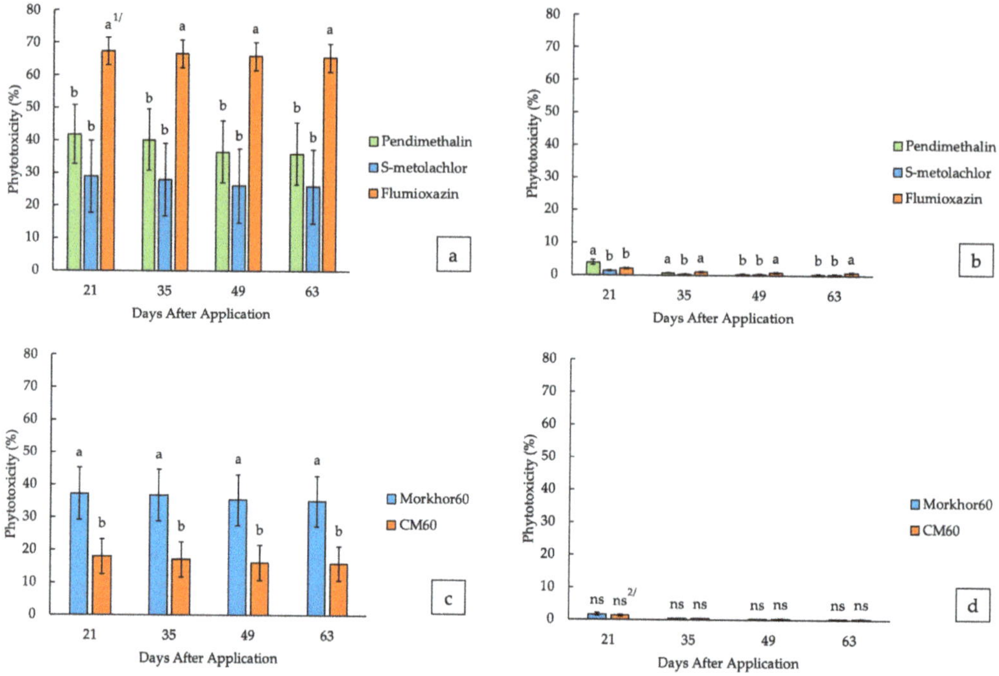

Figure 3. Effects of pre-emergence herbicides and soybean varieties on phytotoxicity (%) at 21, 35, 49, and 63 DAA. (**a**) Pre-emergence herbicides during the rainy season of 2023, (**b**) pre-emergence herbicides during the dry season of 2024/2025, (**c**) soybean varieties during the rainy season of 2023, (**d**) soybean varieties during the dry season of 2024/2025. [1/] Means followed by the same letter are not significantly different according to LSD at $p < 0.05$; [2/] not significant.

3.2.2. Phytotoxicity on Soybean

An analysis of phytotoxicity in soybeans following pre-emergence herbicide application revealed substantial differences among treatments and between seasons, with direct implications for crop safety and the practical use of these herbicides in Northeast Thailand. During the rainy season of 2023, all herbicide treatments induced some level of phytotoxicity, with the magnitude and persistence varying significantly by active ingredient (Figure 3).

Among the herbicides, flumioxazin consistently produced the highest phytotoxicity scores, exceeding 65% at all time points from 21 to 63 DAA (Figure 3a). This high and persistent level of crop injury was significantly greater than that observed with pendimethalin and s-metolachlor ($p < 0.01$), aligning with the marked reduction in soybean germination rate and plant vigour observed during flumioxazin treatment. Pendimethalin and s-metolachlor also caused significant phytotoxicity in the rainy season, but to a lesser extent, with scores ranging from approximately 36% to 42% for pendimethalin and 26% to 29% for s-metolachlor. The statistical analysis confirmed that both weed control treatments and soybean varieties had significant effects on phytotoxicity ($p < 0.05$), and their interaction was also significant, indicating the importance of considering varietal tolerance to herbicide injury.

Notably, the CM60 variety exhibited lower phytotoxicity scores than Morkhor60 across all treatments and time points in the rainy season, suggesting that CM60 possesses greater tolerance to pre-emergence herbicides (Figure 3c). This varietal difference was statistically significant ($p < 0.05$) and consistent with the higher germination rates observed for CM60, affecting plant populations and soybean yields (Table 9).

Table 9. Effect of weed control treatments and soybean varieties on yield and yield components.

Variable	Soybean Yield and Yield Component													
	Rainy Season of 2023							Dry Season of 2024/2025						
	Pods on Branches	Pods on Main Stem	Seed Number/Pod	Weight of 100 Seeds (g)	Grain Yield (kg ha^{-1})	Grain Yield Loss (%)	Plant Population (Plant ha^{-1})	Pods on Branches	Pods on Main Stem	Seed Number/Pod	Weight of 100 Seeds (g)	Grain Yield (kg ha^{-1})	Grain Yield Loss (%)	Plant Population (Plant ha^{-1})
Weed Control (WC)														
Hand weeding	20.21 a	18.31 ab	2.40	17.15 a	1090.50 a	0.00	224,400 a	11.46 a	19.40 a	2.26	17.62 a	1366.50 a	0.00	228,050 ab
Weedy	8.95 d	15.64 b	2.30	16.43 ab	481.30 b	55.86	202,950 ab	2.80 c	8.56 c	2.16	16.94 ab	422.40 d	69.09	230,550 a
Pendimethalin	13.82 c	15.44 b	2.43	15.87 b	365.50 c	66.48	194,850 ab	9.85 b	18.13 a	2.30	16.78 ab	1129.40 bc	17.35	228,350 ab
S-metolachlor	16.50 b [1/]	20.90 a	2.48	16.94 a	1036.30 a	4.97	175,950 b	9.84 b	14.89 b	2.28	15.61 c	954.80 c	30.13	227,750 ab
Flumioxazin	17.59 b	20.43 a	2.50	16.93 a	360.20 c	66.97	82,350 c	9.93 b	17.37 a	2.30	16.24 bc	1242.60 ab	9.07	219,850 b
Variety (Var)														
Morkhor60	18.68 a	19.10 a	2.48 a	15.48 b	603.05 b	-	152,580 b	11.11 a	14.98 b	2.34 a	15.94 b	1028.7	-	219,980 b
CM60	12.15 b	17.18 b	2.36 b	17.85 a	730.48 a	-	199,620 a	6.44 b	16.36 a	2.17 b	17.35 a	1017.5	-	233,840 a
F-tests														
Weed Control (WC)	** [4/]	**	ns [2/]	*	**	-	**	**	**	ns	**	**	-	ns
Variety (Var)	**	*	* [3/]	**	**	-	**	**	*	**	**	ns	-	**
WC × Var	**	**	ns	ns	**	-	ns	**	**	ns	**	ns	-	ns
CV WC	15.53	14.51	6.75	4.21	13.55	-	23.84	13.05	14.39	5.44	4.76	17.70	-	3.48
CV WC × Var	11.75	13.45	5.84	7.57	17.93	-	23.01	13.99	10.96	4.54	4.77	16.5	-	3.35

[1/] Means indicated by the same letters are not significantly different based on LSD at $p < 0.05$, [2/] not significant at $p < 0.05$, [3/] significant at $p < 0.05$, and [4/] significant at $p < 0.01$.

During the dry season of 2024/2025, overall phytotoxicity levels were significantly reduced across all treatments. Flumioxazin, pendimethalin, and s-metolachlor induced only minimal crop injury, with scores below 4% at 21 DAA and declining to less than 1% at later intervals (Figure 3b). No phytotoxicity was observed in hand-weeded or untreated plots. During this season, neither the variety nor the interaction between the variety and weed control had a significant effect on phytotoxicity, as indicated by the non-significant F-tests. This seasonal difference suggests that environmental conditions, such as lower soil moisture and reduced rainfall during the dry season, may mitigate the expression of herbicide-induced crop injury.

3.2.3. Soybean Plant Height

The efficacy of pre-emergence herbicides in soybean cultivation was evaluated using plant height measurements across two growing seasons (rainy season 2023 and dry season 2024/2025) under varying weed control regimes. During the rainy season of 2023, hand weeding consistently produced the tallest soybean plants, which reached 56.15 cm at 63 DAA, significantly outperforming herbicide-treated plots (e.g., pendimethalin: 53.25 cm; flumioxazin: 54.01 cm) ($p < 0.05$) (Figure 4a). This aligns with its 100% WCE (Table 4 and Figure 2), eliminating weed competition and thereby optimising resource availability for soybean growth (Figure 2a). Conversely, the weedy control group exhibited the shortest plants (53.05 cm at 63 DAA), reflecting severe resource competition from uncontrolled weeds.

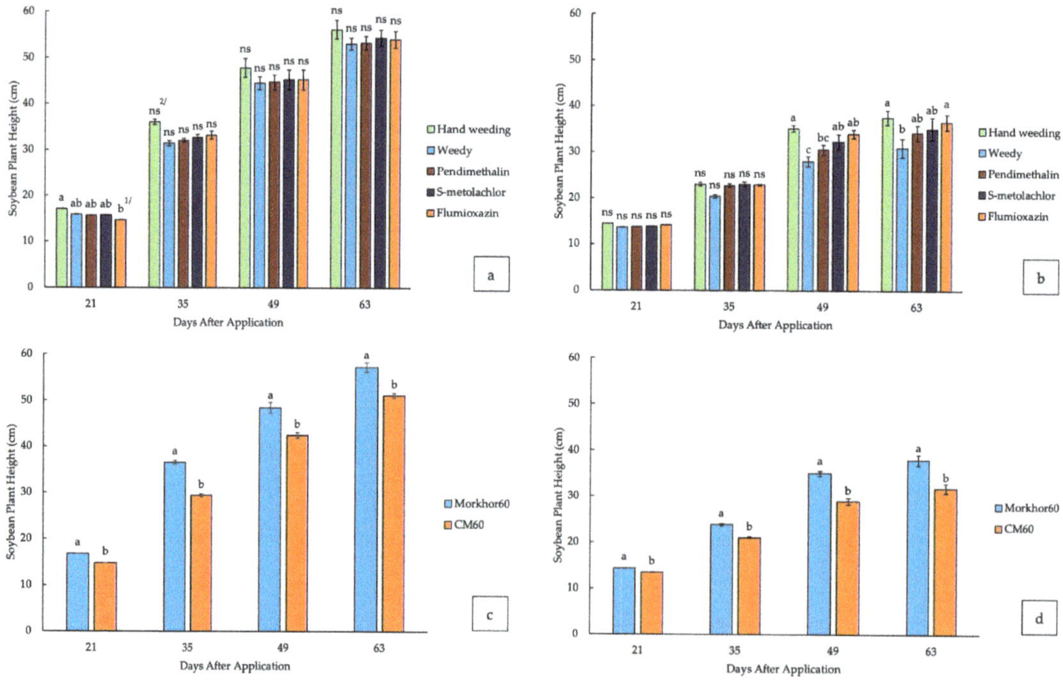

Figure 4. Effects of weed control treatments and soybean varieties on soybean plant height (cm) at 21, 35, 49, and 63 DAA. (**a**) Weed control treatments during the rainy season of 2023, (**b**) weed control treatments during the dry season of 2024/2025, (**c**) soybean varieties during the rainy season of 2023, and (**d**) soybean varieties during the dry season of 2024/2025. [1/] Means followed by the same letters are not significantly different based on LSD at $p < 0.05$; [2/] not significant.

Herbicide-induced phytotoxicity (Figure 3a) further modulated plant growth. For instance, flumioxazin, which caused the highest phytotoxicity during the rainy season (67.35% at 21 DAA), resulted in reduced plant height (14.68 cm at 21 DAA) compared to hand weeding (17.16 cm; Figure 4a). However, by 63 DAA, flumioxazin-treated plants recovered to 54.01 cm, suggesting transient phytotoxicity effects. S-metolachlor, with moderate phytotoxicity (28.85–25.90%), maintained competitive plant heights (54.33 cm at 63 DAA; Figure 4a), likely due to its balanced WCE (48.30–61.54%; Table 4 and Figure 2a), which suppressed weeds without severely impairing soybean growth. Varietal differences were pronounced, with Morkhor60 exhibiting taller plants than CM60 across both seasons (e.g., 57.20 cm versus 51.12 cm at 63 DAA in the rainy season; $p < 0.05$) (Figure 4c). This superiority persisted, even under herbicide stress, highlighting Morkhor60's resilience, possibly linked to its higher germination rate (92.10% in the dry season versus CM60's 97.85%) (Table 8), which may enhance early establishment and resource capture.

The decline in WCE for herbicides over time (e.g., pendimethalin's WCE dropped from 54.54% to 10.03% by 63 DAA in the rainy season; Table 4 and Figure 2a) was correlated with gradual weed resurgence, which likely contributed to reduced plant height gains in the later growth stages. However, during the dry season of 2024/2025, higher WCE values for herbicides (e.g., flumioxazin: 78.50% at 49 DAA) (Table 4 and Figure 2b) corresponded with improved plant heights (36.50 cm at 63 DAA) (Figure 4b), underscoring seasonal variations in herbicide performance and weed pressure.

3.2.4. Number of Nodes

The number of nodes in soybean plants, a critical indicator of vegetative development and yield potential, exhibited significant variations under different pre-emergence herbicide treatments and soybean varieties during the rainy season of 2023 and dry seasons of 2024/2025 (Figure 5). During the rainy season of 2023, weed control (WC) methods significantly influenced the number of nodes at 21 DAA ($p < 0.05$), with flumioxazin yielding the highest early-stage nodes (4.00 nodes/plant), followed by pendimethalin (3.85 nodes/plant) (Figure 5a). Conversely, s-metolachlor produced the lowest node count (3.28 nodes/plant), likely due to its moderate phytotoxicity, which may have temporarily suppressed meristematic activity. Hand weeding, a non-chemical control, showed intermediate node numbers (3.68 nodes/plant), reflecting the absence of herbicide stress but potential early weed competition. By 63 DAA, s-metolachlor surpassed other treatments (10.25 nodes/plant), suggesting recovery from initial phytotoxic effects, while flumioxazin exhibited a decline (9.9 nodes/plant), aligning with its persistent phytotoxicity.

Varietal differences were pronounced, with Morkhor60 consistently outperforming CM60 in node counts ($p < 0.01$ at 21 DAA in 2023). For instance, Morkhor60 recorded 3.89 nodes/plant at 21 DAA compared to CM60's 3.44 nodes/plant (Figure 5a), highlighting the genetic disparities in stress tolerance. This aligns with the germination data (Table 8), where CM60 exhibited higher germination rates (83.00% in the rainy season) but lower subsequent node counts, implying that early vigour may not compensate for herbicide-mediated growth suppression. Morkhor60's resilience was further evident during the dry season, where it maintained superior node numbers (9.51 nodes/plant at 63 DAA versus 7.98 for CM60; Figure 5d), correlating with its lower phytotoxicity susceptibility (1.58% versus 1.26% for CM60 at 21 DAA in 2024/2025; Figure 3d).

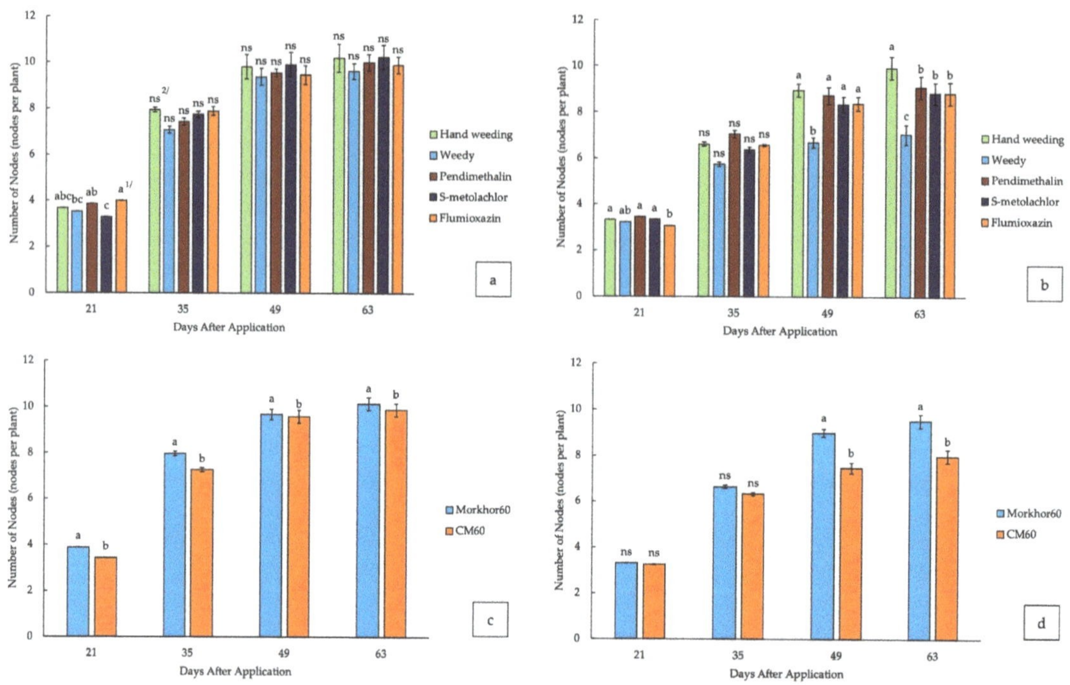

Figure 5. Effects of weed control treatments and soybean varieties on the number of nodes at 21, 35, 49, and 63 DAA. (**a**) Weed control treatments during the rainy season of 2023, (**b**) weed control treatments during the dry season of 2024/2025, (**c**) soybean varieties during the rainy season of 2023, and (**d**) soybean varieties during the dry season of 2024/2024. [1/] Means followed by the same letters are not significantly different based on LSD at $p < 0.05$; [2/] not significant.

During the dry season of 2024/2025, WC treatments significantly affected node counts at the later growth stages (49 and 63 DAA, $p < 0.05$) (Figure 5b). Hand weeding and pendimethalin demonstrated the highest node numbers (8.96–9.93 nodes/plant at 63 DAA), attributable to their balanced WCE (77.41–85.32%; Figure 2b) and minimal phytotoxicity (≤3.83% in 2024/2025; Figure 3b). Conversely, weedy plots exhibited the lowest node counts (6.70–7.05 nodes/plant), underscoring the detrimental impact of uncontrolled weed competition. Flumioxazin, despite its high WCE (78.50% at 49 DAA in the dry season; Figure 2b), showed reduced node counts (8.83 nodes/plant at 63 DAA), likely due to cumulative phytotoxic stress (0.85% at 63 DAA; Figure 3b) and residual effects on meristem development.

3.3. *Effects of Pre-Emergence Herbicides on Yield and Yield Components*

The evaluation of soybean yield and its components under various pre-emergence herbicide treatments in Northeastern Thailand revealed that across both the rainy and dry seasons, hand weeding consistently resulted in the highest seed yields, with values reaching 1090.5 kg ha^{-1} during the rainy season and 1366.5 kg ha^{-1} during the dry season (Table 9). This treatment also supported the greatest number of pods on both branches and the main stem, as well as a superior 100-seed weight, reflecting optimal crop establishment and minimal interference from weeds. These outcomes were closely linked to the observance of the highest plant population densities, which were maintainable due to the negligible phytotoxicity effects and effective weed suppression.

Among the herbicide treatments, s-metolachlor demonstrated strong performance, producing yields comparable to those of hand weeding, particularly during the rainy season, when seed yield reached 1036.30 kg ha^{-1} with only 4.97% yield loss. This treatment also maintained relatively high pod numbers and seed counts, indicating its capacity to balance effective weed control with acceptable crop safety. Flumioxazin, despite inducing the highest phytotoxicity during the rainy season (reaching up to 67.35% at 21 DAA), still supported moderate pod formation and seed development. However, its overall yield was substantially lower (360.2 kg ha^{-1} in the rainy season), largely due to a significantly reduced plant population caused by early-season crop injury and lower germination rates. This pattern underscores the trade-off between herbicide phytotoxicity and weed control efficacy, where severe early injury can limit stand establishment and consequently reduce yield potential despite effective weed suppression. Pendimethalin exhibited moderate phytotoxicity and declining weed control efficiency over time, as reflected in its intermediate yield levels (365.5 kg ha^{-1} in the rainy season and 1129.4 kg ha^{-1} in the dry season). The reduced weed suppression capacity allowed for increased weed competition, which negatively impacted pod formation and seed filling. The weedy control plots consistently produced the lowest yields and yield components across seasons, highlighting the detrimental effect of unchecked weed pressure on soybean reproductive development. Compared to hand weeding, the absence of weed management resulted in a yield loss of 55.86% during the rainy season and 69.09% during the dry season, underscoring the critical importance of effective weed control for maximising soybean productivity.

The relationships between yield components and other agronomic parameters further elucidate the mechanisms underlying these results. Treatments supporting higher germination rates and plant populations (such as hand weeding and s-metolachlor) also promoted greater plant height and node number, both of which are critical for maximising photosynthetic capacity and reproductive sites. Conversely, high phytotoxicity during the rainy season of 2023 in flumioxazin-treated plants suppressed germination and early growth, limiting node development and ultimately reducing yield, despite relatively good pod and seed sets on surviving plants.

Seasonal variations also influenced herbicide performance and crop response. The dry season generally exhibited higher germination rates and yields across treatments, possibly due to more favourable environmental conditions that mitigated herbicide injury and enhanced crop recovery. In contrast, the rainy season's higher moisture levels may have exacerbated herbicide phytotoxicity, particularly for flumioxazin, thereby reducing stand establishment and yield potential. Varietal differences were apparent, with CM60 achieving higher seed yields and plant populations, while Morkhor60 tended to produce more pods and seeds per pod. These findings suggest that varietal selection should be integrated with herbicide choice to optimise soybean productivity under local conditions.

4. Discussion

This study highlights the importance of effective weed management in optimising soybean yield and its components within the rainfed agroecosystems of Northeast Thailand. The weed flora predominantly consisted of broad-leaved species (62.5%), followed by grasses (20.8%) and sedges (16.7%). Notably, *T. portulacastrum*, *O. corymbosa*, *D. ciliaris*, *D. aegyptium*, and *C. rotundus* constituted the majority of weed biomass and density. This observed weed spectrum aligns with previous findings from upland fields in the region [30], reinforcing the necessity for comprehensive weed management strategies.

Hand weeding consistently yielded high WCE, completely suppressing weed growth throughout the rainy and dry seasons (Figure 2). This treatment produced the highest grain yields, reaching 1090.5 kg ha^{-1} during the rainy season and 1366.5 kg ha^{-1} during the dry season, as well as the most significant number of pods per branch and main stem, 100-seed weight, and optimal plant population densities (Table 9). These results align with previous studies indicating that manual weed removal remains the benchmark for maximising crop productivity in smallholder systems [36,37]. The superior performance exhibited by hand weeding reflects the effective suppression of weed competition, and the use of this conventional method avoids damage to plants caused by herbicides.

In contrast, the absence of weed control (weedy plots) dramatically reduced yield and yield components. Grain yield declined to 481.3 kg ha^{-1} during the rainy season and 422.4 kg ha^{-1} during the dry season, corresponding to a yield loss of 55.86% and 69.09%, respectively, relative to hand weeding. This substantial yield penalty underscores the severity of weed interference, which is consistent with the findings of Jha et al. [38] and Chauhan and Johnson [39], who reported that uncontrolled weed growth can reduce soybean yield by more than 50% in tropical environments. The reduced yield in weedy plots was accompanied by lower pod numbers, seed weight, and plant populations, indicating that weed competition limits resource availability and impairs crop establishment and reproductive development.

Among the herbicides tested, s-metolachlor 96% EC at 900 g a.i. ha^{-1} provided the most consistent and sustainable weed control, with WCE values of 87.2% at 21 DAA and 61.5% at 63 DAA during the rainy season and similarly high WCE throughout the dry season (Figure 2). This translates to grain yields (1036.3 kg ha^{-1} during the rainy season; 954.8 kg ha^{-1} during the dry season) comparable to hand weeding, indicating that s-metolachlor effectively balances weed suppression and crop safety. The correlation between increased grain yield and higher WCE (Figure S1) highlights that early weed suppression was pivotal to achieving yields comparable to hand weeding. This aligns with global studies in which s-metolachlor-based treatments boosted yields by 80–97% [35,40,41]. The superior performance of s-metolachlor is likely due to its longer soil residual activity (15–50 days; [42]) and broad-spectrum efficacy, as reflected in the low weed density and biomass observed in treated plots (Tables 6 and 7). These findings are consistent with those of Meseldžija et al. [35] and Qadeer et al. [43], who highlighted the yield benefits of s-metolachlor in various cropping systems.

Pendimethalin and flumioxazin exhibited moderate to high initial WCE. However, their effectiveness declined over time, especially against sedge species such as *C. rotundus*, which demonstrated biological resilience due to its deep tuber system and staggered emergence [44]. This observation is evident in the increased weed density and biomass during the later growth stages, particularly during the rainy season. The limited residual activity of these herbicides, along with their lipophilic nature and heightened leaching during periods of heavy rainfall [45], resulted in diminished weed suppression and moderate yield outcomes (pendimethalin: 365.5–1129.4 kg ha^{-1}; flumioxazin: 360.2–1242.6 kg ha^{-1}). Notably, flumioxazin induced severe phytotoxicity during the rainy season (up to 67.35% at 21 DAA), resulting in poor germination (33.65%) and drastically reduced plant populations (82,350 plants ha^{-1}), which ultimately limited yield despite moderate weed control (Table 8 and Figure 3). Consistent with previous findings, Taylor-Lovell et al. [20] observed that elevated soil moisture levels intensify herbicide-induced damage in soybean plants.

Soil moisture is essential for activating pre-emergence herbicides and ensuring their bioavailability for weed suppression, particularly during the rainy season, when prolonged exposure to cool, wet conditions during crop emergence can hinder soybeans' capacity to detoxify these herbicides, resulting in heightened phytotoxicity [46–49]. In addition, during the "soil cracking" stage of emergence, precipitation can cause the splashing of higher concentrations of PPO-inhibitor herbicides onto sensitive parts of the soybean plant, such as the hypocotyl and cotyledons. This can lead to tissue necrosis, further increasing the risk of injury when soil moisture levels are not optimal [46,50].

Weed management practices also influenced soybean growth parameters. Treatments with higher WCE, such as hand weeding and s-metolachlor, supported greater plant height and node number at maturity (Figures 4 and 5). These traits have been associated with yield components, as increased node number provides more sites for pod development [51]. Seasonal differences were evident, with higher plant height and node numbers generally observed during the rainy season, likely due to higher temperatures (28.6 °C versus 26.4 °C) and greater soil moisture, which can enhance vegetative growth [52]. Research by Madhu and Hatfield [53] supports this observation, demonstrating that adequate soil moisture levels substantially improve crops' photosynthetic rates and total dry matter production. The growth response to seasonal conditions appears complex, as elevated CO_2 levels associated with warmer temperatures may initially slow early vegetative growth before promoting increased leaf number, leaf area expansion, and dry matter accumulation [54]. However, as demonstrated in plots treated with flumioxazin, severe phytotoxicity can counteract these beneficial environmental effects, significantly suppressing early growth and node formation. This suppression ultimately leads to reduced yield potential despite otherwise favourable seasonal growing conditions.

Variations among soybean cultivars significantly influenced yield responses to weed management. The CM60 variety demonstrated higher herbicide tolerance, resulting in greater germination rates and plant populations than Morkhor60, ultimately achieving superior yields (730.5 kg ha^{-1} versus 603.1 kg ha^{-1} during the rainy season). The enhanced performance of CM60 was linked to a more effective canopy closure and lower weed biomass in *C. rotundus* species (42.6 g m^{-2} compared to 70.8 g m^{-2}), reinforcing the idea that cultivar selection can improve soybeans' competitive ability against weeds [42,45]. Reinforced by a statement from Bianchi et al. [55] and Bastiani et al. [56], more stem dry matter, greater shoot dry matter, and increased soil coverage by the crop canopy were associated with superior competitiveness. Plant height also played a role in competition, with barnyardgrass showing greater competitiveness compared to short-height soybean cultivars. Conversely, although Morkhor60 exhibited greater height and a higher number of nodes, it experienced more significant yield reductions due to its increased vulnerability to herbicide phytotoxicity.

These findings highlight the critical importance of integrated weed management strategies that combine effective chemical control with the selection of tolerant cultivars to maximise soybean productivity in tropical rainfed systems. The dramatic yield losses observed in the absence of weed control emphasise that timely and effective weed management is indispensable for sustainable soybean production in Northeast Thailand.

5. Conclusions

This study demonstrates that the efficacy of pre-emergence herbicides for weed management in soybean cultivation in Northeast Thailand is season-dependent. S-metolachlor (96% EC, 900 g a.i. ha^{-1}) is optimal for the rainy season, offering weed control efficiency, low phytotoxicity, and high yields. For the dry season, flumioxazin (50% WP, 125 g a.i. ha^{-1}) is recommended, with significantly reduced phytotoxicity compared to rainy season applications and high yields. Both herbicides effectively controlled the most dominant weed species, although *Cyperus rotundus* L. was less affected. The CM60 variety consistently showed greater herbicide tolerance and yield stability, especially under rainy season conditions. However, these findings are limited to two growing seasons, and multi-year trials are recommended to validate the consistency of herbicide performance and varietal responses. These insights underscore the importance of integrating season-specific herbicide programs with appropriate variety selection to optimize sustainable soybean production in Northeast Thailand.

Supplementary Materials: The following supporting information can be downloaded at https://www.mdpi.com/article/10.3390/agronomy15071725/s1, Table S1. Homogeneity of variance for each parameter and season based on Levene's test. Figure S1. Correlation between weed control efficiency (%) and grain yield of soybean at different days after application (DAA) in each weed control treatment in 2023 (a) and 2023/2024 (b).

Author Contributions: Conceptualization, S.G. and S.C.; methodology, S.G. and U.R.R.P.; software, U.R.R.P. and T.M.; validation, U.R.R.P., S.G., S.C., N.J., and T.M.; formal analysis, S.G.; investigation, S.G.; resources, S.G.; data curation, U.R.R.P.; writing—original draft preparation, U.R.R.P.; writing—review and editing, S.G.; visualization, S.G.; supervision, S.G.; project administration, S.G.; funding acquisition, S.G. All authors have read and agreed to the published version of the manuscript.

Funding: This research was partially funded by the government of Thailand's grants to Khon Kaen University (KKU), (Project no. FY2014: 572603 and FY2015: 581803) through the Research Scholar project of Santimaitree Gonkhamdee. This study was funded by Khon Kaen University (KKU) Scholarships for the Association of Southeast Asian Nations (ASEAN) and Greater Mekong Subregion (GMS) countries, with the main funding in 2023.

Data Availability Statement: Data is contained within the article or Supplementary Materials.

Acknowledgments: We would like to express our gratitude to Khon Kaen University (KKU) Scholarships for the Association of Southeast Asian Nations (ASEAN) and Greater Mekong Subregion (GMS) countries for funding this study, along with the partial funding support from the Northeast Thailand Cane and Sugar Research Center (NECS) and Research Program Funding of the Research and Innovation Department at Khon Kaen University. The authors are highly grateful to the Department of Agronomy, Faculty of Agriculture, Khon Kaen University, Thailand, for providing plant materials, research facilities, and other technical support.

Conflicts of Interest: The authors declare no conflicts of interest.

References

1. Ciampitti, I.A.; Schapaugh, W.T.; Shoup, D.; Duncan, S.; Ruiz Diaz, D.; Peterson, D.; Rogers, D.H.; Whitworth, J.; Schwarting, H.; Jardine, D.; et al. *Soybean Production Handbook*; Kansas State University Agricultural Experiment Station and Cooperative Extension Service: Manhattan, KS, USA, 2016.
2. European Commission. *EU Agricultural Outlook for Markets and Income 2019–2030*; Directorate-General for Agriculture and Rural Development, Publications Office: Luxembourg, 2019.
3. Office of Agricultural Economics. *The Study of Soybean Supply Chain in Chiang Mai, Thailand*; Ministry of Agriculture and Cooperatives: Bangkok, Thailand, 2018.
4. Sritongtae, C.; Monkham, T.; Sanitchon, J.; Lodthong, S.; Srisawangwong, S.; Chankaew, S. Identification of Superior Soybean Cultivars through the Indication of Specific Adaptabilities within Duo-Environments for Year-Round Soybean Production in Northeast Thailand. *Agronomy* **2021**, *11*, 585. [CrossRef]
5. Prasertsri, P. *Oilseeds and Products Annual: Thailand*; USDA Foreign Agricultural Service: Bangkok, Thailand, 2025.
6. Pacanoski, Z.; Mehmeti, A. Worldwide Complexity of Weeds. *Acta Herbol.* **2021**, *30*, 79–89. [CrossRef]
7. Merga, B.; Alemu, N. Integrated Weed Management in Chickpea (*Cicer arietinum* L.). *Cogent Food Agric.* **2019**, *5*, 1620152. [CrossRef]
8. Suryanto, P.T.; Sulistyani, E.; Putra, E.T.S.; Kastono, D.; Alam, T. Estimation of Critical Period for Weed Control in Soybean on Agro-Forestry System with Kayu Putih. *Asian J. Crop Sci.* **2017**, *9*, 82–91. [CrossRef]
9. Vitorino, H.S.; da Silva Junior, A.C.; Gonçalves, C.G.; Martins, D. Interference of a Weed Community in the Soybean Crop in Functions of Sowing Spacing. *Rev. Ciênc. Agronômica* **2017**, *48*, 605–613. [CrossRef]
10. Gesimba, R.M.; Langat, M.C. A Review on Weeds and Weed Control in Oil Crops with Special Reference to Soybeans (*Glycine max* L.) in Kenya. *Agric. Trop. Subtrop.* **2005**, *38*, 56–62.
11. Jadhav, V.T.; Kashid, N.V. Integrated Weed Management in Soybean. *Indian J. Weed Sci.* **2019**, *51*, 81–82. [CrossRef]
12. Ribeiro, V.H.V.; Oliveira, M.C.; Smith, D.H.; Santos, J.B.; Werle, R. Evaluating Efficacy of Preemergence Soybean Herbicides Using Field Treated Soil in Greenhouse Bioassays. *Weed Technol.* **2021**, *35*, 830–837. [CrossRef]

13. Kniss, A.R. Genetically Engineered Herbicide-Resistant Crops and Herbicide-Resistant Weed Evolution in the United States. *Weed Sci.* **2018**, *66*, 260–273. [CrossRef]
14. Norsworthy, J.K.; Ward, S.M.; Shaw, D.R.; Llewellyn, R.S.; Nichols, R.L.; Webster, T.M.; Bradley, K.W.; Frisvold, G.; Powles, S.B.; Burgos, N.R.; et al. Reducing the Risks of Herbicide Resistance: Best Management Practices and Recommendations. *Weed Sci.* **2012**, *60*, 31–62. [CrossRef]
15. Zain, S.; Dafaallah, A.; Zaroug, M. Efficacy and Selectivity of Pendimethalin for Weed Control in Soybean (*Glycine max* (L.) Merr.), Gezira State, Sudan. *Agric. Sci. Pract.* **2020**, *7*, 59–68. [CrossRef]
16. Sridhara, S.; Nandini, R.; Gopakkali, P.; Somavanshi, A.V. Weed Control Efficiency and Weed Index in Soybean as Influenced by Flumioxazin and Its Effect on Succeeding Green Gram. *Int. J. Chem. Stud.* **2019**, *7*, 872–875.
17. Schelter, M.L.; Prates, A.A.; Fruet, D.L.; de Souza, M.P.; Guerra, N.; de Oliveira Neto, A.M. Response of Soybean Cultivars with Different Maturation Times to Pre-Emergence Herbicides. *Semina Ciênc. Agrár.* **2023**, *44*, 841–858. [CrossRef]
18. Alonso, D.G.; Constantin, J.; Oliveira, R.S., Jr.; Biffe, D.F.; Raimondi, M.A.; Gemelli, A.; Blainski, E.; Carneiro, J.C. Selectivity of Glyphosate in Tank Mixtures for RR Soybean in Sequential Applications with Mixtures Only in the First or Second Application. *Planta Daninha* **2010**, *28*, 865–875. [CrossRef]
19. Belfry, K.D.; Soltani, N.; Brown, L.R.; Sikkema, P.H. Tolerance of Identity Preserved Soybean Cultivars to Preemergence Herbicides. *Can. J. Plant Sci.* **2015**, *95*, 719–726. [CrossRef]
20. Taylor-Lovell, S.; Wax, L.M.; Nelson, R. Phytotoxic Response and Yield of Soybean (*Glycine max*) Varieties Treated with Sulfentrazone or Flumioxazin. *Weed Technol.* **2001**, *15*, 95–102. [CrossRef]
21. Fornazza, F.G.F.; Constantin, J.; Machado, F.G.; Oliveira, R.S., Jr.; Silva, G.D.; Rios, F.A. Selectivity of pre- and post-emergence herbicides to very-early maturing soybean cultivars. *Comun. Sci.* **2018**, *9*, 649–658. [CrossRef]
22. Lamego, F.P.; Fleck, N.G.; Bianchi, M.A.; Schaedler, C.E. Tolerância à interferência de plantas competidoras e habilidade de supressão por genótipos de soja: II. Resposta de variáveis de produtividade. *Planta Daninha* **2004**, *22*, 491–498. [CrossRef]
23. Nordby, D.E.; Alderks, D.L.; Nafziger, E.D. Competitiveness with Weeds of Soybean Cultivars with Different Maturity and Canopy Width Characteristics. *Weed Technol.* **2007**, *21*, 1082–1088. [CrossRef]
24. Raimondi, M.A.; Oliveira, R.S.; Constantin, J.; Franchini, L.H.M.; Blainski, É.; Raimondi, R.T. Weed interference in cotton plants grown with reduced spacing in the second harvest season. *Rev. Caatinga* **2017**, *30*, 1–12. [CrossRef]
25. Zobiole, L.H.S.; Oliveira, R.S., Jr.; Huber, D.M.; Constantin, J.; Castro, C.; Oliveira, F.A.; Oliveira, A., Jr. Glyphosate reduces shoot concentrations of mineral nutrients in glyphosate-resistant soybeans. *Plant Soil* **2010**, *328*, 57–69. [CrossRef]
26. Bell, R.W.; Seng, V. Rainfed lowland rice-growing soils of Cambodia, Laos, and North-East Thailand. In Proceedings of the CARDI International Conference on Research on Water in Agricultural Production in Asia for the 21st Century, Phnom Penh, Cambodia, 25–28 November 2003.
27. Oshunsanya, S.O. Introductory Chapter: Relevance of Soil pH to Agriculture. In *Soil pH for Nutrient Availability and Crop Performance*; IntechOpen: London, UK, 2019. [CrossRef]
28. Jandong, E.A. Evaluation of some soybean [*Glycine max* (L.) Merrill] genotypes for yield and agronomic performance under low pH stressed soils. *J. Agric. Sci.* **2021**, *12*, 45–56.
29. Uguru, M.I. Determination of yield stability of seven soybean (*Glycine max*) genotypes across diverse soil pH levels using GGE biplot analysis. *Afr. J. Biotechnol.* **2011**, *10*, 12345–12356.
30. Aekrathok, P.; Songsri, P.; Jongrungklang, N.; Gonkhamdee, S. Efficacy of post-emergence herbicides against important weeds of sugarcane in North-East Thailand. *Agronomy* **2021**, *11*, 429. [CrossRef]
31. Janiya, J.D.; Moody, K. Weed populations in transplanted and wet-seeded rice as affected by weed control method. *Trop. Pest Manag.* **1989**, *35*, 8–11. [CrossRef]
32. Hasan, M.; Mokhtar, A.S.; Rosli, A.M.; Hamdan, H.; Motmainna, M.; Ahmad-Hamdani, M.S. Weed control efficacy and crop-weed selectivity of a new bioherbicide WeedLock. *Agronomy* **2021**, *11*, 1488. [CrossRef]
33. Mani, V.S.; Malla, M.L.; Gautam, K.C.; Bhagwandas. Weed-killing chemicals in potato cultivation. *Indian Farming* **1973**, *23*, 17–18.
34. Li, Q.; Lu, Y.; Shi, Y.; Wang, T.; Ni, K.; Xu, L.; Liu, S.; Wang, L.; Xiong, Q.; Giesy, J.P. Combined effects of cadmium and fluoranthene on germination, growth and photosynthesis of soybean seedlings. *J. Environ. Sci.* **2013**, *25*, 1936–1946. [CrossRef] [PubMed]
35. Meseldžija, M.; Rajković, M.; Dudić, M.; Vranešević, M.; Bezdan, A.; Jurišić, A.; Ljevnaić-Mašić, B. Economic feasibility of chemical weed control in soybean production in Serbia. *Agronomy* **2020**, *10*, 291. [CrossRef]
36. Chauhan, B.S.; Awan, T.H.; Abugho, S.B.; Evengelista, G.; Sudhir-Yadav. Effect of crop establishment methods and weed control treatments on weed management, and rice yield. *Field Crops Res.* **2015**, *172*, 72–84. [CrossRef]
37. Rao, A.N.; Wani, S.P.; Ahmed, S.; Haider Ali, H.; Marambe, B. An overview of weeds and weed management in rice of South Asia. In *Weed Management in Rice in the Asian-Pacific Region*; Asian-Pacific Weed Science Society: Hyderabad, India, 2017; pp. 247–281.
38. Jha, P.; Kumar, V.; Godara, R.K.; Chauhan, B.S. Weed management using crop competition in the United States: A review. *Crop Prot.* **2017**, *95*, 31–37. [CrossRef]

39. Chauhan, B.S.; Johnson, D.E. The role of seed ecology in improving weed management strategies in the tropics. *Adv. Agron.* **2010**, *105*, 221–262. [CrossRef]
40. Oliveira, M.C.; Feist, D.; Eskelsen, S.; Scott, J.E.; Knezevic, S.Z. Weed control in soybean with preemergence- and postemergence-applied herbicides. *Crop Forage Turfgrass Manag.* **2017**, *3*, 1–7. [CrossRef]
41. Safdar, M.E.; Nadeem, M.A.; Rehman, A.; Ali, A.; Iqbal, N.; Mumtaz, Q.; Javed, A. The screening of herbicides for effective control of weeds in soybean (*Glycine max* L.). *J. Weed Sci. Res.* **2020**, *27*, 251–266. [CrossRef]
42. Gazola, T.; Gomes, D.M.; Belapart, D.; Dias, M.F.; Carbonari, C.A.; Velini, E.D. Selectivity and residual weed control of pre-emergent herbicides in soybean crop. *Rev. Ceres* **2021**, *68*, 219–229. [CrossRef]
43. Qadeer, A.; Ali, Z.; Ahmad, H.; Qasam, M.; Toor, S. Invasion of different weeds on gladiolus and their control by herbicides. *Plant Gene Trait* **2016**, *7*, 1–9. [CrossRef]
44. Lakra, K. Effect of irrigation and herbicides on most tenacious weed Cyperus rotundus in wheat. *Int. J. Environ. Clim. Chang.* **2021**, *11*, 29–37. [CrossRef]
45. Kaur, P.; Bhullar, M.S. Effect of repeated application of pendimethalin on its persistence and dissipation kinetics in soil under field and laboratory conditions. *Environ. Technol.* **2017**, *40*, 997–1005. [CrossRef] [PubMed]
46. Arsenijevic, N.; De Avellar, M.; Butts, L.; Arneson, N.J.; Werle, R. Influence of sulfentrazone and metribuzin applied preemergence on soybean development and yield. *Weed Technol.* **2021**, *35*, 210–215. [CrossRef]
47. Moomaw, R.S.; Martin, A.R. Interaction of metribuzin and trifluralin with soil type on soybean (*Glycine max*) growth. *Weed Sci.* **1978**, *26*, 327–331. [CrossRef]
48. Niekamp, J.W.; Johnson, W.G.; Smeda, R.J. Broadleaf weed control with sulfentrazone and flumioxazin in no-tillage soybean (*Glycine max*). *Weed Technol.* **1999**, *13*, 233–238. [CrossRef]
49. Osborne, B.T.; Shaw, D.R.; Ratliff, R.L. Soybean (*Glycine max*) cultivar tolerance to SAN 582H and metolachlor as influenced by soil moisture. *Weed Sci.* **1995**, *43*, 288–292. [CrossRef]
50. Wise, K.; Mueller, D.; Kandel, Y.; Young, B.; Johnson, B.; Legleiter, T. Soybean Seedling Damage: Is There an Interaction Between the ILeVO Seed Treatment and Pre-Emergence Herbicides? Available online: https://crops.extension.iastate.edu/cropnews/2015/05/soybean-seedling-damage-there-interaction-between-ilevo-seed-treatment-and-pre (accessed on 16 July 2025).
51. Board, J.E.; Kahlon, C.S. Soybean yield formation: What controls it and how it can be improved. In *Soybean Physiology and Biochemistry*; InTech: Rijeka, Croatia, 2011; pp. 1–36.
52. Thanacharoenchanaphas, K.; Rugchati, O. Simulation of climate variability for assessing impacts on yield and genetic change of Thai soybean. *World Acad. Sci. Eng. Technol.* **2011**, *59*, 1484–1487.
53. Madhu, M.; Hatfield, J.L. Interaction of carbon dioxide enrichment and soil moisture on photosynthesis, transpiration, and water use efficiency of soybean. *Agric. Sci.* **2014**, *5*, 410–429. [CrossRef]
54. Madhu, M.; Hatfield, J.L. Elevated carbon dioxide and soil moisture on early growth response of soybean. *Agric. Sci.* **2015**, *6*, 263–278. [CrossRef]
55. Bianchi, M.A.; Fleck, N.G.; Federizzi, L.C. Características de plantas de soja que conferem habilidade competitiva com plantas daninhas. *Bragantia* **2006**, *65*, 623–632. [CrossRef]
56. Bastiani, M.O.; Lamego, F.P.; Agostinetto, D.; Langaro, A.C.; Silva, D.C. Relative competitiveness of soybean cultivars with barnyardgrass. *Bragantia* **2016**, *75*, 435–445. [CrossRef]

Disclaimer/Publisher's Note: The statements, opinions and data contained in all publications are solely those of the individual author(s) and contributor(s) and not of MDPI and/or the editor(s). MDPI and/or the editor(s) disclaim responsibility for any injury to people or property resulting from any ideas, methods, instructions or products referred to in the content.

MDPI AG
Grosspeteranlage 5
4052 Basel
Switzerland
Tel.: +41 61 683 77 34

Agronomy Editorial Office
E-mail: agronomy@mdpi.com
www.mdpi.com/journal/agronomy

Disclaimer/Publisher's Note: The title and front matter of this reprint are at the discretion of the Guest Editors. The publisher is not responsible for their content or any associated concerns. The statements, opinions and data contained in all individual articles are solely those of the individual Editors and contributors and not of MDPI. MDPI disclaims responsibility for any injury to people or property resulting from any ideas, methods, instructions or products referred to in the content.

www.ingramcontent.com/pod-product-compliance
Lightning Source LLC
Chambersburg PA
CBHW061442210926
51850CB00008B/1582